KELVIN: LIFE, LABOURS AND LEGACY

KELVIN: LIFE, LABOURS AND LEGACY

EDITED BY

RAYMOND FLOOD,

MARK MCCARTNEY,

and

ANDREW WHITAKER

Great Clarendon Street, Oxford OX2 6DP

Oxford University Press is a department of the University of Oxford.
It furthers the University's objective of excellence in research, scholarship,
and education by publishing worldwide in

Oxford New York

Auckland Cape Town Dar es Salaam Hong Kong Karachi
Kuala Lumpur Madrid Melbourne Mexico City Nairobi
New Delhi Shanghai Taipei Toronto

With offices in

Argentina Austria Brazil Chile Czech Republic France Greece
Guatemala Hungary Italy Japan Poland Portugal Singapore
South Korea Switzerland Thailand Turkey Ukraine Vietnam

Oxford is a registered trade mark of Oxford University Press
in the UK and in certain other countries

Published in the United States
by Oxford University Press Inc., New York

First published 2008

British Library Cataloguing in Publication Data
Data available

Library of Congress Cataloging in Publication Data
Data available

Typeset by Newgen Imaging Systems (P) Ltd., Chennai, India
Printed in Great Britain
on acid-free paper by
Biddles Ltd., King's Lynn, Norfolk

ISBN 978–0–19–923125–6

1 3 5 7 9 10 8 6 4 2

FOREWORD

Brian Pippard

In 1870, when Cambridge University were looking for their first Cavendish Professor of Physics, they naturally thought first of Sir William Thomson who had held the chair of Natural Philosophy in Glasgow since 1846. He had been elected when only 22 years old, and had no wish to leave his personal creation, the first significant physics department in a British university. Indeed, he stayed there for 53 years, and for another 8 after his retirement, still hard at work up to the time of his death at 83. Created Baron Kelvin in 1892, he was the first British scientist to be ennobled for his contribution to learning and to industry. As for Cambridge, they had to make do with James Clerk Maxwell, a much less prominent figure at the time.

Nowadays Maxwell's researches are revered and form an essential part of every physics student's lecture course, while Kelvin's hardly get a mention. To be sure, we still have the Kelvin temperature scale, but to most physicists this is little more than a token, his pioneering ideas on thermodynamics having been overshadowed by those of his friend James Joule and his German contemporary Rudolph Clausius. How is it that in this, and the other fields where his innovations were so important, the memory of one who had been the unquestioned leader of science and technology, the versatile and prolific inventor, should fade so soon after his death? He is not alone in this—his mentor and lifelong friend Sir George Stokes has fallen into similar obscurity—but Kelvin was the most prominent of the British scientists who contributed to the great advances towards the end of the ninteenth century.

Part of the explanation can be traced to his precocity. Although he outlived the nineteenth century the young Thomson began publishing learned work as early as 1840, still only 16 but having already spent several years at Glasgow University. His first paper (for the sake of propriety published under a pseudonym) corrected the Edinburgh Professor Kelland's misunderstanding of Fourier's great book, which Thomson had needed but a fortnight to read and appreciate in the original French. He learnt his mathematics from Fourier and other French mathematicians, and something of precise experimentation from Regnault; but the more immediate inspiration, after graduating from Cambridge, came from Faraday's ideas on electromagnetism which Thomson expressed in a mathematical form that was foreign to Faraday's creative genius. Early in his life of research he became convinced of the perfection of the electromagnetic world and its intelligibility as a mechanism—if you could represent its behaviour by a working model, this was some guarantee that your picture of what was going on was not mere fantasy.

Maxwell belonged to the same tradition, and began his systematic development of Thomson's ideas on electromagnetism also by imagining models. But, having expressed his final conclusions in mathematical form he discarded his models. The version we now know as Maxwell's equations was a second translation, after his death, which introduced vector electric and magnetic fields, visualizable (if you were so inclined) as stresses and strains in the pervasive, but always enigmatic, ether. With Einstein's 1905 theory of relativity the existence of the ether became highly dubious, but the equations have remained the basis of electromagnetic calculations and the ideal of a field theory for the next century of theorists. Thomson's work may be overlaid and forgotten, but he should still be remembered as an important pioneer.

With the development of electromagnetic theory came industrial applications, and Thomson was at the forefront in devising instruments, from the most delicate (in my student days there were still some who remembered using his quadrant electrometer and cursing its temperamental quirks) to the robust voltmeters and ammeters of power engineering. Without his zeal and inventive brilliance the Atlantic cable would have remained the failure it was before he took charge and made Britain the leader of long-distance communication. Yet here again he was overtaken by the invention, during his lifetime, of the diode and triode valves. They developed only slowly until the stimulus of the First World War brought about a major industry that eventually replaced cables by radio communication, and the beautiful old galvanometers by the black boxes of electronics.

For physicists, however, it was the discoveries made around 1900 that created the dichotomy between Thomson's 'classical' physics and the 'modern' physics of electrons and radioactivity and the quantum. Energy, central to classical mechanics, is one concept that has remained central while the determinism of the classical picture has evaporated with the rise of quantum mechanics since 1925; indeterminism has replaced the comfortable assurance of visualizable models, so that if chemists and solid-state physicists continue to imagine electrons as little billiard balls, they nevertheless recognize that such fancies are merely a preliminary to serious model-free calculation. In this they do not greatly differ from Thomson, who never deluded himself his models were exact representations of reality.

The vast expansion of science has inevitably led to much classical work disappearing from students' courses—history, even more recent history than Thomson's, must be excluded to leave room for more immediately relevant concepts. But science does not progress by simply disregarding older views; however novel a theory, its origins always show. The most revolutionary of innovations, quantum theory, began with Planck's attempt to patch up a flaw in classical mechanics; and Newton himself did not create classical mechanics from scratch, but built on foundations laid by Galileo, Kepler, and millennia of devoted astronomers. We cannot truly appreciate the methods and meaning of science without looking into its history, and one of the most agreeable approaches to history is through its great men, of which William Thomson, Baron Kelvin of Largs, was certainly one.

PREFACE

The year 2007 sees the centenary of William Thomson's death. On 23 of December 1907, with all the pomp that the British Empire could muster, William Thomson, Baron Kelvin of Largs was buried in Westminster Abbey beside the nation's most venerated scientist, Sir Isaac Newton. In his lifetime Thomson was seen as the most important natural philosopher of the Victorian age, but the waxing and waning of the twentieth century and the supplanting of classical physics have eroded his reputation, so that for many scientists he is remembered as little more than a unit of temperature.

This book of collected essays seeks to re-evaluate and rehabilitate Lord Kelvin, setting both the man and his work in historical and scientific context. It is not of course the first book to do so, and the debt to previous scholarship, and in particular to the thorough and encyclopaedic scholarship of Crosbie Smith and Norton Wise in their *Energy and Empire: A Biographical Study of Lord Kelvin*, published by Cambridge University Press in 1989, is acknowledged by many of the authors in this volume.

The current work is distinctive in bringing the expertise of a range of authors, historians of science and of mathematics, physicists, mathematicians and engineers, to bear on Kelvin's life, labours, and legacy. This range of contributors in itself testifies to the extent of Kelvin's work, and it is hoped that it will also enable the reader to view the subject from a variety of perspectives, each of which complements and enriches the others. Each essay is self contained and can be read independently of the others. This has inevitably introduced some examples of repetition across the book, but as editors we feel this small price is worth paying, as it allows readers to move directly to the aspects of Kelvin's life and work which interest them most. Of course we hope that such a reader, their appetite thus whetted, will be led to many of the other essays!

We would like to thank the authors, who have each brought their distinctive enthusiasm for Kelvin to their contribution, and have graciously submitted to our editorial requests. We are also grateful that many of them have taken great pains in providing suitable illustrations for their contributions.

Raymond Flood
Mark McCartney
Andrew Whitaker

April 2007

CONTENTS

1
William Thomson: An Introductory Biography

Mark McCartney

AN IRISH CHILDHOOD

It was a family joke that William Thomson had two birthdays, and once while signing his name in a birthday book, as William Thomson on the 25 June, and Kelvin on 26 June he quipped 'It is convenient to have two birthdays when you have two names!'[1] It is not quite clear how the confusion arose, and though in childhood his birthday was celebrated on 25th we have it on the authority of his father that William Thomson was born on the 26 June 1824 at 5 a.m., and on the authority of his eldest sister Elizabeth that the day of his birth was one of beautiful summer sunshine.

Thanks to the labours and love of their father, Dr James Thomson, William and his brothers and sisters enjoyed a happy childhood, steeped in knowledge and learning, which their father took every opportunity to impart.

Dr James Thomson, born in 1786 in a farmhouse outside Ballynahinch, was the youngest—by ten years—of five children. His two older sisters taught him to read and he taught himself arithmetic from a copy of John Bonnycastle's *The Scholar's Guide to Arithmetic*, but also attended a local day school run by a Dr Edgar. He went on to study at Glasgow University, coming home each summer to work in Dr Edgar's school, and by 1815 he was Professor of Mathematics at the newly opened Belfast Academical Institution (today a school, but in the early days of its existence a school and college combined, with the college functioning like a small Scottish university).

James was hard working and ambitious. Opposite the Academical Institution he built two houses, one to raise his family in, and one to rent out. He wrote a range of successful textbooks—on topics such as arithmetic, calculus, trigonometry, and geography (rising at 4 a.m. in the morning to work on them) and he never missed an opportunity to teach his children. William's sister Elizabeth's earliest memory is of a storm in Belfast in the winter of 1821. The wind from the north blew so hard that every window in the front of the house was shattered and the tall kitchen chimney crashed through the roof. James, ever the teacher, took his three year old round the house and showed her the damage done, and the storm still raging across the hole in the kitchen roof and then, using the house barometer, gave her a simple lesson in meteorology.

Elizabeth (born 1818) was the first of James and his wife Margaret's seven children, followed by Anna (1820), James (1822), William (1824), John (1826), Margaret (1827), and Robert (1829). In her reminiscences Elizabeth describes a happy and affectionate childhood in Belfast, filled with books to be read, and books being read to them, and lessons given to the children in the evenings by their father, and family friends and long summer holidays.

James's brother Robert, who still farmed the land outside Ballynahinch wrote to his sister of their Belfast nephews and nieces stating that they were 'wonderfully apt in learning; but I don't think it strange, as both father and mother are drilling them'[2].

A GLASGOW YOUTH

In 1830 the family circle was fractured when their mother, who had never recovered from Robert's birth, died. Two years after this the family moved to Glasgow, where James had been appointed Professor of Mathematics at the university.

Joining Professor Thomson in his Junior Mathematics class in the 1832–33 academic year were his two elder boys James and William. They attended the class, but were not formally enrolled, and did not take examinations. However, in the 1834–35 session, with James now aged 12 and William 10, they did enroll. This is remarkable, but not quite as remarkable as it might first appear, as the usual minimum age for matriculation at Glasgow was 14. A great aunt promised the boys a guinea each if they each won two university prizes and a guinea between them if they each won one. The money was easily won by the boys. They took prizes in a range of classes throughout their time at Glasgow. However, most consistent was their performance in the mathematics and natural philosophy classes, where William regularly took first prize and James second.

Dr James Thomson's hard work had brought financial success, and by 1839 he was wealthy enough to take the family, and a servant, to the continent for two months during the summer. As part of their journey they travelled from Glasgow to Liverpool by boat and from Liverpool to London by train;[3] 'Going at a tremendous rate—no less than thirty-six miles an hour!' wrote Elizabeth in her diary. They then spent a month in London, and went from there to France. The family spent two weeks in Paris together and then James took Anna and Elizabeth to Switzerland, leaving the boys in Paris with the family servant to have daily lessons in French. While there William (or Thomson, as we shall refer to him from now on) went to the Bibliothèque Royale to read Laplace's *Mécanique Céleste* in preparation for an 85 page and mathematically sophisticated essay *On the Figure of the Earth* which won him a University medal in his next academic year.

The next year James took his family to Germany, and in preparation the family all took German lessons, with a teacher coming to the family home. While in Germany James engaged another teacher to come each day and help them polish their conversational German. For Dr James Thomson all opportunities were opportunities to learn. Thomson inherited his father's love and unquenchable thirst for knowledge, and though on holiday in Germany, and under his father's express command to leave all other work behind and concentrate on learning German, Thomson, by now just 16, had packed a little French reading: Fourier's *Théorie Analytique de la Chaleur*. Thomson continues the story:

'Going that summer to Germany with my father and my brothers and sisters, I took Fourier with me. My father took us to Germany and insisted that all work should be left behind, so that the whole of our time could be given to learning German. We went to Frankfort, where my father took a house for two months . . . Now just two days before leaving Glasgow I had got Kelland's book (*Theory of Heat*, 1837), and was shocked to be told that Fourier was mostly wrong. So I put Fourier in my box, and used in Frankfort to go down to the cellar surreptitiously every day to read a bit of Fourier. When my father discovered it he was not very severe upon me.'[4]

James Thomson's lack of severity is probably explained by the fact that his son claimed that Philip Kelland, the Professor of Mathematics at Edinburgh was wrong in his criticisms of Fourier. Although initially incredulous of his son's claims, a closer examination showed that they were accurate, and Thomson wrote the matter up in what was to be his first published paper, which appeared in May 1841 in the Cambridge Mathematical Journal, with the author simply designated as 'P. Q. R.' Ivor Grattan Guinness gives more detail on Thomson's interaction with Fourier's work in Chapter 3.

A CAMBRIDGE STUDENT

Thomson's strong abilities in mathematics were clear to all, and the obvious place for him to study after finishing Glasgow was Cambridge. However, there were concerns that graduating from Glasgow might disadvantage his prospects at Cambridge, and so, although both Thomson and his brother James passed the

BA examinations in 1839, and the MA examinations in 1840, on each occasion only James graduated. Around this time Thomson would designate himself as William Thomson BATAIAP (Bachelor of Arts to all intents and purposes).

October 1841 saw Thomson, aged 17, enter St Peter's College (Peterhouse) Cambridge as a pensioner, that is, a student who paid his own way. During his time at Cambridge the Cambridge Mathematical Journal published a further 10 papers from William Thomson—all pseudonymous, and all but one penned under the initials P. Q. R. Indeed, Thomson was being tipped as Senior Wrangler virtually as soon as he arrived in Cambridge.

While at Cambridge there was regular correspondence with his family. His father, who was footing the considerable bills, wrote to him of the wise use of time and money, and care in the friendships he cultivated:

'Recollect my invaluable maxim never to quarrel with a man (but to waive the subject) about religion.'[5]

'Use *all economy* consistent with respectability. Be most circumspect about your conduct and about what acquaintance you form. You are young: take care you be not led to what is wrong.'[6]

Dr James Thomson's worries were not unreasonable, many a young man came to Cambridge and filled his time with nothing but boating and wine parties. Thomson began his career at Cambridge with whole-hearted promises to his father: he had given no wine parties, he had not joined the rowing club. His proposed schedule for the working day included rising at 5 a.m. to light the fire; reading until 8:15 a.m.; attending his daily lecture; reading until 1:15 p.m.; exercise until 4:15 p.m.; chapel until 7 p.m., reading until 8:30 p.m. and finally, to bed at 9.p.m.

Alas, Thomson's good intentions began to slip. Initially he claimed that the college was divided between rowing men and reading men and that 'rowing for the races is too hard work for getting on well with reading'[7], but by February 1842 he had bought a secondhand boat, a bargain at £7 and, he claimed, much cheaper than renting a boat. His father was unimpressed, but sent money to cover the bill. In spring of the next year his diary records late nights with a wide circle of friends, bathing, skating, walking, reading, and serious rowing. He was also one of the founding members of the Cambridge University Musical Society, which gave its first public performance in December 1843 (with Thomson playing French horn and cornopean). Letters to his father are breezy and relaxed:

'My Dear Father, I have again to write to you on the same pleasant business that I had to write to you so lately, which is to say that my money is again all gone.'[8]

To soften his father up when asking for more money he would sometimes include a mathematical problem for use in the exams at Glasgow, or mention the receipt of another scholarship or prize (of which he received over a dozen during his student days at St Peter's). However, nearly all these prizes and scholarships were for small amounts (the most substantial being £30) and over the three years from autumn 1841 to autumn 1844 Thomson's education had cost over £770. On sending his son a letter containing a summary of this cost Dr Thomson writes, with exasperation:

'How is this to be accounted for? Have you lost money or been defrauded of it? . . . you must exercise the strictest economy . . . not spending a penny unnecessarily.'[9]

To put this in context an upper middle class income at this period would have been around £900–1000 per annum.

Dr James Thomson may have been exasperated at his son's full life and alarming expenditure, but his studies were not suffering. He was working hard under the sharp eye of his private tutor William Hopkins. Hopkins reputation as 'the senior Wrangler maker' was well earned, by 1849 he had coached some 200 Wranglers—a fifth of whom had landed one of the top three places. He was an excellent teacher, and a hard task master. The Mathematical Tripos was the most gruelling academic race course in the country, and Hopkins trained and honed his candidates well. Thomson's examinations commenced on New Year's Day 1845, and finished on the 7th of January. There were twelve papers, with morning papers being two and a half hours long, and afternoon papers three hours long. There was a high proportion of 'bookwork' based questions and the papers required fluent recall, a good knowledge of mathematical tricks and shortcuts, and fast penmanship.

To universal surprise, when the results were read out Thomson came second, with one Stephen Parkinson, a student at St. John's coming first. The family were disappointed, indeed Aunt Agnes, James's headstrong sister-in-law was indignant:

'I am most *desperately* disappointed . . . I am not consoled to learn that so and so, and so and so, stood second. I expected him to stand first, and the only thing that reconciles me is that *we* all needed this mortification.'[10]

One explanation for Parkinson's victory is that, quite simply, he wrote much faster that Thomson. Indeed one of the examiners commented that if he had not seen Parkinson's speed in the examination hall he would have struggled to believe that the sheer volume of material could have been written by him in the time limit. However, when it came to the Smith's Prize examinations later in January Thomson easily took first place. These papers were more heavily weighted towards problem solving, and Thomson outperformed Parkinson in each of the four papers, and in two of the papers the marks were in the proportion of three to two.

William Thomson's failure to be Senior Wrangler has generated at least two, possibly apocryphal, stories. The first, which has at least the provenance of being told by Joseph Larmor and being recorded in Thomson's official biography by Silvanus

P. Thompson[11], is that in the Smith's Prize papers the two top candidates presented solutions to a question which were so similar that the matter was investigated further. When questioned, the Senior Wrangler said that the solution he had given came from a paper he had read in the *Cambridge Mathematical Journal*. The paper, which had been published some years earlier, had been authored by someone who signed himself simply as 'P. Q. R.'.

The second anecdote, which certainly does not appear in any of the early biographies of Thomson, is that on the day the results of the Tripos were to be announced at the Senate House he dispatched his college servant with the words 'Oh, just run down to the Senate House, will you, and see who is Second Wrangler'. The servant returned and announced 'You sir!'[12] Clearly it could be argued that if true, such an uncomplimentary story would probably have been suppressed by an early biographer, but to a reader of Thomson's life it does not appear to ring quite true. Thomson had faults, indeed he could be very quick tempered, but he does not appear, in relationships or letters, to have been conceited.

Thomson's undergraduate success brought an election to a Fellowship of Peterhouse in June 1845: he had just turned 21.

While Thomson was still a Cambridge undergraduate events were taking place at Glasgow which would help determine his future career. William Meikleham had been the Professor of Natural Philosophy at Glasgow University since 1803, and during the academic session of 1838–39 his health became poor and his classes were covered by colleagues for the remainder of that academic year and the next. By 1841 it was realized that Meikleham was unlikely ever to return to his classes, and the thoughts of Dr James Thomson began to turn to his colleague's eventual replacement. He wanted academic excellence, of the Cambridge calibre—a member of the intellectual elite, but not an elitist. The new member of staff needed to be in sympathy with the broad and non-hierarchical Scottish university education system and above all he had to be a good teacher.

There was a range of possible candidates, but by perhaps as early as Christmas 1842 Dr Thomson had realized that his son William, then only 18, could be in the running for the job. Certainly by March 1843 he was encouraging his son in a letter to cultivate particular friendships which could prove useful 'in case of a certain event coming round'[13] and the next month he was gently testing his ideas out on a colleague, Dr William Thomson, the Professor of Materia Medica. Dr William Thomson pointed out the young Thomson's lack of experience in experimental work, and his father advised him to get all the experience he could at Cambridge. Thus he attended lecture courses on experimental natural philosophy twice (once in 1843 and again in 1844), and in 1844 he attended lectures on practical astronomy and astronomical instruments. During the summer of 1843 he spent a month working in the chemistry labs at Glasgow.

Part of the manoeuvering also included a four and a half month trip to Paris after graduation in 1845. In Paris he attended lectures on chemistry and physics at the Sorbonne, and on the further advice of Dr William Thomson, bought and studied French texts. Through introductions provided by Scottish academics like J. D. Forbes of Edinburgh and Sir David Brewster of St Andrews he met eminent men like Cauchy and Biot, and through Biot he was introduced to Victor Regnault, who was the Professor of Natural Philosophy at the Collège de France. Regnault was happy for Thomson to assist in his laboratory, which at times amounted to mundane tasks such as working the air pump, or stirring water in a calorimeter. However, Thomson made the most of the opportunity, spending from eight in the morning to five or six at night in the evening helping with experiments, making use of the library of books, and observing what he described much later as Regnault's 'faultless technique, a love of precision in all things, and the highest virtue of the experimenter—patience.'[14]

Back in Glasgow, Dr James Thomson was pleased to hear about his son's progress. Not only was he gaining practical experience, but he was meeting men who may well be able to write him influential testimonials in the near future. Not all, however, were necessarily looking forward to the prospect of William Thomson joining his father as a professor. James was a man who sought change at Glasgow, and that change irked some of his colleagues; appointing Thomson would mean one more professorial vote in support of it. In a remark to one of his children Dr James Thomson commented that at least one of his colleagues would rather see Satan appointed to the Chair of Natural Philosophy than have William take the post.

With due, though doubtless unintentional, timing Professor William Meikleham died on 6 May 1846, and from that moment James Thomson's covert manoeuverings turned to overt action. He wrote immediately to his son 'The enclosed notice [of Meikleham's death] must put you into active and energetic motion without delay'[15]. On the 26 May, Thomson wrote letters to each of the electors announcing his wish to be a candidate for the post, and then a long list of testimonials were collected to swamp and impress the Glasgow professors, and impressed they could not fail to be. The list reads like a selection from a *Who's Who* of nineteenth century science including George Boole, Arthur Cayley, James D. Forbes, Sir William Rowan Hamilton, Joseph Liouville, Agustus De Morgan, Victor Regnault, George Gabriel Stokes, James Joseph Sylvester, and William Whewell.

When it came to the election none of Dr. James Thomson's fears were realized—possible candidates who James had fretted over, and who would have been genuine competitors to his son, simply did not apply. On the 11 September 1846, William Thomson, then aged 22, was unanimously elected to the Chair of Natural Philosophy at Glasgow. He held the post until 1899, and could not be induced to leave even by the Cavendish Chair at Cambridge, which was offered to him three times, on its creation in 1870, again on Maxwell's death in 1879, and once more when Lord Rayleigh vacated it in 1884.

THE GLASGOW PROFESSOR

Thomson was clearly unhappy with his initial performance at Glasgow. To be confirmed in post he had to read an essay in Latin on the prescribed title *De caloris distributione per terrae corpus* before the faculty. The faculty 'unanimously expressed their satisfaction with the trial essay'[16] when it was read to them on 13 October, but Thomson burnt the manuscript of the essay on the very day he delivered it. Then when he gave his first lecture to students on 4 November he wrote the next day to Stokes saying that it was a failure: it was all written down and he'd read it too fast. The letter which contained this comment is one of the first in a correspondence[17] between Stokes and Thomson which spans over 50 years, and the relationship between the two men is discussed by Alastair Wood elsewhere in this book.

However faltering Thomson felt his start was, his first year of teaching seems to have gone well enough and he became a very popular lecturer, with his sister Elizabeth commenting on his reception at the College prize giving ceremony on 1 May 1850:

'The prizes were distributed to-day, and William shewed to very great advantage. He was received with deafening cheers, beyond anything I ever witnessed on the noisy first of May. 'The young Professor' was cheered, and cheered again, and the students seemed never to tire of shouting 'Three cheers more for the young Professor!'[18]

Behind the boisterous students' cheers there is hard evidence for this early popularity. A set of lecture notes taken by a student in Thomson's 1849–50 Junior Natural Philosophy class shows well structured teaching.[19] Recalling his own student days in the early 1860s, David Murray[20] remembers Thomson as 'an enthusiastic and inspiring teacher' who 'dispensed the professorial gown after his introductory lecture and stood when lecturing, eager, alert and animated like a runner waiting for the starting signal'. As Thomson's recent biographers, Smith and Wise[21] note, his reputation for frequent digressions and lectures, which ranging far and wide from the subject in hand left students floundering, comes perhaps from a slightly later time when his rising profile in science and engineering made more and more calls on his time.

Perhaps one genuine weakness was that, according to Murray:

One of his peculiarities was that he assumed that the person he was talking to knew quite as much as he did.[22]

Or, more alarmingly for some, another old student, John Hutchison, recalled 'Sir William unfortunately believed that everybody could learn mathematics'.[23]

However, the fact that Thomson was a dynamic communicator with an infectious sense of enthusiasm is clear. During the 1882 British Association meeting at Southampton Thomson gave an evening discourse, and one member of the audience that night recalled:

> 'Sir William Thomson's lecture ... which was given to a large audience, was good for all who understood it. But Thomson himself was splendid; he danced about the platform in all directions, with a huge pointer in his hand; he shook in every fibre with delightful excitement, and the audience were as delighted as he.'[24]

It captures Thomson as a teacher perfectly—not everyone may have understood, but it was hard not to be carried along with the lecturer's enthusiasm. The point is underlined even further by the fact that the title of Thomson's talk that night was what one feels to be the somewhat unpromising topic of the tides.

Thomson would bring the excitement of new discovery to lectures at Glasgow. Thus, for example, his verification of his brother's prediction of the lowering of the freezing point of water under pressure was announced to the class in 1850. Or again, recalling an experience in 1859 Thomson writes:

> One Friday morning I had been telling my students that we must expect the definite discovery of other metals in the sun besides sodium, by the comparison of Fraunhofer's solar dark lines with artificial bright lines. The next Friday morning I brought Helmholtz's letter with me into my lecture and read it, by which they were told that the thing had actually been done with splendid success by Kirchhoff.[25]

CREATION OF THE PHYSICAL LABORATORY

On opening the Physical and Chemical Laboratories of University College North Wales in 1885 Thomson recalled that:

> When I entered upon the professorship of Natural Philosophy at Glasgow, I found apparatus of a very old-fashioned kind. Much of it was more than a hundred years old, little of it was less than fifty years old, and most of it was worm eaten. Still, with such appliances, year after year, students of natural philosophy had been brought together and taught as well as possible ... But there was absolutely no provision of any kind for experimental investigation, still less idea, even, for any kind of students' practical work. Students' laboratories for physical science were not then thought of.[26]

Thus, shortly after his appointment, Thomson approached the faculty about the need for new equipment, and found himself knocking at an open door. J. P. Nichol, the Professor of Astronomy (who in covering Meikleham's Natural Philosophy classes in 1839–40 had taught Thomson), had already recommended to the faculty that there was a need to bring the apparatus up to date. The result was that Thomson was given a grant of £100 and a grandly named *Natural Philosophy Class-Room and Instrument Committee* was set up to oversee expenditure. In the summer of 1847, during a visit to Paris, Thomson ordered apparatus, and while in London he went to various instrument makers with the hope of ordering even more. By either happy accident, or shrewd design, he had spent only £80 of the allocated £100 when the Class-Room and Instrument Committee reported in November 1847 and they recorded their pleasure glowingly at 'the caution and ceremony with which the purchase and selection of the valuable instruments was carried into effect'.[27]

The result was that the committee recommended that the Professor be given a further £150—adding (presumably by use of 'creative accounting') that this could be considered as equivalent to only £130 given the previous year's under-spend. A further £150 was duly given and over the first five years of his appointment the Faculty gave Thomson a total of £550 for new apparatus. The apparatus bought was for lecture demonstrations, that is, standard teaching.

However in the realm of experimentation, it was in the creation of a physical laboratory for students to work in where Thomson's proved himself to be highly innovative. The creation of the laboratory can be dated to around 1850. In 1846 Thomson had begun with a fiefdom which was made up of a classroom, an apparatus room, and use of the college clock tower, but around 1850 the Faculty gave him use of an old wine cellar and then around 1858 he annexed an examination room which had been left unused, asking Faculty to sanction the action after the event. Eventually students could be found working in all of these spaces, from clock tower to cellar. Although the gaining of the old wine cellar dates to around 1850, it seems that Thomson was involving students in experimental work before that date.

Again from his address on the opening of the laboratories at University College North Wales Thomson recalls:

Soon after I entered my present chair in the University of Glasgow in 1846 I had occasion to undertake some investigations of electrodynamic qualities of matter, to answer questions suggested by the results of mathematical theory ... The labour of observing proved too heavy, much of it could scarcely be carried out without two or more persons working together. I therefore invited students to aid in the work. They willingly accepted the invitation, and lent me most cheerful and able help. Soon after, other students, hearing that their class-fellows had got experimental work to do, came to me and volunteered to assist in the investigation.

I could not give them all work in the particular investigation with which I had commenced—'the electric convection of heat'—for want of means and time and

Fig. 1.1. The north-west corner of Old College of Glasgow became Thomson's kingdom. Along the ground floor to the right of the leftmost (angular) turret are three windows onto a large part of his laboratory (which had previously been an examination room), from there rounding the corner to the next turret was his classroom. Above the laboratory was an apparatus room. The college clock tower was also used for experiments. (From Andrew Gray, *Lord Kelvin: An Account of His Scientific Life and Work*, J. M. Dent & Co., London (1908), opposite p.70.)

possibilities of arrangement, but I did all in my power to find work for them on allied subjects (Electrodynamic Properties of Metals, Moduluses of Elasticity of Metals, Elastic Fatigue, Atmospheric Electricity, etc.). I then had an ordinary class of a hundred students, of whom some attended lectures in natural philosophy two hours a day, and had nothing more to do from morning till night. These were the palmy days of natural philosophy in the University of Glasgow.[28]

Thus began pragmatically what was to be, as Thomson would proudly tell his students in later years, the first laboratory of its kind in Great Britain. Thomson involved the best students in more serious work and often employed a student over the summer months. He also employed a fulltime assistant, Donald MacFarlane. By at least the early 1860s MacFarlane had effective control over the day to day running of the laboratory, with Thomson giving him instructions as to what investigations he wanted carried out and MacFarlane distributing the work amongst the students. Thomson, however, was always keen to know how work was proceeding, and in later years when he was increasingly away from Glasgow he would keep in contact by post, giving directions and expecting updates to be sent to him by return, indeed sometimes if impatience got the better of him he would communicate by telegraph.

EARLY DAYS OF THE YOUNG PROFESSOR

In 1847 Thomson came across the Mancunian, James Prescott Joule, at the meeting of the British Association in Oxford. Joule had spoken at the British Association in Cork in 1843, giving an experimental value for the amount of mechanical work required to raise one pound of water through one degree Fahrenheit. He spoke again in Cambridge in 1845, updating the accuracy of his previous work and suggesting that water at the bottom of Niagara falls should be 0.2°F warmer than the water at the top, and also suggesting a value for the zero of temperature to be 480°F below the freezing point of water (–248°C). The British Association, however, greeted Joule's work with the silence of disbelief and apathy. Thomson recalled one scientist saying 'he did not believe in Joule because he had nothing but hundredths of a degree to prove his case by'.[29]

But in 1847, in Oxford, William Thomson was in the audience, and although incredulous of some of Joule's work, apathetic he was not. He was very interested. Joule recalls that Thomson asked questions from the floor and provoked debate, Thomson recalls the matter slightly differently, saying he waited until afterwards to question Joule. Whatever the truth of the matter, the meeting was a significant one. Within a few days Thomson was writing to his father about how enjoyable the Oxford BA had been and remarking:

'Joule is, I am sure, wrong in many of his ideas, but he seems to have discovered some facts of extreme importance.'[30]

But Joule was not wrong, and through careful thought Thomson came to agree with him. Along the way he reconciled Joule's work with that of Carnot on heat engines; he devised a more fundamental way to define the absolute zero of temperature; saw the idea of energy as a great unifying principle; and introduced the ideas of statical and dynamical (or what is now called potential and kinetic) energy. Iwan Rhys Morus (chapter 8) tells the story of Thomson and Joule's interaction and the evolution of Thomson's ideas.

However it should not be thought that Thomson stands as sole and conquering developer of thermodynamics. As Andrew Whitaker (chapter 16) points out in his retrospective on Thomson's enduring significance, many scientists were involved in the creation of the new science, with Joule, Rudolf Clausius, Justus von Liebig, Hermann von Helmholtz, Julias R Mayer, and Thomson's colleague at Glasgow, W. J. Macquorn Rankine all playing important roles. All contributed to the development of thermodymanics, all deserve honourable mention. Thomson can certainly be seen as performing an important role in unifying the emerging subject, and along with Peter Guthrie Tait in their *Treatise on Natural Philosophy*[31]—discussed in Raymond Flood's chapter—he can be seen as one who aimed to re-centre natural philosophy around the principle of conservation of energy.

Thomson had been elected a Fellow of the Royal Society of Edinburgh in February 1847, but on 6 June 1851, just before his 27th birthday, his rising scientific star was recognized by his election to the more august Royal Society of London. His election came on the same day as that of his lifelong friend G. G. Stokes and future adversary T. H. Huxley.

Chronology

A brief list of Kelvin's life, achievements and awards	
26th June 1824	William Thomson born
October 1834	Matriculates at Glasgow (aged 10)
May 1840	Reads Fourier's *Théorie Analytique de la Chaleur* in a fortnight (aged 15)
October 1840	Enters St Peter's College, Cambridge (aged 16)
May 1841	First paper published in the Cambridge Mathematical Journal (aged 16)
January 1845	Second wrangler and first Smith's Prizeman (aged 20)
June 1845	Elected Fellow of St Peter's College, Cambridge (aged 21)
September 1846	Unanimously elected to Chair of Natural Philosophy in Glasgow (aged 22)
February 1847	Elected Fellow of the Royal Society of Edinburgh Served as President 1873–8, 1886–90, 1895–1907. RSE was one of the first of over 100 learned societies across the world to elect Thomson to their ranks.
June 1851	Elected Fellow of the Royal Society of London (aged 26) Elected on the same day as Huxley and Stokes. Served as PRS 1890–5.
1857	Honorary Doctor of Laws, Dublin University. This is the first of 21 honorary doctorates which Thomson was awarded.
November 1866	Knighted by Queen Victoria (aged 42) Takes as his motto–Honesty is the best Policy
January 1892	Elevated to Peerage Takes as his motto–Honesty without Fear
June 1902	Awarded the Order of Merit Thomson was an inaugural member of this order and was awarded the honour on his 78th birthday.
August 1902	Made Privy Councillor
April 1904	Elected Chancellor of Glasgow University
17th December 1907	Dies. Buried beside Newton in Westminster Abbey on 23rd

In September 1852, aged 28, Thomson married his second cousin Margaret Crum. Thomson's official biographer, Silvanus P. Thompson,[32] weaves Margaret gently into his early narrative so that it comes as no huge surprise when we find them engaged. However, in Glasgow at the time there was at least one person who was extremely surprised to hear of the engagement: Sabina Smith had rejected Thomson's proposals of marriage on no less than three occasions—just before New Year 1850, in April 1851, and for the last time in April 1852, but by 13 of July 1852 Thomson was engaged to Margaret. In a sorrowful sidelight on history, Smith and Wise[33] record Sabina's lifelong regret at having rejected her suitor once too often. Writing to her sister in December 1891, nearly 40 years after the event, the regret sounds as fresh as if she had been writing in the summer of 1852 when she discovered he was to marry another instead:

> I suppose you must know that I have regretted all my life having refused him? . . . it was the extremity of folly to think I c[d] go on refusing a man, & yet have him at my disposal whenever I choose! . . . certainly *he* was not to blame.[34]

CABLES AND CONTROVERSY

By the 1850s Thomson had made his mark in pure science, and by the end of 1854 he had over 80 publications to his name on topics ranging wide over the broad areas of electricity, magnetism, and heat[35]. A paper read before the Royal Society in 1855, entitled 'On the theory of the electric telegraph', however, gives an indication of a new direction in his work. By 1850 telegraphy over land was already established, and in 1851 a submarine cable had been laid between Dover and Calais. By the early 1850s the possibility of a connection across the Atlantic was being thought of, but the prospect of laying a cable from Britain to America was recognized as a huge engineering project.

Thomson initially became interested in telegraphy via the problem of how an electrical signal was distorted as it was transmitted along a long telegraphic cable (essentially as cable length increased short sharp pulses were smoothed out into longer pulses rising to and falling from a gentle maximum), but by December 1856 he had effectively launched into a completely new area of work when it was announced that he was to be one of the directors of the newly formed Atlantic Telegraph Company. Although Thomson was not employed in any technical role, he increasingly became involved in technical issues, and when, at the last minute the chief electrician Mr. O. E. W. Whitehouse claimed that due to ill health he could not join the ship on the first cable laying expedition, it was Thomson who went on board instead. This first attempt to lay a transatlantic cable was a failure, and it was only in 1866, after multiple failures, that a successful connection was finally made. By the time the cable had been laid Thomson, now aged 42, had spent many months at sea, and had become intimately and enthusiastically involved with the project. For his labours he, along with others who had been involved in the cable project, was knighted in November 1866.

The telegraph work gave rise to a number of patents, one of which was for his mirror galvanometer, which was capable of detecting the small currents associated with signal transmission along the cable. His patents, his involvement in other cable laying projects—including the laying of the French Atlantic cable in 1869—and his formation of a firm of consulting engineers made him a very wealthy man. His engineering activities are discussed further in Sir Bernard Crossland's chapter.

Thomson's views on the age of the Earth brought him into conflict with some of his scientific contemporaries and in particular with Thomas Henry Huxley. Thomson's interest in the topic went back as far as work he published as a Cambridge undergraduate in 1844; indeed in a conversation with Silvanus P. Thompson in 1906 he claimed it went back further still. Pointing to line in a paper he published in 1842 he said:

> It was this argument from Fourier that made me think there must have been a beginning... Trace back the past, and one comes to a beginning—to a time zero beyond which all values are impossible. It's all in Fourier.[36]

Thomson used Fourier's techniques to estimate the Earth's age by considering it to be a cooling sphere, but as Patrick Wyse Jackson points out (chapter 10), this was only one of three approaches to the problem of the age of the Earth which he took.

Fig. 1.2. In the prime of his age for invention: a picture of Thomson taken by his brother-in-law, David King, with one of the earliest cameras. The note at the bottom of the picture, written by Thomson in 1892, reads *W Thomson reading a letter or letters from Fleeming Jenkin, about experiments on sub-marine cables probably about March 1859.* (From Agnes Gardiner King, *Kelvin the Man*, Hodder & Stoughton Ltd., London, 1925, frontispiece.)

He also used the effect of tidal friction slowing the rate of rotation of the planet, and estimates for the age of the Sun to help set limits. He published more than 20 papers on these matters during his life, with about half of them appearing in the 1860s. In Darwin's words, uniformitarian geology gave a past which was[37] 'incomprehensibly vast'. Thomson's conflict with geologists, and to a lesser degree with supporters of Darwin's theory of evolution, centred on his contention that uniformitarianism was patently false, and his calculations set limits on the age of the earth, of about 100 million years, which were much too low for their liking.

In February 1869 Huxley gave an address to the Geological Society of London where he sought to neutralize Thomson's attacks. He parried the issue of uniformitarianism by claiming that:

> I do not suppose that, at the present day, any geologist would be found to maintain absolute uniformitarianism ... If Hutton and Playfair declare the course of the world to have been always the same, point out the fallacy by all means, but in so doing, do not imagine that you are proving modern geology to be in opposition to natural philosophy.[38]

When it came to Thomson's calculations he retorted:

> Mathematics may be compared to a mill of exquisite workmanship, which grinds you stuff of any degree of fineness; but, nevertheless, what you get out depends on what you put in; and as the grandest mill in the world will not extract wheat-flour from peascods, so pages of formulae will not get a definite result out of loose data.[39]

Huxley had spoken with typical bravado and wit, but he was no match for Thomson. He could not refute Thomson's 'pages of formulae' and less than two months later in an address to the Glasgow Geological Society Thomson weighed and dissected Huxley's wit and arguments, and found them wanting.

It is tempting, with the imperialism of modern physics and geology, to smile upon the debate now. This is a case study from history where the story could all too simply be concluded with 'Of course now we know that Thomson was wrong'. But that would be as foolish as criticizing his work on the Atlantic telegraph cable on the grounds that we now possess mobile phones. Thomson used the best science and data available to him at the time, and more importantly his work on the age of the Earth shows how he creatively and enthusiastically brought the power of mathematics and natural philosophy to bear upon new disciplines.

SAILING IN NEW DIRECTIONS

On June 17th 1870, Thomson's wife Margaret died. She had been ill for virtually the whole of their married life. Three months after her death Thomson bought, with some

Fig. 1.3. With the increasing wealth which came from his involvement in the laying of submarine telegraph cables, Thomson bought a 17 year old, 126 ton, oak built yacht—the *Lalla Rookh*. (From Silvanus P. Thompson, *The Life of William Thomson Baron Kelvin of Largs*, 2 volumes, Macmillan & Co., London, 1910, opposite p.616.)

of the new wealth that his cable work was bringing him, a 17 year old, 126 ton, oak built yacht—the *Lalla Rookh*. The long months on board ships when laying telegraph cables had clearly given Thomson a love of the sea. In a letter to Tait in 1873 he writes:

> My desk at the NPL [Natural Philosophy Laboratory] and the L. R. [*Lalla Rookh*] are the only places in the world for which I am fit.[40]

Thomson would spend large parts of the six months between Glasgow University teaching sessions (which ended on the 1st of May, and recommenced in November) on his yacht.

In spring 1871 Thomson attempted to convince Tait, Maxwell, Helmholtz, Huxley, and Tyndall to come cruising with him that summer, in the Western Isles of Scotland, after the British Association meeting in Edinburgh. It seems that all but Helmholtz demurred, with Tait claiming that he did not mind sailing, but would much rather be playing golf. A few days before the trip Helmholtz met Tait in St Andrews and in a letter home to his wife described golf as:

> a kind of ball game, which is played on the green sward with great vehemence by every male visitor and by some of the ladies: a sort of ball game in which the ball lies on the ground and is continually struck by special clubs until it is driven, with the fewest possible blows, into a hole, marked by a flag ... Mr. Tait knows of nothing else here but golfing ... W. Thomson must be now just as much absorbed in yachting as Mr. Tait in golfing.[41]

Indeed he was—but as ever Thomson's new interest was an opportunity for new discovery, new work, new enthusiasm. In the same year that Thomson bought *Lalla Rookh* he began to publish and speak on nautical matters, and over the 1870s and 1880s he gave over 50 papers and presentations on matters such as the tides, compass design, lighthouse signalling, depth sounding, and the behaviour of water waves in deep sea, with further papers not directly on seamanship, but on allied issues such as the magnetization of iron (which was relevant for nautical compasses on board iron ships). A sequence of patents was also taken out, for the mariner's compass, and depth sounding apparatus (which allowed accurate soundings to be taken quickly and without having to stop the ship), and improvements thereon.

Thomson's compass was more stable than existing ones, compensated for the effects of modern ironclad ships, and performed well in tests. However even though by 1878 German, Russian, and Italian Admiralties had all ordered Thomson's compass, the British Admiralty were only interested in letting it even be tested on one of Her Majesty's Ships if Thomson would pay for it. It was not until 1889 that the British Admiralty agreed to its adoption.

His work on tides resulted in a mechanical tide predictor which was capable of graphing the tides for any port for a full year in advance. It was based on a mechanical integrator which had been developed by his brother James (who by 1873 had joined Thomson back in Glasgow as Professor of Engineering, having previously held the chair of Civil Engineering at Queen's College, Belfast from 1854).

In 1873 Thomson's work on laying submarine telegraph cables took him to Madeira. Due to a fault in one of the cables, the ship had to remain off the coast of the island for some 16 days, and during this time Professor Sir William Thomson was entertained by the island's wealthiest landowner, Mr. Blandy. The next year on 2 May, the day after the end of Glasgow's teaching session, Thomson set sail again for Madeira, this time in the *Lalla Rookh*, and by 11 May he was engaged to Mr. Blandy's second daughter, Fanny. They were married in the British Consular Chapel on the island on the 24 June, just before Thomson's 50th birthday. Fanny was 36.

He had departed from Glasgow on 2 May leaving instructions that a new heating stove should be put in the house to make it warmer and more like Madeira, but soon after he returned in the autumn he bought land and built a new home, Netherhall, at Largs about 25 miles west of Glasgow. Thomson, aided by his brother, oversaw much of the design and building. It was an impressive baronial style home, complete with peacocks for the grounds supplied by James Clerk Maxwell. Fanny was a very different wife to Margaret. Margaret, almost permanently ill, languishing at home, writing poetry about sadness and death, was replaced by Fanny, who enjoyed sailing and loved to entertain and see the house filled with guests. And filled with guests it often was.

Margaret Ethel Gladstone, Thomson's grand niece writes of staying at Netherhall in 1892:

generally the house was full of company; 24 people stayed in it while I was there, and endless people came to dinner and tea and lunch. Aunt Fanny likes company very much, and as for Uncle William it doesn't seem to make much difference to him what happens; he works away at mathematics just the same, and in the intervals holds animated conversations with whomever is near . . . mathematics went on vigorously in the 'green book.' That 'green book' is a great institution; there are a series of 'green books'—really note books made specially for Uncle William, which he uses up at a rate of five or six a year, and which are his inseparable companions. They generally go upstairs, downstairs, out of doors and indoors, wherever he goes, and he writes in his 'green book' under any circumstances. Looking through them is quite amusing; one entry will be in the train, another in the garden, a third in bed before he gets up; and so they go on at all hours of the day or night. He always puts the place and exact minute of beginning an entry. It gives one a rather common-sense way, or scientific way, of looking at things to stay with Uncle William: you begin to feel that everything has a reason, and that that reason may be found out; and things should not be slurred over or left to chance when you can direct them by taking a little trouble and using a little thought.[42]

Thomson's 'green books' were indeed an institution. He had started using them after a fall on ice in December 1860. A fracture was not diagnosed until a week after the fall, and he spent many weeks in bed, and was left with a limp for the rest of his life. While bedridden he started to use a green-backed notebook. The books (of which there are over 100) became his constant companion, with him having a special pocket sewn into his coats to keep them in, and his habit of apparently submerging himself in work in the middle of a social gathering was one which he had cultivated for a long time.

Back during the cruise in *Lalla Rookh* in the summer of 1871, Helmholtz wrote to his wife, of a visit to meet Professor Blackburn (who had been a contemporary of Thomson's at Cambridge and had replaced Dr James Thomson as the Professor of Mathematics at Glasgow after his death in 1849) and his family at their holiday home in the Western Highlands:

W. Thomson presumed so far on the freedom of his surroundings that he always carried his mathematical note-book about with him, and as soon as anything occurred to him, in the midst of the company, he would begin to calculate, which was treated with a certain awe by the party.[43]

In 1884 Thomson, who was now aged 60, went with his wife to North America. The British Association was meeting on Montreal, and from there they went to Johns Hopkins University in Baltimore, where Thomson was to deliver a set of lectures. Johns Hopkins were eager to have such a giant of physics in their midst: Thomson could lecture on whatever he wished, and the university would not only cover travel expenses, but also give him £400. The lectures commenced on 1 October and the 20th lecture finished the series on 17 October. Thomson, it appeared, had not prepared in advance; he certainly did not have any notes. He lectured apparently *ex tempore* and with an *ex cathedra* authority and command. Lord Rayleigh, who along with men like Albert A. Michelson and Edward W. Morley made up the regular and intimate audience of 21 people, remarked:

> What an extraordinary performance that was! I often recognized the morning's lecture was founded on the questions that had cropped up when we were talking at breakfast.[44]

Thomson spoke on light and molecules. The ether was like Scottish shoemaker's wax—which is brittle enough to be shaped into a tuning fork and made to vibrate, but fluid enough to let a massive object like a lead bullet gradually fall through two inches of it over a period of several months. Mechanical models constructed from springs and rods and concentric metal spheres were investigated to explain various aspects of molecular behaviour. It was a tour de force. The lectures were seen as a master class for the elite attending academics, but Andrew Gray, Thomson's eventual successor to the chair of Natural Philosophy at Glasgow, commenting on the published form of the lectures[45] notes:

> The book reads in many places like a report of some of the higher mathematical lectures which were given every session at Glasgow; and on that account, if on no other, it will be read by the old students of the higher class with affectionate interest.[46]

John Roche and Elizabeth Garber discuss Thomson's modelling activity in the areas of atoms and the magnetic field in their respective chapters.

THE ESTABLISHED SCIENTIST

On New Year's day 1892 Sir William Thomson was raised to the peerage. Within a few days it was decided he would take the name Kelvin, after the Kelvin river which ran by the new buildings of Glasgow University. The British Medical Journal wrote[47] 'A peerage has at length been conferred upon a scientific man because he is a scientific man', but Thomson's political activity in support of Liberal Unionism had also played a role. Indeed, Thomson's recent biographers, Smith and Wise[48], argue that his political activity over the six years previous to 1892 were crucial to the offer of the peerage being made. Thomson was by now fulfilling the many roles of an elder statesman of science. He opened new electric light works, factories, and science labs: at the opening of a new school science laboratory in 1893 he delighted pupils, and doubtless irked classics masters, by stating that translations should be used more freely to allow more time for teaching science.

The year 1896 saw Glasgow University celebrate the 50th Jubilee of its most distinguished professor. There were 2500 guests and the celebrations ran from 15 to the 17 June. There was an exhibition of not only all of Thomson's inventions, but also of the medals he'd been awarded[49] and certificates of membership of the 80[50] or so learned societies across the world which had elected Thomson to their ranks over the years. A temporary telegraph station was set up and congratulatory messages, whether by telegraph or as written addresses, came from as far away as Moscow and Toronto, Bombay, and Tokyo.

In 1899 Thomson, aged 75, finally retired from his chair at Glasgow. He had been Professor of Natural Philosophy for 53 years, and true to his love of the university and unfailing enthusiasm to be always learning, he insisted on enrolling as a research student.

The new century brought more honours. In 1902, on his 78th birthday, he was one of the inaugural group of people to be awarded the newly instituted Order of Merit by King Edward VII. He was made a Privy Councillor in the same year. In 1904 he was elected as Chancellor of Glasgow University. Up to the end of his life he was active and busy, still attending the British Association meetings, still writing and speaking, but finding himself increasingly detached and in disagreement with

Fig. 1.4. Lord Kelvin's last lecture as professor of Natural Philosophy at Glasgow in 1899. (From Silvanus P. Thompson, *The Life of William Thomson Baron Kelvin of Largs*, 2 volumes, Macmillan & Co., London, 1910, opposite p.1011.)

the new scientific hegemony. A recent biographer of Thomson, alluding to this, subtitled his work[51] 'The Genius and Tragedy of William Thomson'. But there was no tragedy; it was just old age. The fresh brilliance of genius, and Thomson was most assuredly that, can only burn for so long.

After a short illness Thomson died, at Netherhall, on 17 December 1907 and on a foggy 23 December he was buried in Westminster Abbey beside Isaac Newton. The choir sang, the candles in the abbey flickered, and William Thomson, Baron Kelvin of Largs, was gone.

2

Educating William: Belfast, Glasgow, and Cambridge

Alex D. D. Craik

EARLY EDUCATION

To appreciate the education of William Thomson, it is necessary to know something of his father, James Thomson, whose influence was immense. James Thomson was born in 1786, the son of a farmer, and the youngest of five children. The farm, called Annaghmore, was near Ballynahinch, Co. Down, in the north of Ireland. The family was descended from seventeenth-century Scottish settlers. Retaining their Presbyterian religion, this large contingent of immigrants, sometimes known as 'Ulster Scots', formed a distinct strand of Irish society. Few were rich, and many later emigrated to the USA.

James Thomson had little formal schooling, receiving his early education at home from his elder sisters, and at a day school run by the Reverend Samuel Edgar. A dilapidated copy of Bonnycastle's *Arithmetic* gave him his first taste for mathematics. With Edgar, he studied mathematics and classics in preparation for entering Glasgow University. This he did in 1810 at the age of 24. The academic session began on 1 November and ended on 30 April, when most students returned home to work to help finance their studies. James' employment was then as an assistant at Dr. Edgar's school, where he gained valuable teaching experience. After graduating at Glasgow as M.A. in 1812 he attended two further sessions, to study medicine and theology, with the intention of becoming a minister of the Presbyterian church.

Instead, he was appointed in 1814 as a teacher of arithmetic, mathematics, and geography at the newly-founded non-sectarian Belfast Academical Institution. This soon evolved into a college for more advanced study with attached preparatory schools; and in 1815 James was appointed as the college's Professor of Mathematics while retaining his school mastership. The Institution received the appellation 'Royal' in 1831.

Fig. 2.1. Professor James Thomson, William Thomson's father, aged about 61, drawn by his daughter Elizabeth. (Courtesy of the National Portrait Gallery, London.)

In 1817, James married Margaret Gardner, daughter of a well-to-do Glasgow merchant. They had seven children: Elizabeth (born 1818), Anna (1820), James (1822), William (June 26th, 1824), John (1826), Margaret (1827), and Robert (1829). Both Margaret and Robert were sickly and seemed unlikely to survive. Margaret became mentally impaired and died at the age of three; but Robert, despite bouts of severe pain in childhood, recovered and lived until 1905, having emigrated to Australia around 1850. Their mother, Margaret, was very ill following Robert's birth, and remained an invalid until her death in May 1830.

The family as a whole was not blessed with good health, but William was the exception. James (Jr.) was forced several times to interrupt his engineering career to recuperate at home for long periods. He seems to have suffered from a range of complaints. John, following success at Glasgow College (as the university was then popularly known), was well set on a medical career, but died aged just 20, of fever caught when working in hospital. Anna died in 1857, aged just 37. Elizabeth, too, had indifferent health although she lived until 1896. In 1848 her health broke down; and, to aid her recovery, she and her husband, the Rev. David King, went to Jamaica for the winter.[1]

Margaret Gardner's only sister Agnes had helped to look after the children since Robert's birth, and apart from one brief period, she remained as the family's housekeeper in Belfast and later in Glasgow, until William himself married in 1852. Agnes had made a disastrous marriage to one Arthur Gall: she returned to him during 1834–36, but the reconciliation did not last for long.[2]

In December 1831, James Thomson was appointed Professor of Mathematics at Glasgow College in succession to his old teacher James Millar. Millar had been professor there since 1789, but had become ineffectual and unable to maintain discipline in his classes: he resigned when charged with incompetence in the Scottish

Fig. 2.2. James Thomson built the two creeper-clad houses to the left of the drawing, in College Square East, Belfast. The one with the prominent arched doorway was William Thomson's birthplace. The drawing was made by William's niece Agnes Gardner King in 1909. The site of the house, now demolished, is today marked by a plaque. (From Elizabeth Thomson King (ed.) *Lord Kelvin's Early Home*, Macmillan & Co. Ltd., London (1909), opposite p.21.)

University Commission's report of 1830. But, unknown to James Thomson when he accepted appointment, Millar had negotiated to retain his professorial salary in lieu of a pension. Though Thomson, like the other professors, was provided with a house, he discovered that his only income was students' fees, then at a low ebb as his classes were small. He responded to the challenge by offering afternoon lectures on geography and astronomy for young ladies, and he soon built up the numbers taking his regular mathematics classes. Later, he received his full professorial salary; but it has been suggested that his decision to educate James (Jr.) and William at home, rather than send them to school, was on account of this temporary financial difficulty. In contrast, both the younger sons, John and Robert, attended Glasgow Academy after receiving elementary instruction from their elder sisters.[3]

While still in Belfast, James Thomson wrote several successful textbooks on arithmetic, trigonometry, differential and integral calculus, geography, and Euclid's *Elements* of geometry. Several were reissued in later editions: his *Treatise on Arithmetic, In Theory and Practice* was particularly popular, and his *An Introduction to the Differential and Integral Calculus*, which first appeared in 1831, was among the first such texts in English to adopt the continental 'Leibnitzian' notation.[4]

James Thomson's education of his family was thorough, and his sons James (Jr.) and William were particularly apt and precocious pupils who spurred each other on. Within the close-knit family, William was his father's special favourite. In the words of his sister Elizabeth, published after her death in *Lord Kelvin's Early Home*:

> He [James] was indeed both father and mother to us, and watched over us continually. William was a great pet with him—partly, perhaps, on account of his extreme beauty, partly on account of his wonderful quickness of apprehension, but most of all, I think, on account of his coaxing, fascinating ways, and the caresses he lavished on his 'darling papa'.... Sometimes the others thought that there was a little affectation in this, especially when he used baby language after he could speak quite well; and we laughed at him, but he never heeded...
>
> I do not remember that any of us were ever in the slightest degree jealous of William on account of our father's making him a little more of a pet than the rest of us...[5]

Though James undertook much of their instruction, during 1830–31 James (Jr.) and William attended some classes at the Belfast Academical Institution. There, seven-year old William was awarded a first prize and nine-year-old James came second.[6] When the family moved to Glasgow a year later, the same pattern continued. William, James, and their older sisters studied Latin at home with their father; and the boys unofficially attended, as listeners, his Junior mathematical class in the university. Education was woven into family life. Elizabeth studied painting as well as Latin, and became a very proficient artist: her drawings of the family and of landscapes made during their holidays are vivid records of this time. She also taught her younger brother John, while Anna learned to play the piano and instructed young

Robert. When James returned from his morning duties in the university, the family dined together:

> After dinner our father gave us a short mathematical lesson, and after that he read aloud to us. During this winter he thus read the whole of Pope's *Iliad* and *Odyssey*, several plays of Shakespeare, those of Goldsmith and Sheridan, besides selections from the old poets. William had the strongest sense of humour of any of us, and not only enjoyed it himself, but set all the little party laughing mirthfully whenever a humourous passage occurred.[7]

In their second Glasgow year, James and William attended a class in Natural History, given by a Dr. Cooper, and each evening they passed on what they had learned to Elizabeth and Anna.[8]

AT GLASGOW COLLEGE

In November 1834, James and William together matriculated at Glasgow College: James was aged 12 and William just 10. Though young, they were not as precocious as this seems today: many entrant students were then aged 13 or 14, while some older entrants had worked for several years to finance their belated education. Their father, James Thomson, had been one such, finally entering university at the age of 24. Compared with his early struggles, the educational path of his sons was made smooth indeed. They first attended the humanity [Latin] class of Professor Ramsay; during 1835–36 they studied natural history and Greek; in 1836–37 they took their father's junior mathematical class, followed the next year by his senior mathematical class and by classes in logic. Finally, in their fifth session of 1837–38, they studied junior natural philosophy under William Meikleham, and chemistry under Thomas Thomson (no relative). This rather leisurely programme entitled them to the B.A. degree, just as their younger brother John was about to begin his studies at the college. Two years later, in 1840, James graduated M.A. in mathematics and natural philosophy, having taken additional classes.

During 1839–40, William took the courses in moral philosophy and senior natural philosophy, the latter mostly taught by the Professor of Astronomy, John Pringle Nichol, owing to Meikleham's illness. Then, in 1840–41, he took Ramsay's senior humanity class. In that session, he also assisted David Thomson, a new substitute lecturer for Meikleham, by preparing the experimental demonstrations that accompanied the lectures. During their time at the college, both William and James were awarded several prizes.[9]

William had attended Glasgow College for seven consecutive sessions, excluding previous unofficial attendances, and was still some months short of his

17th birthday. Doubtless, he could have completed these studies far more quickly; but it seems certain that his father did not wish him to leave home too early. Though entitled to the M.A. degree, William never graduated, perhaps believing that this might adversely affect his acceptance as an undergraduate in Cambridge.[10]

During the long summer breaks from May to October, James Thomson and his family moved away from the grimy city to the countryside of the Clyde valley. Their favoured destinations were a cottage at Knock Castle, near Largs, and another on the Isle of Arran. In these near-idyllic surroundings, James imparted his knowledge of geology, geography, and natural history, Elizabeth made sketches, and they went fishing, boating, and hill walking. Doubtless it was then that William acquired his love of sailing and the sea, that in later life became his main recreation and was characteristically integrated into his researches.

But these summer vacations did not separate the Thomsons from university society. Professor Nichol's family visited Arran, and his son became a close friend of the Thomson boys. The Meiklehams, too, were family friends, and had a cottage close to the Thomsons'. Professor William Meikleham, the Professor of Natural Philosophy, was remembered by Elizabeth Thomson a little unkindly:

> we had much friendly intercourse with him and his family. He was a good-natured, fat, little hunchback, with a very red face; and he had a fat, little, curly-haired black dog called Jura, that always toddled beside him ... His daughter was a short, stout, benevolent lady, who was extremely kind to us.[11]

His three sons were a little older than the Thomson boys, and they went boating together. In 1840 Professor Meikleham's son Edward and James Thomson (Jr.) together joined the Dublin office of a firm of civil engineers. Typically, Professor James Thomson had made careful enquiries before arranging for this, their first employment.

Clearly, with such family ties, the Thomson boys would have received more than the usual attention from the professors who taught them; but it is also clear that their talent was such that they required no special treatment. William Meikleham fell ill around 1840 and never recovered sufficiently to resume teaching. He did the Thomson family his greatest favour by living on until 1846: for, by that time, William Thomson was fully qualified to take over his post, and his father James had thoroughly prepared the ground for his election.

James Thomson was not only an effective teacher; he was also a shrewd participant in college politics. When he was first appointed, the Glasgow College hierarchy were mostly Tories (or, in Church of Scotland terms, 'Moderates'). But James was a confirmed Whig, and supported the 'Evangelical' wing of the Scottish Kirk. As early as 1818, the reform-minded Regius Professor of Chemistry, Thomas Thomson (no relative), had been responsible for establishing a chemical laboratory to train practical chemists; but his efforts to achieve more general improvements had largely failed, and he was nearing retirement just when James Thomson was most active.

The arrival of J. P. Nichol as Professor of Astronomy in 1836 brought to Glasgow an inspiring teacher and a political radical. A friend and correspondent of James Mill and John Stuart Mill, Nichol had high educational and philosophical ideals that struck a chord with James Thomson. In 1841, they were joined by (another) William Thomson, Professor of Medicine and a committed Whig. These three gradually obtained dominance over the Tory faction, as old professors were replaced by new.[12] By 1846, when his son William became a candidate for the Professorship of Natural Philosophy, James Thomson could rest assured that no Tory political influence could obstruct his well-laid plans.

In 1842, not one but two Glasgow professors faced financial ruin. The Professor of Logic, Robert Buchanan, had to sell most of his possessions following the collapse of the Renfrewshire Bank; and J. P. Nichol amassed such debts that he was declared bankrupt. It seems that the altruistic Nichol had overspent his own fortune to buy instruments for the Glasgow observatory, and as a result lost his own books and other belongings. Regarding Nichol's downfall, Anna Thomson wrote to her brother William:

> I am sure he must have many a heavy thought when he remembers the ruin he has involved himself and his family in ... I do feel very sorry both for him & Mrs Nichol and for his children. It will make a sad change for he will not be able to give them the advantages they might have had. How very differently papa has done for all of us![13]

James Thomson had himself known poverty, and was anxious to avoid the fate of his friend and colleague. He paid close attention to his own expenditure and that of his family. After William went to Cambridge, the two exchanged many letters about money: James urging his son to exert economies whenever possible and to avoid debt (unless to his college tutor), and William trying to account for and to justify all his expenditure. These financial lessons would have instilled in the young William an appreciation of the benefits of a steady income, and may well have helped to inspire the entrepreneurial business activities of his later life.

If James exerted the main influence on his son's education, and was particularly responsible for his thorough grounding in mathematics, both William Meikleham and John Pringle Nichol were crucial in exciting William's love of natural philosophy. Meikleham was the first teacher at Glasgow to introduce his students to Lagrange's *Mécanique Analytique* and Laplace's *Mécanique Céleste*. It is likely that only his best students would have tried to read these difficult works, and they probably did not progress far; but William Thomson certainly progressed further than most.[14]

After attending Meikleham's class of 1838–39, William continued his reading during the summer. In preparation for an essay on the 'Figure of the Earth', he studied French works by Lagrange, Laplace, Legendre, and recent English treatises by G. B. Airy and J. H. Pratt. It is no surprise that his essay won a medal. The theory of the 'Figure of the Earth' concerned the Earth's shape under the joint action of gravitation and the centrifugal force of rotation, assuming that the whole Earth is fluid.[15] This problem afforded a key illustration of the power of applied mathematics: the theory had originated with Newton, Maclaurin, and Clairaut, and was progressively improved by Legendre, Laplace, and Ivory. It formed a section of George Biddell Airy's *Mathematical Tracts* and was dealt with at length in John Henry Pratt's *Mathematical Principles of Mechanical Philosophy and Their Application to the Theory of Universal Gravitation* of 1836. Both these works were then studied by advanced students for the Mathematical Tripos at Cambridge. William's reading for his essay was an excellent preparation for his time at Cambridge.

Whereas Meikleham had promoted the older French mathematicians Lagrange and Laplace, J. P. Nichol espoused the more modern works of Fresnel on light and Fourier on heat. It was William Thomson's good fortune that Nichol replaced the ill Meikleham for the natural philosophy class of 1839–40. Though Nichol apparently did not claim to have mastered these works' mathematical technicalities, he was clearly aware of their importance to physics, and imparted his enthusiasms to his brilliant student. As soon as the session ended, William Thomson read Fourier's *Théorie Analytique de la Chaleur* from beginning to end: this work was to exert a major influence on him throughout his research career.

William's association with David Thomson during 1840–41 was also scientifically fruitful. David Thomson, no relation but a cousin of Michael Faraday, had been an outstanding Glasgow graduate. He went on to Trinity College, Cambridge, where in 1839 he gained only a lowly Tripos place as 21st Senior Optime: a disappointing performance attributed to illness. Among the books recommended by David Thomson for advanced reading were Poisson's *Mécanique* and Samuel Earnshaw's *Mechanics*, then used at Cambridge.

William later described himself as 'inoculated with Faraday fire', on learning from David Thomson about Faraday's ideas of lines of electrical and magnetic forces. As David Thomson would have included little of such advanced material in his first lecture course, the two must surely have discussed the topic privately. However, Thomson was at first sceptical of Faraday's ideas and critical of Faraday's lack of mathematics: only gradually did he come to appreciate the value of Faraday's insights, that later provided a major stimulus for his own researches on electricity and magnetism.[16]

The ready availability to William of advanced mathematical works, and his ability to read French at this stage, were advantages of his privileged situation.

In the summer of 1839, James had taken all his children on an extended trip to London and the continent. The stay in London included sight-seeing, theatre visits, and socializing with acquaintances; but also had the more serious purpose of a successful surgical operation for Robert (in the days before chloroform). When the patient had recovered, the family left for Paris on the first of July. After two weeks, James took his two daughters on a visit to Switzerland, leaving the four boys in Paris in the care of their 'old servant Jane' and a Madame Putois, who was employed to provide instruction in French. A 'Mr. Smith, a cultivated young Englishman' whom they had met on the way, also promised to keep an eye on them. The family finally returned to Glasgow on 21 September, spending another week in London en route. Their sister Elizabeth reported that the boys had made 'excellent progress in French'.[17]

William's knowledge of Fourier's work was soon put to good use when he encountered Philip Kelland's *Theory of Heat* (1837). Kelland was a fellow of Queens' College, Cambridge and had recently been appointed as Professor of Mathematics at Edinburgh University. During the summer of 1840, William took the book with him on another educational trip, taken by the Nichol and Thomson families, this

Fig. 2.3. Pencil drawing of James Jr. at the age of 16, by his sister Elizabeth. (Courtesy of the National Portrait Gallery, London.)

Fig. 2.4. Pencil drawing of William at the age of 16, by his sister Elizabeth. (Courtesy of the National Portrait Gallery, London.)

time to Germany. When he was supposed to be studying German, William was discovering that Kelland had failed properly to understand Fourier's methods, and had made unwarranted criticisms of many of Fourier's results. Though James was annoyed that his son had been failing to make the most of his opportunity to learn German, he was sufficiently impressed to suggest that William write a paper about the matter.

This paper, with the author unnamed, was sent by James to Kelland, and also to Duncan F. Gregory, who was editor of the recently-founded *Cambridge Mathematical Journal*. Kelland was naturally less than overjoyed to find himself upstaged; but he admitted his error, and a revised version was accepted for publication by Gregory. The publication of this paper under the pseudonym 'P. Q. R.' did not conceal William's identity for long. Kelland realized that he had been outdone by the young son of his Glasgow counterpart; while Gregory was doubtless happy to accept a paper critical of the man who, two years previously, had controversially defeated him for the Edinburgh chair. As for William, it announced his credentials in Cambridge mathematical circles, just as he was about to enrol as an undergraduate.

On their way back from Germany, the Thomsons stopped briefly in London, where James negotiated the above-mentioned engineering apprenticeship for James (Jr.) at the Dublin office of the firm of John McNeil. But young James soon had to return home, disabled by an injured knee and a general deterioration of health that kept him an invalid for several years.[18]

A STUDENT AT CAMBRIDGE

Much of the summer of 1841 was spent on Arran, where, according to Elizabeth, 'all the boys did a great deal of work under their father's supervision, and were very busy also working out their own ideas'.[19] In late October, William and his father set out for Cambridge, accompanied part of the way by Elizabeth, on her way to visit a cousin in Lincolnshire. This was still before the days of railways. The journey involved taking the mail coach from Glasgow to Carlisle, where they stayed overnight, then another coach to Hull. Disliking the look of a boat that would have taken them to Ely, they opted instead for more coach journeys to Cambridge.

James lost no time in visiting his contacts. He had arranged that William should join the small St Peter's College (now called Peterhouse), for he knew of the high reputation of William Hopkins as a 'Wrangler maker'. Looking ahead, he was also doubtless aware that this college admitted Scots to the fellowship, whereas fellows at some other colleges had to be residents of particular English counties. (In fact, Hopkins was not a fellow, being married; but he was a Peterhouse graduate and the college's mathematical lecturer, as well as Cambridge's leading private tutor.) Within a few years, Peterhouse was to become the preferred college of many able Scots, among them Peter Guthrie Tait and (for just a year) James Clerk Maxwell.[20]

Henry Wilkinson Cookson, the tutor and later the Master, personally showed William his apartments. He also invited James to dine with him at the Fellows' table, and introduced William to a student who took him to the students' table. William wrote to his sister of his good fortune 'in getting comfortable rooms at once, as most of the students do not get them till the second year'. Visits were made to Hopkins, James Challis (the Plumian Professor of Astronomy), and Duncan F. Gregory. Though Hopkins did not take first-year students as private pupils, William wrote to Elizabeth that 'he will examine me now and then, and supply me with a tutor when I need one'.[21] His first-year college tutors were Cookson and Frederick Fuller, who later became Professor of Mathematics at Aberdeen.

Before long, William became a friend of Duncan F. Gregory and Archibald Smith, both fellows of Trinity and both Scots. These two were running the *Cambridge Mathematical Journal*, to which William had already submitted papers. It was most unusual for college fellows to seek out first-year undergraduates, but William was no ordinary student: within just a few days of his arrival, it was predicted in his college that he would become the Senior Wrangler.[22]

His first letters extol the beauties of Cambridge. They also describe his refusal of services of various tradesmen touting for business, including a 'very cheap' offer of regular haircuts at much greater cost than he was used to. Soon, he tried out rowing on the Cam in a 'funny' for one or two oarsmen; but he did not at first join the college boat club, for 'the men connected with the club are generally a rather idle set'.[23]

Cookson's opening lecture, on Euclid, first outlined the university's ideas on education, which William immediately appreciated were directly opposed to those of the Whig-led Society for Diffusion of Useful Knowledge (SDUK). While the university sought to provide a liberal education for the sons of a social élite, who would in due course become leaders of society, the SDUK was trying to make education available to those classes who could not afford to attend university.[24]

James lost no time in writing to his son, urging him to exercise economy: 'You must keep up a gentlemanly appearance, and live like others—keeping, however, rather behind than in advance'. This was a regular theme, and he kept William's expenditure under close scrutiny. He also offered more general advice, such as 'never to quarrel with a man (but to waive the subject) about religion and politics'. William, for his part, often had to write to his father to ask for more funds, telling him that discounts could be secured by prompt payment.[25]

William's letters to his father and sisters are full of information of his doings. He reassured his father that 'I have gone to as few wine-parties as I possibly could, and at any to which I have gone there has not been the least approach to excess . . . I have given no wine parties, or indeed any parties yet, but I suppose I must return some of the invitations next term'.[26] He details one eminently respectable party, held by William Hopkins for his private students and prospective students:

> A few wrangling-looking men soon began to drop in, and a great many freshmen, or raw materials for manufacture. Any to whom I spoke said that they were going to read with Hopkins if, or as soon as, he would take them . . . Mr. and Mrs. Hopkins and a young lady sang some glees, and Mr. Hopkins asked all of us whether we performed on any instrument; and when he heard that we did not, he said he was very glad to hear it. After music, conversation, and looking at a great many beautiful prints, we adjourned into another room for supper, which was in very splendid style.[27]

In fact, Thomson *did* play a wind instrument, the cornopean or cornet, and he later became a founding member of the Cambridge University Musical Society (which was begun by Peterhouse students), playing the cornopean and French horn in its orchestra.[28] He had also started to wear spectacles to correct short-sightedness.

Though James comes across in his letters as over-protective and controlling, he would have been well aware of the bad reputation of some Cambridge students for extravagance, self-indulgence, gambling, and vice. Over the years, quite frequent published broadsides had accused the university and colleges of neglecting their responsibilities. For instance, a former student, Robert Mackenzie Beverley, in 1833 addressed the Chancellor of the University, complaining of the extravagance and immorality of students and fellows alike. An American student, Charles Astor Bristed, was an undergraduate at Trinity College during 1840–44, and so overlapped with William Thomson. His book *Five Years in an English University* (1852) confirms the financial and moral dangers of student life in Cambridge at that time.[29]

Early in 1842, James sent bankers' drafts to William, questioning him closely about his expenditure, which did not tally with the overall total. Exhorting him to adopt accurate business habits, he recounted the financial disaster that had just overtaken J. P. Nichol. William dutifully replied with more details, and requested that his father send on several books that he had left in Glasgow, including his Fourier, Poisson's *Mécanique*, George Peacock's *Examples*[30], also his own prize-winning essay on the 'Figure of the Earth', 'and as many books of the lighter kind as you choose, as my library is so very scanty that I shall almost be obliged to buy books to fill the shelves'.[31]

Just a month later, James was shocked to get a letter from his son telling him that he had just bought a second-hand single-seater boat, a 'wonderful bargain' at seven pounds. This, he claimed, would be a great saving on renting one. Doubtless to placate his father, he ends: 'I have been going on reading steadily, about eight hours a day, and getting up perfectly regularly a little before six o'clock'. James sought reassurance from Cookson that his son had not acted foolishly, and was eventually convinced, but not before sending further moralizing advice:

> Use all economy consistent with respectability. Be most circumspect about your conduct and about what acquaintance you form. You are young: take care you are not led to what is wrong. A false step now, or the acquiring of an improper habit or propensity, might ruin your life. Frequently look back on your conduct and thence learn wisdom for the future.[32]

Such financial exchanges extended throughout William's university career. It is enough to record William's overall expenditure, as detailed by James in October 1844:

The following are the amounts of your three years' expenditure:—

Up to Sept. 26, 1842 . . .	£240 : 11 : 7
From that date till Nov. 1, 1843 .	238 : 15 : 0
" " Oct. 12, 1844 .	295 : 0 : 0
	£774 : 6 : 7

James was unhappy about the increase in the third year: 'Have you lost money or been defrauded of it, or have you lived on a more expensive scale?'.[33] These sums included all fees to the university and the college and those for private tuition by Hopkins, who then charged about £72 per annum.

Compared with most students, William was certainly prudent in his expenditure. In addition, he was awarded several small sums by his college, as scholarships and prizes. In total, these amounted to £36 in 1842, £44 - 7/- in 1843, £41 - 10/- in 1844, and about £35 for the part-year of 1845. He was also awarded a special college prize of £21 on completing the Tripos examinations. The prize money was intended to be spent on books; but his most lucrative award, the Gisborne Scholarship of £30 p.a., helped significantly with fees and living expenses.[34]

William had arrived already well-prepared for the Cambridge curriculum, and though he studied hard at classics and mathematics, he had enough spare time both for rowing, or rather sculling, and for writing several original papers for the *Cambridge Mathematical Journal*. Though his tutor Cookson approved of the former recreation, he had doubts about the propriety of the latter, for undergraduates were not expected to publish such work. But the *Journal*'s editor, Archibald Smith, encouraged William to continue.

William continued rowing throughout most of his time at Cambridge. He regularly exercised in his own boat with 'the fleet', an early-morning group that included some other able mathematicians.[35] In the spring of 1843, he at last joined the college boat club and took his place in an 'eight'. Being fairly small and light, he rowed at 'bow', and his boat managed to win most of its races. Though he found the experience exhilarating: 'It was three weeks clean cut out of my time for working at Cambridge; so I determined to do no more rowing . . . [but] six months afterwards I won the silver sculls'.[36]

William's biographers all note his indefatigable energy. A typical day began at six or seven in the morning and continued until midnight or later. Hours of intense study were interspersed by boating, swimming, walking, skating, playing his cornopean, reading, and discussing with friends. Much of this is recorded in his personal diary, though some passages (perhaps alluding to female acquaintances) have been deliberately removed.[37]

His first summer vacation was spent with the family at Largs, where William drafted more papers for publication in the *Journal*. On his return to Cambridge in October 1842, he began to study with William Hopkins in a small class of about five hand-picked students. Though at first apprehensive that the others in his class might be 'extremely formidable', he soon found that he could more than hold his own. Two of the class were the German-born Franz W. L. Fischer, who had previously studied at universities in Germany and France, and Hugh Blackburn, a Scot who had attended Eton College and then Glasgow University: both were later to become professors in Scottish universities, Blackburn as James Thomson's successor in Glasgow and Fischer in St Andrews.

Hopkins gave lectures, and provided much additional information in manuscript notebooks, which the class members were required to transcribe. (Learning to write rapidly with a quill pen was one of the skills needed for the Tripos examinations.) He set questions for the students to answer both in writing and orally. Reporting to his father, William wrote:

What we have had already approximates very much to the plan w^{h} you pursue with your class. He asked us all questions on various points in the diff^{l} calculus, in the order of his manuscript, w^{h} he has given us to transcribe, and gave us exercises on the different subjects discussed, w^{h} we are to bring with us tomorrow. He says he never can be quite satisfied that a man has got correct ideas on any math[l] subject till he has questioned him viva voce.[38]

Though none of Hopkins' original manuscripts seem to have survived, several students' transcripts exist, including some made by G. G. Stokes, William Thomson, and later James Clerk Maxwell and Edward J. Routh. Hopkins' one published mathematical textbook, his *Elements of Trigonometry* (1833), also gives a clear idea of his teaching style. The standard theory is clearly and succinctly described, and many worked examples are set out, with others left for the student to try.[39]

Surprisingly, Hopkins did not restrict his teaching to mathematics alone, but conducted experimental demonstrations for the interest of his pupils. In 1839, he ordered a thermo-multiplier from J. D. Forbes, Edinburgh's Professor of Natural Philosophy. This device, which detected small quantities of radiant heat, consisted of a thermopile battery and a sensitive galvanometer. When placing his order, Hopkins rather imperiously advised Forbes that:

your own self interest is perhaps in some measure involved in the matter as there are probably few individuals in the country who have the same opportunity as myself of communicating a knowledge of your experiments to those who can thoroughly understand them... Therefore send me the best apparatus you can, or incur the peril of a certain quantity of scepticism which would be the natural result of unsuccessful experiments...[40]

The apparatus sustained some slight damage in transit, and several months later Forbes was still advising Hopkins on its use. Presumably it was demonstrated to William Thomson's class.

The study of heat, originally inspired by J. P. Nichol's lectures and by Fourier's treatise, was to remain a lifelong interest of Thomson; and he was later to invent his own far more sensitive galvanometers for use in telegraphy. Hopkins' own researches were mainly attempts (not entirely successful) to apply mathematics to geology and glaciology. These included estimates of the age of the Earth by calculating its heat loss from a supposed uniformly-hot initial state. This, too, was to become one of Thomson's later preoccupations.[41]

Attendance at professorial lectures at Cambridge was seldom encouraged by the tutors, as these were perceived not to 'pay' in the examinations. As a result, attendance at them was small and sometimes non-existent. But: 'Hopkins was strongly opposed to the ordinary idea of cramming for the Senate House ... He set before his pupils as their first object a clear understanding of the principles of what they were doing, and he urged them to leave all questions of success to take care of themselves', and he recommended attendance at some professorial lectures.[42] William Thomson is known to have attended the lectures of James Challis on experimental natural philosophy in both 1843 and 1844, and, in the latter year, Challis's lectures on astronomy and astronomical instruments. Few such opportunities were then available at Cambridge to gain experimental training.

Early 1844 brought some distractions. Returning from Glasgow to Cambridge in January, he visited his brother James in London, where he was now re-embarked on his engineering apprenticeship. He also called to visit his cousin Margaret Crum at her school nearby. (William and Margaret had known each other since childhood, and the two were to marry in September 1852.)[43]

During the Easter vacation, William again visited London. He met up with James in Millwall; and with his sister Anna and her new husband (presumably then on honeymoon), having been unable to attend their wedding. There, they together attended performances of Rossini's opera *William Tell* and Handel's *Messiah*. He also accompanied Archibald Smith, now a trainee barrister, to meet Col. Edward Sabine (later President of the Royal Society).[44]

A traditional part of the training for the Tripos examinations was the summer reading party, when the tutors took their students to seaside or mountain retreats for two months' study. In the summer of 1844, Hopkins took his charges to Cromer in Norfolk. There, each morning and evening, Thomson, Fischer, Blackburn, and the rest were given intensive tuition and took mock examination papers, with relaxing breaks in the afternoons. From Cromer, Hopkins wrote to James Thomson that William:

> has given us entire satisfaction. His *style* is much improved, and though still perhaps somewhat too *redundant* for examinations in which the time allowed is strictly limited, it is very excellent as exhibiting the copiousness of his knowledge as well as its accuracy. I consider his place as quite certain at the tip-top, but I am anxious that he should recollect that he has his own reputation to contend against.[45]

After more individual coaching back in Cambridge, William took the Tripos Examinations in the Senate House during 1st–7th January 1845. He was confidently expected to be the Senior Wrangler, at the head of the list of candidates. The 12 examinations covered a range of mathematics, pure and applied, some devoted to 'bookwork' that could be memorized, and some to 'problems' that could not. The 'problems', however, were mostly of a standard predictable sort, and part of the tutor's task was to train his students to recognize and tackle the various kinds that might occur.

The first two 'bookwork' papers concerned comparatively elementary topics in geometry, algebra, trigonometry, and conic sections; and applications were limited to aspects of mechanics, hydrostatics, optics, and astronomy for which the differential and integral calculus was not required. There followed a 'problems' paper; then papers four and five, respectively on 'mixed' (i.e. applied) mathematics and 'pure' mathematics (including calculus); then another 'problems' paper. The remaining six papers carried more advanced questions on both pure and 'mixed' mathematics, including partial differential equations, rigid-body dynamics, hydrodynamics and the wave theory of light.[46]

Contrary to expectations, William was pushed into second place by Stephen Parkinson, a student of St John's College tutored by John Hymers. Though Robert L. Ellis is said to have commented to his fellow-examiner Harvey Goodwin that 'You and I are just about fit to mend his [Thomson's] pens', Parkinson's superior memory, concise expression, and astonishing speed of writing enabled him to gather more marks.

William does not seem to have been much cast down by his second place. Writing during the Smith's Prize examinations, he assured his sister Elizabeth that:

> The principal thing that I care about in the result of the [Tripos] examination is the disappointment which I am afraid Papa must feel, as I am afraid he had rather raised his hopes about it; though I tried to keep him from expecting too much before the examination, as I knew the uncertainty. Will you write to me and tell me what he has said? I suppose he will write, but I should like to hear from you what he thinks.[47]

Knowing that James Thomson would be bitterly disappointed by his son's second place, Hopkins quickly wrote to reassure him that, despite Parkinson's talent for rapid exposition, William far exceeded him in promise of future distinction. In similar terms, William Whewell informed Edinburgh's J. D. Forbes that 'Thomson of Glasgow is much the greatest mathematical genius: the Senior Wrangler was better drilled'.[48] Further consolation soon followed in the Smith's Prize examinations, when William easily won the first prize ahead of Parkinson.

In June of that year, William was elected to a Foundation Fellowship at Peterhouse, worth £200 per annum, and in October he was appointed college lecturer in mathematics. During 1845–46, he earned a further £80 by college teaching and examining, and he also took some private pupils.[49] At last, at the age of 21, he had achieved financial independence.

PREPARING FOR THE GLASGOW CHAIR

As early as 1841, James Thomson had begun to look for a likely successor to the aged and ailing Professor Meikleham. David Thomson, then Meikleham's substitute, was one possible candidate, but James sought others. He considered three more Scots: Duncan F. Gregory and Archibald Smith, who had both studied at Cambridge as well as in Scotland, and the well-established James D. Forbes of Edinburgh.[50]

Forbes, an aristocratic Tory much influenced by the educational philosophy of Cambridge's William Whewell, was deemed by James to be too out of sympathy with the democratic traditions of Glasgow University. In any case, Forbes had no wish for the job. William knew both Gregory and Smith through his involvement with the *Cambridge Mathematical Journal*, and James sought his advice. In early 1842, William assured his father that Gregory 'would make a splendid professor for you . . . He is a considerable Whig (both in mathematics and politics) but that need not be known. I think he is undoubtedly the best and most original math^{n} in Cam[bridge], and it is also said that he has a great knowledge of experimental and physical subjects'.[51] Sadly, just a year later Gregory's health failed and he died in 1844, aged only 30.

Archibald Smith, on the other hand, had in 1836 been an unsuccessful candidate for Glasgow's Chair of Practical Astronomy that had been awarded to J. P. Nichol. But Smith was now embarked on a law career and was thought unlikely to canvass energetically for the Glasgow post. In the event, with typical indecision, Smith considered applying for the Natural Philosophy Chair when it eventually fell vacant—thereby causing James Thomson some disquiet—but did not do so. James Thomson also approached William Hopkins to see whether he might apply, but Hopkins was content with his current situation.[52]

By 1843, James had come to the view that his own 18-year-old son would soon be a strong candidate. But James planned to enhance William's credentials by advocating further diversification of his educational experience. The widely unpopular report of the 1826 Royal Commission into the Scottish universities had engendered a climate of hostility towards perceived Anglicization. Accordingly, a top degree from Cambridge was not uniformly perceived as a desirable qualification. James Thomson had no doubts about the strengths of the Cambridge Tripos, and J. D. Forbes in Edinburgh favoured, for different reasons, a measure of reform towards the Cambridge educational model. But many Scottish educators held the view that a Cambridge education ill-qualified a man to teach the wide range of classes, starting at a very elementary level, that existed in the Scottish universities.

Accordingly, if William was to succeed in his candidature, he had to show evidence of teaching ability, and also a familiarity with laboratory work. Though he had successfully tutored some Peterhouse students for the Tripos, William had little experience of elementary teaching. His familiarity with laboratory experiments came mainly from his time as a student in Glasgow and during summer vacations at home. At Cambridge, as we have seen, he took such limited experimental opportunities as were available, attending James Challis's lectures on experimental natural philosophy and on practical astronomy and astronomical instruments.

James Thomson sought the advice of his colleagues and political associates Dr. William Thomson (Professor of Medicine at Glasgow) and J. P. Nichol, and both recommended that William spend some time in Paris to study experimental techniques. James was forthright in his advice to his son: he wrote that Dr. William Thomson:

> speaks emphatically about the necessity of your giving very great attention to the experimental part as soon as you can; as he says no one will have any doubt as to your mathematical attainments, but that some may even think them to be such as to make you neglect the popular part of Natural Philosophy.[53]

Accordingly, soon after his graduation in 1845, William set out for Paris, furnished with letters of introduction from J. D. Forbes, David Brewster and, more surprisingly, from the theologian Thomas Chalmers.[54] Doubtless it was James Thomson who arranged for these letters; and he also sent on detailed advice from Dr. William Thomson, who recommended the purchase of French texts on *la physique expérimentale* from prize money recently awarded by St Peter's College. For this visit to Paris, William was accompanied by his good friend Hugh Blackburn.[55]

Soon, William was acting as an unpaid assistant in the laboratory of Victor Regnault at the Collège de France, then considered the best physical laboratory in France.[56] William also made contact with the leading Parisian mathematicians, whose ready acceptance of him owed something to his letters of introduction, but more to his own persona and intellect.

Shortly before William Thomson left Cambridge for Paris, William Hopkins had given him two copies of the little-known 72-page *Essay on the Application of Mathematical Analysis to the Theories of Electricity and Magnetism* by George Green, that had been privately published in 1828. William quickly recognized its importance and showed it to the French *savants* Joseph Liouville, Charles-François Sturm, and Michel Chasles, who were surprised to find that it had anticipated some of their own results as well as Thomson's. The German August Crelle, then visiting Paris, immediately offered to republish the essay in his *Journal für die Reine und Angewandte Mathematik*. It duly appeared in three parts during 1850–54 with a short introduction by Thomson.[57]

Back in Cambridge, William Thomson took over the editorship of the *Cambridge Mathematical Journal* from R. L. Ellis, who had kept it going in the year following D. F. Gregory's death. He soon enlisted George Gabriel Stokes to join him in writing a series of didactic 'Notes on Hydrodynamics' that were a useful aid to students at a time when no adequate textbook existed. At that time, Stokes was making fundamental advances in his researches on water waves and on viscous flows. The friendship of Thomson and Stokes then established was to be a lifelong one. Their extensive scientific correspondence has survived intact, and it charts one of science's most remarkable collaborations: in it, new discoveries are described, speculations are raised and disputed, questions asked and answered. Stokes was the more careful and precise, and Thomson the more imaginative and speculative: each used the other as a sounding board and confidant, to their mutual benefit.[58]

In the same year of 1845 a complication arose, for it seemed that James Thomson would be appointed as Principal of a new non-sectarian Northern Irish College (later Queen's University, Belfast). James was attracted by the prospect, and he thought also that it would double William's chances of securing a Glasgow chair as his chair of mathematics would thereby fall vacant. But William did not want his father to leave Glasgow, and James (Jr.) was sure that William would not want to take up the mathematics chair as he was too committed to natural philosophy. In the event, sectarian politics denied James the appointment: a deal had apparently been done by the Government that the first Principal should be a Presbyterian minister acceptable to most parties. As a consolation, James was offered, and refused, the post of Vice-President of the new college.

By the time that Meikleham finally died in 1846, James Thomson's well-laid plan was in place. William had his high Cambridge degree, now enhanced by further experimental credentials. All that was now necessary was to convince the appointing committee: though several were political friends and associates of James, some others were in the Tory camp. James overruled the advice of William's old Cambridge tutors H. W. Cookson and W. Hopkins that a few carefully-chosen references would suffice; instead, William's application was accompanied by over thirty printed testimonials. In the words of Smith and Wise (1989):

> The list, which reads like a roll of honour of mid-nineteenth-century scientific worthies, included Augustus de Morgan, Arthur Cayley, Sir William Rowan Hamilton, George Boole, G.G. Stokes, Victor Regnault, J.D. Forbes, and Thomas Thomson, in addition to William's Cambridge examiners.[59]

The illustrious list also included the Master and Resident Fellows of St. Peter's College, William Hopkins, Henry Cookson, Frederick Fuller, William Whewell, George Peacock, James Challis, James Joseph Sylvester, and David Thomson. Two letters of support from Joseph Liouville arrived too late to be printed but were forwarded to the committee. Archibald Smith wrote no reference, but neither did he come forward as a rival candidate, as had been feared.[60]

Only one brief extract from these testimonials need be reproduced here, that from William Hopkins, the person best acquainted with Thomson's work:

> It is scarcely necessary to offer individual testimony with respect to Mr. Thomson's mathematical talents and acquirements, which are best attested by the academic distinction which he has gained, and the original papers which he has written. But I would here remark, that his power as an *analyst* is not unaccompanied (as is not unfrequently the case) by a taste for physical investigations. I have known other young men as good analysts as himself; but I doubt whether, in the course of my long experience, I have ever met with any one of his own age who combines such a knowledge of abstract mathematics with such an almost intuitive perception of physical truths, so accurate a knowledge of physical

principles, and such enlarged and matured views of the great physical problems which Nature presents to us . . .[61]

But James was still worried, for a rumour had reached Glasgow that William's college teaching at St. Peter's college had proved too advanced for the 'ordinary students'. To counter it, some referees were asked specifically to write favourably of William's teaching abilities.[62]

In the event, no serious rival candidate came forward and, on 11 September, 1846, William was elected unanimously at the age of 22.[63] His father's extreme delight is recorded in a letter to Elizabeth from her husband, the Rev. David King:

When I came up to Glasgow yesterday I was just in time to receive the joyful tidings of William's unanimous and cordial appointment to the Chair of Natural Philosophy! The first announcement I had on the subject was your father's face as he came out of the Hall where the election had been conducted. A countenance more expressive of delight was never witnessed. The emotion was so marked and strong that I only fear it may have done him injury . . .[64]

Elizabeth later confirmed her father's 'happy expression', remarking that, in contrast, 'William does not look in the slightest degree elated. He is perfectly composed'.[65]

William moved back into the family home and took up his position on 13 October. But less happy events were soon to follow. Next February, his younger brother John died of fever, probably typhus, contracted when working as a trainee doctor at the infirmary. James (Jr.) had been living at home since 1845, having given up another engineering post because of illness; but he at first showed little improvement despite, or perhaps because of, regular medical attention. Around 1844 Anna moved to Belfast with her husband William Bottomley. There, they witnessed the misery of the potato famine and typhus outbreaks of 1847, and Anna was herself to die ten years later. In 1848 Elizabeth's health broke down, and she was recuperating in Jamaica during the winter of 1848–49, when Glasgow experienced an outbreak of cholera in which over 3500 people died. One of its victims was James Thomson, who died on 12 January, 1849 at the age of 62. He and William had been Glasgow colleagues for only two years. He did not live to enjoy William's major successes, nor to see his eldest son, James, occupy the chairs of engineering at Queen's College Belfast (1857–73) and then Glasgow University (1873–89).[66]

3

On the Early Work of William Thomson: Mathematical Physics and Methodology in the 1840s

I. Grattan-Guinness

INTRODUCTION

William Thomson began to publish exceptionally early, in 1841 while in his late teens; and a steady stream of papers came from him thereafter. Most of them in the 1840s concerned heat diffusion, electricity, or magnetism, and related mathematical methods such as Fourier series and potential theory. In addition to their contents, they manifest a procedure of working by analogy in which these methods were taken from one topic and adapted for use in another one. In this paper I shall review this work of his first decade, especially the papers published by 1845. He also produced a few rather unoriginal papers in mechanics, which I shall not treat.

Many of these papers appeared in the new *Cambridge Mathematical Journal*, which Thomson was to edit (with '*and Dublin*' added to the title) from 1846 to 1854[1] Some papers came out in French in the *Journal des Mathématiques Pures et Appliquées* edited by Joseph Liouville (1809—1882); they became acquainted in 1845 when Thomson spent several months in Paris. Later in life he reprinted almost all of these papers: those in *Electrostatics and Magnetism* in a volume of 1872 (with a slightly revised edition in 1884),[2] and on heat and related topics in the first volume (1882) of his *Mathematical and Physical Papers*[3] To these reprints he added comments on errors and on contemporary and later developments; I quote a few of these. I shall cite these books in the text as 'Thomson *Elec Mag*' and 'Thomson *Papers 1*' respectively, using the page numbers in these re-appearances (from the first edition of the 1872 book) while also indicating the dates of original publications of the papers.

As usual at that time, Thomson built upon a foundation of previous work; much of it emanated from France, which had been the dominant country for pure and applied mathematics from the 1780s to the 1820s and was still a major centre[4] Thus I include summaries of their contributions most pertinent to Thomson, starting, as he did as an author, with heat diffusion.

ON THE FRENCH BACKGROUND, ESPECIALLY FOURIER

The main founder of heat diffusion was Joseph Fourier (1768–1830), whose work started in the 1800s and was very much on the agenda by the 1820s. The interest of others lay partly in heat theory itself: to consider the diffusion equation, the linear partial differential equation representing the phenomenon. For Fourier heat was heat, to be exchanged with cold, and we do not need to delve more deeply into the nature of either. It is fair to call him a positivist, for when Auguste Comte (1798–1857) introduced the word in the late 1820s he saw Fourier's work as philosophically exemplary.

Of broader concern was the form of solution that Fourier chose to revive and greatly enhance in reputation: the representation of a mathematical function in an infinite series of sines and cosines, now named after him:

$$f(x)=\sum_{r=0}^{\infty}(a_r \cos rx + b_r \sin rx), \quad 0 \le x \le 2\pi \tag{1}$$

where the coefficients were determined by certain integral formulae. Not only were they usable elsewhere in science; they posed difficult problems in their own right, such as the interpretation of (1) outside the interval of definition, the convergence of the series, the range of functions so expressible, and the definability of the attendant integrals. From the 1810s onwards Fourier was rivalled by S. D. Poisson (1781–1840), who used different methods both in the physics and in the mathematics.

These series (and some related solutions) applied only to finite physical bodies; what about infinite ones? P. S. Laplace (1749–1827) showed the way ahead in 1809 with an integral solution (not what is now called the 'Laplace transform') of the diffusion equation in such circumstances; and this led soon Fourier to his own such solution, now also named after him. A. L. Cauchy (1789–1857) found it a few years later and studied it rather more systematically. Several other French mathematicians who emerged in 1810s and 1820s took up heat diffusion or Fourier analysis, or both.

THOMSON'S EARLY WORK ON HEAT THEORY

Thomson started out publishing on heat theory because when he went on holiday in 1840 with his family he took with him Fourier's book *Théorie Analytique de la Chaleur* (1822) on heat diffusion for light reading. He had also read the first wide-ranging account of heat theory in English, the *Theory of Heat* (1837) by Philip Kelland (1808–79). On heat diffusion Kelland was no British answer to Fourier, and his criticisms of aspects of Fourier's theory deserved better treatment, as the young man realized. His first paper dealt with the legitimacy of (1) itself (1841: *Papers 1*, 1–6); in declaring that f(x) was 'completely arbitrary', he obviously had not read the beautiful paper by the German mathematician J. P. G. Dirichlet (1805–59) in 1829 (extended in 1837) laying sufficient conditions on the function for convergence actually to take place.

Thomson did make some useful, though not novel, remarks on the reading of (1) outside its interval of definition. In a sequel paper he studied non-harmonic series (as they were to become known, partly though the influence of his own later work on acoustics), where the coefficients *r* of the argument variable in the sines and cosines were replaced by the non-integral roots of certain transcendental equations. It was essential that all these roots be real; Fourier had argued quite well for this, and Poisson much better. Thomson's paper (1841: *Papers 1*, 7–9) really added nothing.

In a two-part paper 'On the linear motion of heat' (his favourite phrase for connoting diffusion) Thomson explored Laplace's integral solution for infinite bodies (1842: *Papers 1*, 10–15, including his later rejection of much of his analysis). Then he applied the results to a problem that both Fourier and Laplace had tackled around 1820: the cooling of the Earth (1843: *Papers 1*, 16–21); later this problem was to loom large in his work. Unlike Fourier he allowed himself to talk of two 'series of waves of heat' moving in opposite directions (p. 17), as if each trigonometric term was talking heat. He also considered an important property in heat theory, its external diffusion from the surface of an extended body into the environment; equalling internal temperature gradient there and the external temperature difference produced a first-order partial differential equation that led to the transcendental equations mentioned above.

In his next papers on heat Thomson moved on to other authors. In a more substantial piece, seemingly inspired by recent work by Gabriel Lamé (1795–1870), he analysed heat diffusion in more general coordinate systems (1843: *Papers 1*, 22–24). He followed up quickly with a similar analysis of the diffusion equation itself in three dimensions, drawing upon not only Lamé but also some differential geometry due to Charles Dupin (1784–1873) (1843: *Papers 1*, 25–35), which he reproved in

THÉORIE

ANALYTIQUE

DE LA CHALEUR,

PAR M. FOURIER.

A PARIS,

CHEZ FIRMIN DIDOT, PÈRE ET FILS,

LIBRAIRES POUR LES MATHÉMATIQUES, L'ARCHITECTURE HYDRAULIQUE

ET LA MARINE, RUE JACOB, N° 24.

1822.

Fig. 3.1. Both William Thomson and James Clerk Maxwell viewed Fourier's treatise as 'a great mathematical poem'.

a succeeding note (1844: *Papers 1*, 36–38). This kind of mathematics was soon to rise in importance for Thomson, as we shall see in later.

In his next paper Thomson entertained the interpretation of solutions when time was taken as negative, as if retrodicting the previous thermal behaviour of the body in question (1844: *Papers 1*, 39–47, including a long note added later). The time terms would now take positive exponents, so endangering the convergence of the series. More fundamental, however, was the physics; the impossibility of these anterior solutions convinced him not only of the direction of time but especially that the universe had an origin.[5]

In 1846 Thomson was appointed Professor of Natural Philosophy at the University of Glasgow, thanks to support from various parties (including Liouville). He succeeded his own teacher William Meikleham, who had been one of those to encourage him to study French authors. He underlined the importance of his work on heat theory by choosing it as the subject of his inaugural dissertation; unfortunately he destroyed his text at once (p. 187). While some of his contributions to heat theory were marginal, they displayed some distinctive features, and we shall soon see their effect on other topics. It is a pity that he was ignored in a useful history of heat diffusion.[6]

ON THE FRENCH BACKGROUND, ESPECIALLY LAPLACE

Even before his holiday Fourier, Thomson had delved into a still more formidable work, in 1839: Laplace's *Traité de Mécanique Céleste*, published in four volumes between 1799 and 1805 with a round-up successor volume on various topics that came out in instalments between 1823 and 1827.[7] Two features attract us here.

The first is potential theory, in the form of analysing the attraction of solid bodies at external points. Assuming inverse square forces, the potential at point P of an extended body B not containing P and composed of elements dm distant r from P was given by $\int_B dm/r$. The main stimulus of the theory was then in planetary mechanics, and the effort was directed toward finding solutions to the linear partial differential equation (now named after Laplace) that the potential satisfied. The main thrust of Laplace's account lay in the first two volumes of his book: the interest for Thomson would have lain in that equation and in the heavy use of Legendre functions and spherical harmonics (again the modern names) in its solutions, and also in the role in general of potentials.

Thomson may not have noticed the second feature, but he reacted in detail to some of its consequences. In the fourth volume, Laplace had analysed atmospheric refraction by assuming a corpuscular theory of light. Soon he was generalizing this approach to an ambitious programme of molecular physics, in which all physical phenomena were to be studied this way.[8] Optics was the branch where most success was gained, while in heat theory progress was limited; our attention falls upon electricity and magnetism, where the molecularism was supported by the further assumption of pairs of electrical and magnetic fluids.

Following Laplacian principles Poisson produced basic studies of electricity and magnetism in the mid-1810s and the mid-1820s respectively. In the former connection he complemented Laplace's equation by finding the equation that determined the potential V of an extended body at an *internal* point I. In a rectangular coordinate system xyz the equation took the form

$$V_{xx} + V_{yy} + V_{zz} = 4\pi\rho, \tag{2}$$

where subscripts denote partial differentiations and ρ was the density at I. Laplace's equation for external points is obtained by setting $\rho = 0$ in (2).

A main task for Poisson was to calculate the thickness of the fluids that lay upon the bodies in question after electrification or magnetization had occurred and equilibrium was achieved. Curiously, he ignored electromagnetism and electrodynamics, which during the 1820s captured the attention especially of A. M. Ampère (1775–1836) among his contemporaries. Overall the French did not display their normal dominance in these areas of physics, where some of the initiative was taken over in the British Isles.

THOMSON'S EARLY WORK ON POTENTIAL THEORY

Poisson's assumption of electrical fluids was also adopted in Britain by Robert Murphy (1806–1843), in his book *Elementary Principles of the Theories of Electricity, Heat and Molecular Actions* (1833); but other approaches were also being followed.[9]

Important for Thomson was the work of Michael Faraday (1791–1867). He had found evidence to cast doubt both on theories of fluids and actions at a distance. In a complicated sequence of reasoning he eventually offered instead a theory in which the effect of one electrified body upon another one depended not only upon them but also upon the environment in which they were situated (including intervening objects such as screens), and that lines of induction could be curved as well as rectilinear. He emphasized the role of materials in which conduction did not happen but within which induction could occur; he named them 'dielectrics'.

Thomson produced his first 'mathematical theory of electricity' by emulating heat diffusion to establish the 'connexion' (1843: *Elec Mag*, 1–14). Considering the diffusion of heat from a single source, he studied isothermal surfaces, exactly the analogous notion to equipotential surfaces in planetary mechanics; and he switched topic easily to examine the electrostatic potential on the surface of an extended body (where he went along with a theory of fluids). Much of the rest of the paper dealt with heat distribution in a homogeneous ellipsoid, but again treated in a manner appropriate also to other physical phenomena—for example, in concentric shells. At the head of his paper he noted, within square brackets, that he had been partly anticipated by Michel Chasles (1793–1880).

The links became more explicit in a follow-up note on 'Propositions in the theory of attraction' (1843: *Elec Mag*, 126–138; also in a version for Liouville), where Thomson stated and proved various theorems of the divergence type concerning the attraction of an extended body at internal and at external points. He then wrote a note showing the equality of the attractions of both 'conducting and non-conducting electrical bodies', which he specified respectively by holding their electricity by the atmosphere 'to a certain extent at least' and by internal friction (1843: *Elec Mag*, 98–99).

Thomson was also aware of the following results. Chasles had proved various theorems on internal and external potentials for an extended body that implied that a given potential on a surface could be produced by an infinity of distributions (claim in 1839, proof in 1842). Meanwhile, C. F. Gauss (1777–1855) asserted in 1840 that an isomorphism held between potentials and distributions. Thomson cited both authors without raising any issue.[10] The issue was a subtle one, not clarified until from the late 1860s onwards; in the end it emerged that on this occasion the mighty Gauss was wrong.

An important source of disaffection for Thomson from theories of fluids came from a theorem proved in 1842 by Samuel Earnshaw (1805–88). Assuming that the aether was punctiform (that is, composed of tiny molecules), he showed that if *only* central inverse square forces obtained between them, then any equilibrium would be unstable. To save the phenomena he mooted the possibility of inverse square attraction but inverse n repulsion, where $n > 2$. The result obtained for any medium consisting of particles that interacted with inverse-square forces—in particular, in the theories of fluids in electrostatics and magnetism.[11] Thomson reproved it in the context of electricity by a *reductio* argument (1845: *Elec Mag*, 100–103), and from it he became convinced that electrical fluids should play no role in theorizing. Instead he adopted a stance broadly similar to Faraday's that induction was propagated through dielectrics and other materials in manners somewhat akin to the conduction of heat.

Although Thomson's adhesion to a theory of fluids had disappeared, no convincing alternative came to the fore. He used names such as 'electric charge' and 'magnetic forces' seemingly in an agnostic spirit concerning the nature of the phenomena in which they were involved. But the analogy with heat theory remained strong; for example, Thomson also used Faraday-like phrases such as 'lines of force' of electricity and magnetism along with his own 'lines of motion of heat'.

THE IMPACT OF GREEN

A major advance in Thomson's understanding of the importance of line and surface integrals in potential theory occurred early in 1845 when at last he was given copies of a book that he had known about for some time: *An Essay on the Application of Mathematical Analysis to Electricity and Magnetism* (1828) by George Green (1797–1841). This remarkable self-taught mathematician had written a masterpiece that he had, however, published in a most obscure manner, by public subscription in his home town of Nottingham.[12] He then produced several important research papers during his remaining years, but hardly even cited his book there.

Green made two great innovations in the book. Firstly, by using integration by parts he proved a theorem that showed the relationship, in terms of potentials, between distributions of material within an extended body and potentials on its

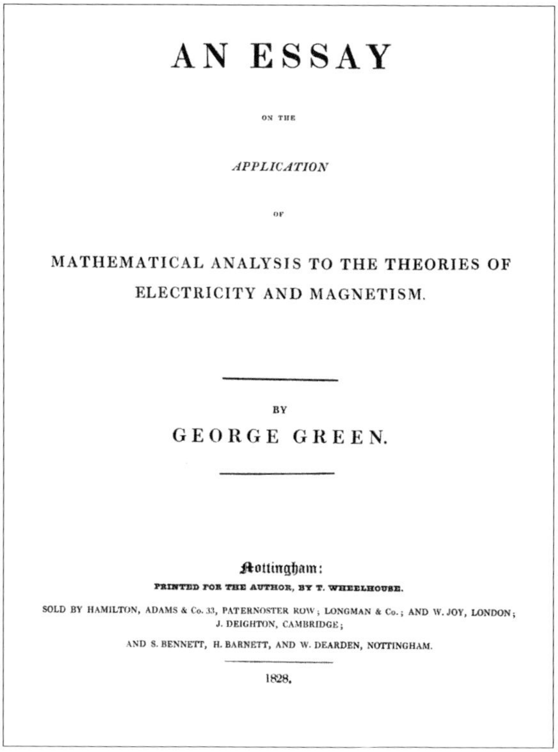

AN ESSAY

ON THE

APPLICATION

OF

MATHEMATICAL ANALYSIS TO THE THEORIES OF ELECTRICITY AND MAGNETISM.

BY

GEORGE GREEN.

Nottingham:

PRINTED FOR THE AUTHOR, BY T. WHEELHOUSE.

SOLD BY HAMILTON, ADAMS & Co. 33, PATERNOSTER ROW; LONGMAN & Co.; AND W. JOY, LONDON; J. DEIGHTON, CAMBRIDGE;

AND S. BENNETT, H. BARNETT, AND W. DEARDEN, NOTTINGHAM.

1828.

Fig. 3.2. The title page of George Green's book which was important in Thomson's understanding of line and surface integrals in potential theory.

surface; of the various versions now known, his was the one called 'symmetric'.[13] Secondly, he recognized the importance of a 'potential function' that satisfied 'Laplace's equation' (both names his) inside the body and equalled a given potential on the surface, and a few other conditions.[14] Poisson had come to the same theorem (but not the function) a few years earlier, in the context of magnetism, but saw it only as a mathematical simplification; it may have been Green's original motivation.

Regarding the physics, Green held magnetism to be less well theorized than electricity; following C. A. Coulomb (1736–1806), he took a magnetic body to be composed of tiny non-communicating 'particles'.[15] Unlike Poisson, he assigned to each category only one fluid; and maybe following Coulomb again, he calculated the 'density' of the electric fluid rather than any thickness.

Thomson had achieved Green's kind of understanding of theorems about potentials by his own efforts, but now he found in Green a body of results that he could use. He first cited Green early in a study of 'electricity in equilibrium'; later here he also cited Faraday on propagation of electricity in dielectrics, and Fourier for the corresponding situation in heat theory (1846: *Elec Mag*, 15–37 (pp. 17–18, 26–29)). He followed with a mathematical exercise for Liouville in 1845, with an English version in (1847: *Elec Mag*, 112–125), evaluating certain multiple integrals of kinds that Green and he found useful in electrostatics. He also reworked and modified Green's analysis of 'induced magnetism in a plate'.[16]

A substantial advance made at this time was Thomson's 'method of images', first published in Paris in the form of a letter to Liouville (1845: *Elec Mag*, 144–146). Adopting a procedure in projective geometry, he considered a sphere of centre C and took any radius CA, extending it to pass through the external point Q that was defined in terms of any given internal point P on that radius by the property

$$\mathrm{CP} \times \mathrm{CQ} = \mathrm{CA} \times \mathrm{CA}. \tag{3}$$

Then he showed that if a charge were placed at Q, it would produce the same equipotential surfaces as those caused by a certain calculable negative charge placed at P; hence a conducting surface was replaced by a point charge. Again Green formed part of the inspiration, especially his considerations of potentials at points inside and outside a sphere, which may have suggested to Thomson that the potential theorem could be reread in reverse, from surface to interior instead of the usual interior to surface. The method was generalizable to a finite (and sometimes even infinite) collection of point charges and to iterated imaging relative to other spheres (and indeed other surfaces); in all cases a main benefit was some simplification of the sums, especially when seeking approximate solutions.

THOMSON FROM 1845 INTO THE 1850s

In the years after 1845 Thomson continued to develop his version of potential theory, with electricity and magnetism as the main applications. Under the influence of Faraday he assumed that the phenomena were taking place in an 'elastic solid', an aether in all but name; in a short paper he presented the similar differential equations involved in 'a mechanical representation of electric, magnetic and galvanic forces' (1847: *Papers 1*, 76–79).

But some uncertainty, or caution, remained. Maybe following Green, in an analysis of 'electricity in equilibrium', Thomson revived Coulomb's name 'electrical density' (defined as the quantity of electricity distributed over some unit surface of area) because he found it 'far more philosophical' than talk of layers of fluids precisely because of its neutrality (1848: *Elec Mag*, 48). In a later paper he gave the definition 'a magnet is a substance which intrinsically possess magnetic properties', not one of his most penetrating thoughts; he then spoke of 'imaginary magnetic matter, northern and southern' (1851: *Elec Mag*, 342, 352). Soon afterwards he began to use Faraday's word 'field' to characterize the environment within which electrical or magnetic forces were acting; and also to draw upon another analogy in emulating dynamics concerning the relationship between energy and work (for example, 1855: *Elec Mag*, 128–132).

On the mathematical side, Thomson had an important idea about the existence of solutions: perhaps inspired by Green's function, he told Liouville in 1847 that a class of linear partial differential equations, including both Fourier's surface diffusion condition and Poisson's equation, admitted a unique solution that went to zero 'for infinite values of the [independent] variables' if the density was zero outside some 'finite closed surface' (1847–48: *Elec Mag*, 139–143; 1848: *Papers 1*, 93–96). His proof relied upon integration by parts and minimizing a certain integral. This proposal was a special case of a principle that came to be named after Dirichlet and became very popular in potential theory until Karl Weierstrass (1815–97) showed in 1870 that the process of minimization was not safe.[17] As for methods of finding solutions, that of images proved fruitful (for example, 1848–49: *Elec Mag*, 60–85).

Among later theorems was the one relating potentials on a surface to conditions on its closed bounding curve; we name it after G. G. Stokes (1819–1903), but in fact Stokes received it in a letter from Thomson in 1850.[18] Several of the researches of these two men were to intertwine in the succeeding decades;[19] for example, in the supposed properties of the aether. At that time Thomson also organized the reprint of Green's book in three parts in the *Journal für die Reine und Angewandte Mathematik* (1850, 1852, 1854), so giving it a much wider, and international, audience.

Heat diffusion rather faded from view, though another emulation occurred in a brief announcement of 'a system of magnetic curves' drawing upon curvilinear coordinates (1847: *Papers 1*, 81–82). By 1854 Thomson saw Faraday's phrase 'conducting power for lines of force' as 'complet[ing] the analogy with the theory the conduction of heat' (*Elec Mag*, 1). Further, heat theory in general played roles from the early 1850s onwards in Thomson's concern with thermodynamics and from a decade later in the various inter-disciplinary (and partly religious) controversies over the age of the Earth.[20]

CONCLUDING COMMENTS: MACROSCOPIC PHYSICS AND ITS ATTENDANT MATHEMATICS

Now the laws of motion for heat which Fourier lays down in *Théorie analytique de la chaleur*, are of that simple elementary kind which constitute a mathematical theory properly so called; and therefore, when we find corresponding laws to be true for the phenomena presented by electrified bodies, we may make them the foundation of the mathematical theory of electricity[.]

Thomson, 1846 (*Elec Mag*, 29)

From the mathematical point of view the most important innovations of Thomson's work were his contributions to potential theory and his republishing of Green's book. He joined Poisson, Ampère, Green, Chasles, Gauss, and some others as pioneers of this subject, and of the use of line and surface integrals in mathematics in general. These topics increased in importance steadily across the whole of mathematical physics (including mechanics) during the rest of the century.[21] The stimulus also involved linear partial differential equations, the flux or gradient expressed in terms of first-order partial derivatives, and the use of curvilinear coordinate systems.

As the quotation at the head of this section exemplifies, for Thomson the analogy was to extend beyond mathematics as far as thinking in the end that electrical and magnetic phenomena were to be treated as mediations through an aether, rather similar to (but not identical with) the propagation of heat. And always he focused on macroscopic aspects of the phenomena, such as internal distributions and surface potentials, and flows in various directions, more than on supposed microscopic constitutions of materials and fluids.

In addition, Thomson played a role in the history of Fourier analysis; for, while his own contributions were not very significant, he helped to bring the theory to the attention of British mathematicians, especially alerting them to its ubiquity. For example, Alexander Freeman cited all of the heat papers in notes adjoined to his English translation of Fourier's book.[22] In an encyclopaedia article of 1880 on heat theory Thomson himself included a 'Compendium of the Fourier mathematics for the conduction of heat in solids, and the mathematically allied physical subject of diffusion of fluids, and the transmission of electric signals through submarine cables'.[23]

Once again we see analogies exhibited in the title of a Thomson paper. Of course analogizing has long been practised in science, especially mathematics; but in his case there may have been a special source of encouragement. From the late eighteenth century there had developed in Scotland a strong philosophical tradition called by its practitioners 'common-sense'. It involved a novel combination of experience (while avoiding empiricism) with a concern with reasoning and understanding (while avoiding psychologism). They paid attention to making analogies from one intellectual domain to another one: founder Thomas Reid (1710–1796) was very cautious of its utility, but Dugald Stewart (1753–1828) saw them as valuable guides in the prosecution of inductivist epistemology. In addition, these philosophers liked geometry and techniques associated with it, of which curves, surfaces, solids, images, and flows can readily be seen as examples.

The consistency of this tradition with Thomson's early science (and indeed also his later work) is evident; but the question of influence is not easy to establish. Several compatriot scientists explicitly drew upon it, but not Thomson.[24] Further, the important role of Fourier was seemingly philosophical as well as technical, and his positivism is consistent with common-sense without drawing upon it. It seems likely that as a student, and maybe also from within his family, Thomson will at least have heard of the tradition; for example, he took courses in moral philosophy from Robert Fleming (1792–1866),[25] who was sympathetic to it. At all events, whether because of explicit influence, cultural absorption or just coincidence of thought, philosophically speaking Thomson's science looks very Scottish.

4

James Thomson and the Culture of a Victorian Engineer

Peter Bowler

Any detailed study of William Thomson needs to take into account his relationship with his brother, James. The two men shared common interests in the intersection of physics and engineering, although they came to the relationship with different priorities. James Thomson was, primarily, an engineer. He was the second Professor of Engineering at Queen's College, Belfast, serving from 1857 to 1873, when he left to take up the Chair of Engineering at Glasgow. In addition to his teaching he was a practicing engineer who made a number of important inventions. He was also active in promoting improvements in the urban environment, especially during his period in Belfast. These wider interests led him to a very broad conception of science and its implications for society, paralleling William's own approach in many respects. My title for this chapter is 'James Thomson and the Culture of a Victorian Engineer' and I want to take it seriously because I am not a historian of technology, but a historian of science with an interest in evolutionism and the environmental sciences.[1] For help with understanding Thomson's contributions to physics and engineering, I would like to record my thanks to Crosbie Smith and to Sir Bernard Crossland. But what attracted me to Thomson was the fact that he was much more than an engineer—he was a scientist too, who did important work in physics, but who also related his science (like his engineering) to the natural environment. James Thomson's culture related engineering to natural philosophy in its widest sense, the whole enterprise being driven by his religious beliefs and his social philosophy.

James and William Thomson collaborated actively in the studies of thermodynamics which made the latter's reputation, and James made important studies in other areas of physics bearing on the engineering problems he encountered in the course of his work. William himself was—as Crosbie Smith has shown—an eminently practical physicist, most of whose work was linked in one way or another to technological developments in areas such as steam power and the electric telegraph.[2] What I want to suggest is that the two brothers stand at opposite ends of a spectrum of interests by which science interacted with technology in the nineteenth century. William was a physicist with a strong interest in engineering, and James was an engineer with a strong interest in physics. They both realized that successful technologies depend on a proper understanding of the physical processes involved. But both also realized that the physics which helps us to understand—and improve—the steam engine or the telegraph also helps us to understand processes going on in the natural world. So Kelvin participated in the great debate over the age of the Earth which racked late nineteenth-century geology (and greatly disturbed the Darwinian evolutionists). James studied how flowing water shaped the beds and banks of rivers, how the properties of ice affected the scouring action of glaciers on the landscape, and how the behaviour of the air as a fluid controlled the great wind systems encircling the globe. He was truly an engineer with a broad vision of the world.

The Thomsons were raised as Presbyterians, and both retained a strong religious faith throughout their lives. After much agonizing with his conscience, James eventually became a Unitarian. His social views were very much those of a liberal Protestant, exemplifying the best aspects of what has been called the Protestant work ethic. Both brothers expected their involvement in science and engineering to yield benefits to themselves—they were always pleased when an idea was patentable. But at the same time they wanted their inventions to benefit the community as a whole by promoting economic activity or public health. James was active in civic reform, working to benefit the community through the provision of a better water supply, and the creation of public parks where the workers of a newly industrialized city such as Belfast could take the air.

Like many nineteenth-century scientists, both brothers saw their investigations of nature as a means of understanding the divine creation. The motivation underlying their work on thermodynamics was both practical and religious. As Crosbie Smith and M. Norton Wise have shown in their study of Kelvin, the brothers' worldview focused on the source of energy which drove all natural processes. The ultimate source of energy was God—He had created just so much energy in the beginning, and the laws of nature He had instituted led to an inevitable decline in the amount of energy that was left available for useful work in natural processes. This was a universe with a built-in trend toward what would later be known as the 'heat death', the point at which all matter was at a uniform temperature. At this point the total amount of energy was still the same as at the creation, but none was available to make anything happen, because useful work can only be obtained if there is a difference of temperature between the source of the energy (such as the steam engine's boiler) and the sink (the environment into which the waste steam and water is exhausted). Small wonder that with a worldview in which the dissipation of useful energy was an inevitable part of the divine plan, the two brothers were driven by a desire to minimize the amount of unnecessary waste in any machine. Throughout their careers, they strove to design machines which extracted as much of the useable energy as possible, losing only what the laws of nature made inevitable.

These views also led them to take up positions on some of the great debates which rocked the scientific world. Kelvin, as is well known, tried to calculate the length of time in which the earth could remain geologically active, before cooling to a dead ball of rock. He came up with figures that reduced the amount of geological time to a level at which Darwin's theory of evolution would have been untenable.[3] In the 1860s James wrote to his brother about the relationship between life and the processes of physical nature. He thought that living things might be the only vehicles that could violate the law of the dissipation of useful energy, in effect creating new energy in addition to that supplied to the universe in the creation.[4] William saw less reason to exempt living things from the laws of physics, suggesting that will-power could only alter the direction of natural processes, switching the consumption of energy into new and unpredictable channels, without actually violating the laws of thermodynamics. It is significant that for both of them, the ability of a living body, including that of a human being, to have a real influence on the world was of paramount concern. Science, religion, and morality went hand in hand.

Having alerted us to the richness of the culture within which James Thomson worked, I want to put a little flesh on the bones of the issues I have just sketched in. Let me begin with an outline of his career.[5] He was born in 1822, two years before William. Their father, also James Thomson, was Professor of Mathematics at the Belfast Academical Institution. The children were educated at home until their father moved to take up the Chair of Mathematics at Glasgow in 1834, after which they studied at Glasgow under their father and the other professors. William eventually went to Cambridge for training in mathematics, while James worked under engineers in various parts of the country, ending up at the Millwall shipbuilding works of William Fairburn, one of the leading figures in the construction of the new ocean-going iron steamships. His health then broke down and he moved back to Glasgow where he collaborated with William in his work on thermodynamics and began his career as an inventive engineer. In 1851 he moved to Belfast where he opened an office as a civil engineer and served as engineer to the Water Commissioners. He married Elizabeth Hancock in 1853. The following year he became Acting Professor of Engineering at Queen's, being appointed to the Chair of Engineering three years later. Thomson replaced James Godwin, the first

Fig. 4.1. The University of Glasgow drawn by James' niece, Agnes Gardiner King. James resigned from his chair at Queen's, Belfast to join his brother in Glasgow, where he held the chair of engineering from 1873 to 1889. (From A. G. King, *Kelvin the Man*, Hodder & Stoughton Ltd. 1925 opposite p.40.)

professor of engineering, who had worked mainly on railways. Although providing much more teaching than Godwin, he retained his practice as an engineer. He lived for some time at No. 17 University Square (three houses removed from my own office) where he redesigned the sewers.[6] He remained at Queen's until moving to take up the Chair of Engineering at Glasgow in 1873. In 1889 he was forced to resign the chair after experiencing severe problems with his eyesight—his papers in the archives at Queen's include some letters from this last part of his life written in enormous handwriting, often on very large sheets of paper (although his daughter also acted as his amanuensis).[7] He died in May 1892, followed within a week by his wife and younger daughter, all succumbing to a 'severe cold', probably pneumonia.

While practicing as an engineer in Belfast, Thomson served as resident engineer to the Belfast Water Commissioners and advised on the introduction of steam engines to improve the city's water supply. He also designed a weir for the river Lagan. He belonged to many of the societies which served as the backbone of the city's social and intellectual life, and frequently read papers at their meetings. A paper read to the Belfast Social Inquiry Society in 1852 advocated the provision of public parks for the benefit of the citizens, and led indirectly to the creation of the Ormeau Park. Thomson's original preference was for a park much closer to the city centre, created on ground surrounding the Blackstaff River which he proposed to drain.[8]

Many of Thomson's most important inventions were derived from his studies of fluid motion. In 1850 he patented a vortex turbine which was designed to minimize the loss of energy by careful control of the manner in which the water entered the turbine wheel.[9] High pressure water was injected from the outside and transferred its energy to specially shaped vanes which kept the water in equilibrium as its pressure dropped. These turbines were immensely successful and were used all over the world. Thomson also designed an improved centrifugal pump which increased efficiency from 50% to 70% and was widely used.[10] He invented a jet pump, which had no moving parts, but relied on a high pressure jet of water to create a suction effect.[11]

Before he returned to Belfast, James had used his experience with marine steam engines to good effect during the collaboration with his brother which led to the creation of the modern science of thermodynamics. Working from the basic principles of the new science, in 1847 he predicted that the freezing point of water would be decreased if the pressure was increased. This effect was demonstrated experimentally by William in the following year.[12] Arising from this work James became interested in the phenomenon of regelation, by which a wire bearing a load can pass through a block of ice by melting the ice beneath, the water then re-freezing on top of the wire where it is no longer subject to pressure. He read a paper on this phenomenon to the Belfast Natural History and Philosophical Society in 1857 and corresponded with Michael Faraday on the subject.[13]

Equally significant was his collaboration with Thomas Andrews, his fellow professor at Queen's, who worked during the 1860s on the relationship between liquids and gases. Andrews applied extremely high pressures and low temperatures to gases which had hitherto resisted all efforts to liquefy them. He argued that for these substances there was a continuous change of state, rather than a sharp transition from gas to liquid. Thomson too published extensively in this area.[14]

Thomson's vortex turbine was inspired by his interest in fluid mechanics, which allowed him to calculate the best way of getting useful work from high pressure water. As with his work on thermodynamics, a major inspiration was the desire to minimize waste. In a world where energy was, by its very nature, always becoming less available for useful purposes, it was vital that no unnecessary waste should occur. Whether the source was heat in a steam engine, or water-flow in a turbine, the laws of physics could be applied to keep waste to the minimum that nature would allow.

I want to turn now to Thomson's interest in what we would today call the environmental sciences. Although primarily an engineer, his interest in the physics underlying the processes he wished to control encouraged him to study the natural phenomena associated with the areas in which he worked. In a surprising number of cases, this led in turn to original scientific work that was not linked directly to his engineering, but which threw light on the processes which shape our environment—the flow of rivers, the circulation of the winds, and the action of ice on the Earth's surface. Thomson's intellectual horizons thus stretched far beyond those that we might expect for someone whose prime enthusiasm was engineering: he was also a natural philosopher in the widest sense of that term—a term still in common use at the start of his career, long before it was replaced by the modern designation of 'scientist'.

As early as 1841, his exploration of the wider implications of the new doctrine of energy led him to speculate about the tides in the oceans. Vast amounts of energy are wasted in the constant rise and fall of the sea-level: where did it all come from? By this time the mechanism governing the tides was well understood, but Thomson realized that the rotation of the Earth beneath the 'humping' of the seas created by the gravitational pull of the Sun and the Moon must imply that the seas exert a frictional effect on the planet's rotation. In effect, the tides are gradually slowing down the Earth's daily rotation, and that is the source of the energy being dissipated so liberally.[15]

More directly related to his practical work as an engineer for the Belfast water commissioners was his interest in the ways in which the water flowing along a river bed actually shape the course of the river. He investigated the effects of the constant flow of water on the curvature of the river's banks, noting how the land surface would be eroded on the outer part of the curve, while on the inner curve the slackening of the flow would lead to the accumulation of sediment. He even built a model river in which lengths of thread attached to pins could be used to illustrate the direction of flow at different points on the bend.[16] Thomson was able to go far beyond the commonsense understanding of the process by which a river flowing along an alluvial plain tends to increase the size of its loops, eventually leaving some to be isolated as 'ox-bow' lakes. Ever the practical man, he also noted that his insights could be used to explain how the curvature of pipes interfered with the smooth flow of water. Similar arguments were used to explain how some rivers accumulate sand and gravel at their mouths to form 'bars' which block access by shipping.[17]

Thomson's studies of the properties of ice, especially its plasticity under pressure, led him to take an interest in the effects of glaciers in shaping the landscape. Victorian scientists made extensive studies of Alpine glaciers, and the concept of an 'ice age' in which much of northern Europe was covered by ice was introduced in the 1840s by Louis Agassiz. A number of British geologists were encouraged to explain the topography of the northern parts of the country by assuming that the land was eroded by ice. James Thomson took an interest in a phenomenon which also attracted the attention of the young Charles Darwin: the so-called parallel roads of Lochaber, especially Glen Roy. These horizontal indentations on the upper levels of the valleys were recognized as the remains of ancient beaches, and Darwin hypothesized that they were produced when the whole of Scotland had been sunk

Fig. 4.2. William, James, and their eldest sister Elizabeth in 1885, drawn by Elizabeth's daughter, Agnes Gardiner King. (Courtesy of the National Portrait Gallery, London.)

beneath the ocean. He later admitted that this was his greatest mistake in science—and it was Thomson who played a major role in establishing the correct explanation. His 1848 paper on the parallel roads of Lochaber argued that the roads were the remains of beaches formed by glacial lakes, created by glaciers damming the lower reaches of the valleys. When the ice melted, the water drained away, leaving the parallel roads to puzzle modern observers.[18]

Thomson's studies of thermodynamics and fluid motion in gasses also led him to take an interest in the circulation of winds in the atmosphere. He gave an address to the British Association in Dublin in 1857 in which he used physical principles to explain why the great wind systems of the globe, including the trade winds, circulate in definite bands at particular latitudes. Here he was engaging with another key problem recognized by environmental scientists in the Victorian era, of concern to meteorologists and oceanographers as well as physicists. Thomson's contributions were significant enough for him to be asked to give the Royal Society's Bakerian Lecture on this topic, which was delivered on 10 March 1892, shortly before his death.[19] Interestingly, Thomson's papers in the Queen's University archives also show that he was trying to apply the same principles to explain the coloured bands on the surface of the planet Jupiter. He was forced to admit, however, that his theory would not work, given the astronomers' arguments that Jupiter's atmosphere is immensely deep. His ideas *would* apply, he insisted, to a Jupiter-sized planet with a thin atmosphere like the Earth's.[20]

I will conclude with another of Thomson's contributions to the environmental sciences, one with a particular connection to Northern Ireland. In 1877 he published a paper on the forces that could have produced the unusual columnar structures found at the Giant's Causeway. It was widely admitted by then that basaltic rocks had cooled from a molten state, but geologists were puzzled as to the nature of the forces that could build up in the solidifying rock to produce such regular fractures. Some thought the sections of the columns had begun as spheres of solidifying rock which expanded until the intersected with one another. Thomson rejected this view and supported a rival theory in which forces analogous to those which create the cracks in drying mud were responsible, but he was able to provide a much more detailed account of the way the forces would build up within a deep, homogeneous mass of cooling rock.[21]

I hope that I have provided enough examples to show that James Thomson was not only an engineer of genius, but also a scientist who thought on the grandest possible scale about the physical forces which govern the processes he wished to control. He was not content to study those forces solely on the small scale needed to design pumps and turbines—he was fascinated by the way the same forces operated on a global scale to create major features of our environment. Given his interest in public health, the supply of clean water, and open spaces within the city, we can see him as a figure whose work not only helped bring together the activities of the scientist and the engineer, but who also saw both science and technology as operating within a global system which it was our duty to understand with a view to benefiting the whole of humankind. If this was the culture of a Victorian engineer, it was a culture with the widest possible dimensions.

5

Fifty-Eight Years of Friendship: Kelvin and Stokes

Alastair Wood

'I always consult my great authority, Stokes, whenever I get a chance'

—William Thomson, Lord Kelvin, Baltimore Lectures, 1884.

INTRODUCTION

The contributions of Sir G. G. Stokes to mathematics and physics were many and varied. A contemporary and close friend of Kelvin, his name has become well known to generations of scientists, mathematicians, and engineers, through its association with various physical laws and mathematical formulae. In standard college textbooks we encounter Stokes drift, Stokes law, Stokes theorem, Stokes phenomenon, Stokes conjecture, and the Navier–Stokes equations. But while the contributions of Stokes to mathematical physics are impressive, many believe that his greater contribution was as a sounding board for his contemporaries, providing sound advice, good judgement, and mathematical rigour. Many great scientists deliberately avoid administration, committees, and editorship of journals on the grounds that such activities would stifle their creativity. But Stokes threw himself into this role and, although a shy man of few words, was regarded with affection by colleagues throughout these islands for his extraordinary generosity in encouraging their work and communicating ideas to others, usually through his extensive correspondence. Kelvin was his principal correspondent over a period of 56 years and we are fortunate to have the largest surviving correspondence between two Victorian physicists preserved, mainly in the Cambridge University Library, and published in a two volume collection by David B. Wilson[1]. These letters give us a valuable insight not only into the research and research methods of two giants of nineteenth-century British science, but also into the day-to-day running of Cambridge and Glasgow Universities and of the Royal Societies of London and Edinburgh at a critical time in their histories.

COMPARISON OF FAMILY BACKGROUNDS AND FORMATIVE INFLUENCES

The family backgrounds and early lives of Stokes and Kelvin at first sight seem similar—both were born in the island of Ireland into families containing academics (as well as ministers of religion and medical doctors in Stokes's case), both received their secondary education in Britain and both were graduates of Cambridge University, Stokes being Senior Wrangler in the mathematical tripos in 1841 and Thomson Second Wrangler in 1845. Both were first Smith's Prize winners and on graduation both took up fellowships in their respective colleges, Thomson in Peterhouse and Stokes in Pembroke. Despite these apparent similarities their social backgrounds were very different.

George Gabriel Stokes was born in Skreen Rectory, County Sligo, on the north-west coast of Ireland, on 13 August 1819. His father, Rev. Gabriel Stokes, is recorded as being of a taciturn nature and was aged 52 when George Gabriel was born. His mother was beautiful, but stern, and the children stood in awe of her[2]. Coupled with his grandfather's known shyness, it is easy to imagine the source of Stokes's 'rich silences'. Stokes was a member of a well-established Anglo-Irish family of Trinity academics (his grandfather, John Stokes, had been Regius Professor of Greek in Dublin University) and clergymen. His three eldest brothers were to be ordained as Anglican priests and throughout his life he clung to the basic evangelical truths that he had learned from his father. This web of Anglican family connections advanced his early education and his career at Cambridge. His family background explains the positions that he was to adopt in later life, namely the maintenance of the established status of the Anglican Church in England and Ireland (he was not successful in the latter), his earnest commitment to the Act of Union between Britain and Ireland and his dedication to the monarch, Queen Victoria, as Defender of the Faith.

The Stokes family had settled in Dublin from England some 200 years earlier. The first of the family to be mentioned in Ireland was Gabriel Stokes, born in 1682, a mathematical instrument-maker in Essex Street, Dublin, who became Deputy Surveyor General of Ireland. At that time the Test Act and the Penal Laws, which placed Catholics under major political and social disadvantage, were still in force.

All the Stokes were Anglican in religious matters. The Church of Ireland was predominantly the church of the Anglo-Irish and English landowners, although in County Sligo the congregation would have been mainly agricultural settlers and artisans who had come over in the wake of Cromwell's army in the 1640s. The largest religious grouping were the Catholics, at that time mainly peasants and tenant farmers in a subsistence economy, their rights removed by the Penal Laws.

A third group consisted of Protestant Dissenters, organized, cohesive, and concentrated mainly in the north of Ireland. It is to this group that the Thomson family, settlers from Scotland in 1641, belonged. The Dissenters were allies of the Church of Ireland in troubled times. Although refused legal toleration (they could not own land), they were prosperous and secure by comparison with the Catholics. But as non-conformists they were denied entry to the older universities in England and Ireland, although not in Scotland where the established church was Presbyterian. Unlike the long academic background of Stokes, Kelvin's father James Thomson was the largely self-taught son of a Ballynahinch farmer who by dint of hard work had in 1810 gained admittance to Glasgow University. He had intended to become a Presbyterian minister, but accepted instead a post as Professor of Mathematics at the Belfast Academical Institution, which he left in 1830 to take the Chair of Mathematics at Glasgow University. It would be fair to say that James Thomson was more concerned with teaching than with original research, unlike his sons William and James.

The social structure in nineteenth century Ireland meant that an Anglo-Irish family enjoyed a considerable advantage in every walk of life. In some respects they formed a separate community, disliked and largely avoided by the Irish, but mistrusted by the British government. Nonetheless they contained a small, closely-knit scientific elite who made a disproportionately high contribution to nineteenth-century physics. This group included, among others, the astronomer William Parsons, 3rd Earl of Rosse (1800–67), the mathematician and physicist Sir William Rowan Hamilton (1805–65), the mathematician James McCullagh (1809–47), the seismologist Robert Mallet (1810–1881), the geometer George Salmon (1819–1904), the physicist Samuel Haughton (1821–97) (a Quaker), the natural philosopher George Francis Fitzgerald (1851–1901), and the statistician Ysidro Edgeworth (1845–1926), nephew of the hydrographer Admiral Francis Beaufort (of wind-scale fame).

By the time of Stokes's birth in 1819, compromise had begun to replace the Penal Laws and relations had become easier. Although his great-uncle Whitley Stokes, a medical fellow of Trinity College Dublin, had a minor involvement with the United Irishmen in the Rebellion of 1798, the Stokes, in common with other Protestant families, were alarmed by the invasion of a French army and the violence that accompanied this movement. Thenceforth they took a pro-unionist stance when the Irish Parliament was abolished through the Act of Union of 1800. But in the case of George Gabriel Stokes, his daughter, Mrs. Laurence Humphrey, records in her memoir[3] 'The late Queen's Jubilees were occasions which he thoroughly enjoyed, for like all Irishmen of his way of thinking, he was a very loyal subject'. There can be little doubt that Stokes regarded himself as British but was greatly influenced throughout his life by his Anglo-Irish upbringing. The same could be said of Thomson, but when it was to his political advantage, as in a speech in Birmingham in 1883 in support of the Liberal Unionists, he was prepared to present himself 'as an Irishman' on the Irish Question. This makes it hard to account for Thomson, writing to Stokes on holiday with his family in Portstewart in the north of Ireland in August 1848: 'Are you trying to re-collect the scattered supporters of Irish Nationality and make another effort for independence? Or do you fraternise with the Saxon, the enemy of our country?' We can only suppose that this forms part of a private joke directed at the cultural nationalism of that time.

Stokes received his early education in Skreen from the parish clerk, George Coulter, and in 1832 he was sent away to Dr. Wall's School in Dublin. His father the Rev. Gabriel Stokes died in 1834, and his widow and two daughters had to leave Skreen Rectory, but money was found to send George Gabriel to continue his education at Bristol College in England. His mathematics teacher, Francis Newman,

Fig. 5.1. George Gabriel Stokes. (From *The Mathematical and Physical Papers of the late G. G. Stokes* volume 4, Cambridge University Press (1904), frontispiece.)

Fig. 5.2. William Thomson. (From Silvanus P. Thompson, *The Life of William Thomson Baron Kelvin of Largs*, 2 volumes, Macmillan & Co., London, 1910, opposite p. 446.)

brother of Cardinal Newman, wrote that Stokes 'did many of the propositions of Euclid as problems, without looking at the book'. Stokes appears to have had a great affection for Newman, whom he records as having 'a very pleasing countenance and kindly manners'. Newman was also responsible for the following anecdote[4]:

> His habit, often remarked in later life, of answering with a plain yes or no, when something more elaborate was expected, is supposed to date from his transference from an Irish to an English school, when his brothers chaffed him and warned him that if he gave long Irish answers he would be laughed at by his school fellows.

William Thomson, on the other hand, was part of a close and affectionate family, but this idyllic upbringing was broken by the early death of his mother in 1830. Shortly after, when Thomson was six, the family moved to Glasgow. Thomson was taught at home by his widowed father before matriculating as a student in his father's university in 1834 at the early age of 10. Thus Thomson lived at home throughout his education until he left for Cambridge in 1841.

KELVIN AND STOKES IN CAMBRIDGE

Stokes entered Pembroke College, Cambridge, as an undergraduate in 1837. Distinguished graduates from Pembroke included the martyr, Bishop Ridley, the poets Spenser and Gray, and the statesman William Pitt. Although a mathematical prodigy at school, Stokes was beaten into second place in his first year at Pembroke by one John Sykes. From second year onwards he studied, as was the custom at that time, for the highly competitive Mathematical Tripos with a private tutor, the 'Wrangler maker' William Hopkins. These private coaches formed a central locus through which the prevailing views of past Wranglers, who had become the examiners and textbook writers of the present, were passed down to the wranglers of the future. So effective were these studies that Stokes was Senior Wrangler (that is, placed first in mathematics in the whole university) and first Smith's Prizeman in 1841 and elected to a Fellowship at Pembroke. Stokes was to remain at Cambridge until his death in 1903, a total of 66 years.

Thomson entered Peterhouse at the age of 17 in 1841, having already published a paper in the *Cambridge Mathematical Journal* defending the work of Fourier. During his time as an undergraduate at Cambridge he published a further 10 research papers in the Journal. Unlike the quiet Stokes, he made the most of social life in College and engaged in various athletic pursuits, including rowing, which his father (who wrote to him frequently on the good use of time and money) feared would bring him into loose company and drinking that would 'ruin him for ever'. Thomson had the same private coach, William Hopkins, as Stokes and great things were expected of him in the Mathematical Tripos, but to general surprise he finished as Second Wrangler. It was not unusual at the time for mathematicians of real creativity to fail to finish first, as the tripos papers tended to be a test of memory and speed. But Thomson came first in the Smith's Prize examination, whose questions emphasized problem solving and were more suited to his abilities, and was appointed Fellow of Peterhouse in 1845.

Brought up with his father's anti-Tory views, Thomson moved easily among the circle of Whig mathematicians of the Journal. This circle did not include Stokes (who was later to sit with the Tories in Parliament) and Thomson only came to know him after taking his degree early in 1845, when he discussed the questions in the Smith's Prize examination with him[5]. But the relationship seems to have prospered, as we find Stokes writing a reference for Thomson's Glasgow University professorial application 'as a personal acquaintance'. Thomson spent only one year teaching at Peterhouse before leaving in 1846 to take up the Chair of Natural Philosophy in Glasgow, which he was to occupy for over half a century. Although Thomson spent only five years in Cambridge, he retained a strong attachment to the university, returning for events at Peterhouse and duties as an examiner, even although he was to turn down the offer of the new Cavendish Chair of Physics on three occasions in the 1870s and 1880s in favour of remaining in Glasgow.

Thomson took over the editorship of the *Cambridge Mathematical Journal* in 1845 and continued in this post during his early years in Glasgow, relying on former colleagues in Cambridge to help with the editorial work and the refereeing of papers. He changed the name to the *Cambridge and Dublin Mathematical Journal*, attracting Irish-based contributors such as Boole and Hamilton: but Thomson became disillusioned with the preoccupation with pure mathematics and sought to change the thrust of the *Journal* to 'mathematical analysis of the material world'. He expressed his frustration in a 1851 letter to Stokes as follows:

> I shall be very glad to get publishing your paper in the Journal, as I am very desirous of getting such papers on physical subjects sometimes in place of the endless algebra and combinatorics which so abound.

After an unsuccessful attempt to get Stokes to take over as editor, he stood down in favour of N. M. Ferrers in 1852.

THE EARLY RESEARCH CORRESPONDENCE

By today's standards the correspondence of Thomson and Stokes covered a surprisingly wide range of topics. Thomson had a steady research output throughout his life and published by far the greater number of research papers on topics ranging from fundamental physics to industrial applications. Stokes enjoyed his most active period from 1845 to 1860, concentrating his activities on his very significant contributions in light and fluid dynamics, thereafter calmly and competently managing the running of British science. Almost all of Stokes 138 published papers (unlike Thomson, only one was jointly authored) appear in the five volume *Mathematical and Physical Papers*[6]. The first three volumes were edited by Stokes personally, the last two by his successor in the Lucasian Chair at Cambridge, Sir Joseph Larmor, also an Irishman. Larmor published the memoirs and scientific correspondence of G. G. Stokes[7] in 1907, an important source for later scholars, as well as the last three volumes of Kelvin's *Mathematical and Physical Papers*.

Thomson, on the other hand, had a tendency to dash off in all directions in pursuit of research topics. There were periods when their correspondence was largely one way, Thomson bombarding Stokes with questions and proposals. Stokes served as a sounding board for Thomson's ideas, keeping him abreast of research that had already been done in the areas he chose to enter. When Thomson was carried away by his new enthusiasms, it was the quiet and cautious Stokes who brought him back down to earth, pointing out flaws in his arguments. Kelvin later acknowledged that 'Stokes gave generously and freely of his treasures to all who were fortunate enough to have opportunity of receiving from him'. The different temperaments of the two men are well illustrated by an anecdote of J. J. Thomson[8], Director of the Cavendish Laboratory in Cambridge:

> When Kelvin was speaking, Stokes would remain silent until Kelvin seemed at any rate to pause. On the other hand, when Stokes was speaking, Kelvin would butt in after almost every sentence with some idea that had just occurred to him and which he could not suppress.

But when Stokes spoke, he did so with authority.

Their correspondence began after Thomson's move to Glasgow. It would be wrong to suppose that their letters were exclusively on research matters. Stokes' first letter to the new Professor Thomson in October 1846 concerns his subscription to the Cambridge Philosophical Society and ends 'I suppose you have hardly begun your lectures yet. I hope you will like your work'. Writing to Stokes on elastic solids in November, Thomson finishes 'Yesterday I gave the introductory lecture which was rather a failure as I had it all written and I read it very fast', a classic mistake of new lecturers in any age. Replying, the more experienced Stokes comforts Thomson with the statement 'When once your first course is completed you will not find it such hard work'.

Stokes's early research was in the area of hydrodynamics, both experimental and theoretical, during which he put forward the concept of 'internal friction' of an incompressible fluid. The fundamental equations for the motion of incompressible fluids were first published in 1822 by the French civil engineer Claude Navier. Modern physicists would find his analysis based on an unacceptable notion of intermolecular forces. Using his concept of internal friction in fluids, it was Stokes in 1845 who put the derivation of these equations on a firm footing. Thus it is by both names, Navier–Stokes, that these equations are known throughout the world today. They are used to describe the wake behind a boat or the turbulence behind a modern aircraft. The equations are employed on a daily basis by aeronautical engineers, ship designers, hydraulic engineers, and meteorologists. While simple examples, such as steady flow in a straight channel, can be solved exactly, and some more complicated cases admit an approximate numerical solution by large-scale computer packages, the mathematical problem of the existence and uniqueness of a general solution to the Navier–Stokes equations remains unsolved today. Stokes's work was independent of the work of Poisson and Saint-Venant that was appearing in the French literature at the same time, but Stokes' methods could also be applied to other continuous media such as elastic solids. He later turned his attention to oscillatory waves in water, producing the subsequently verified conjecture on the wave of greatest height, which now bears his name.

Thomson's researches during this period were mainly on the mathematical theories of heat, electricity, and magnetism, making analogies between them and reconciling earlier research of Coulomb and Faraday in electricity and Carnot and Joule in heat. This led to his sharing with Clausius and Carnot the credit for the second law of thermodynamics which sets limits on the efficiency of engines and is generally agreed to be his main contribution to fundamental physics. Many of the exchanges with Stokes in the late 1840s reflect these interests, but in February of 1849 we find an intensive correspondence of seven letters on a career matter.

A TALE OF TWO PROFESSORSHIPS

The elder James Thomson died on 12 January 1849, creating a vacancy in the Chair of Mathematics in Glasgow University. William Thomson had clearly been pressing Stokes to send in an application: given the relative slowness of communications in early Victorian Britain, there were clear advantages to a research collaboration being conducted in the same university. But a difficulty of religious confession had raised its head. Whereas fellows at Cambridge had to conform to the established Church of England, in Glasgow a statement of conformity to the presbyterian Church of Scotland was required of professorial candidates. Thus on 12 February we find Stokes writing to Thomson from the Rectory of his eldest brother the Rev. John Whitley Stokes in Aughnacloy, County Tyrone, Ireland.

> I arrived here safe on Saturday, after a rather rough passage. It was so rainy that I saw but little of the scenery of the Clyde [this suggests he had been on a visit to Thomson in Glasgow, as Stokes would normally have crossed by the Liverpool route on his regular visits to his family] ... Now as to the [religious] tests. My brother is decidedly of the opinion that the straightforward course is to decline to take them unless I am prepared to become a thorough Presbyterian, which I certainly do not mean to become ... I have determined to back out of it in the most polite way I can ... I suppose my best plan is to write apologies to those whom I have asked to give me testimonials.

Thomson, however, was not prepared to let matters rest there. We find him writing on 14 February to Stokes:

> Your letter which I received this morning has put me quite into a state of agitation ... I thought as you seemed to be convinced about the tests, you would not have felt any difficulty, after consulting with your friends, about coming forward as a candidate.

Thomson went on to explain at length that it should be possible to conform in a lax sense, while maintaining his attendance at the Episcopal Church (as the Church of England was called in Scotland). Thomson himself was liberal in his religious practice and attended both the Episcopalian and Presbyterian churches. Stokes was not moved by this powerfully written letter, and after a further exchange wrote on 22 February to confirm 'the uselessness of standing in the way I spoke of', which was to make a statement at the beginning of his letter of application that he was not prepared to conform to the Church of Scotland. There had been a precedent for this in 1838 when another Cambridge Senior Wrangler, Philip Kelland, had been appointed Professor of Mathematics in Edinburgh University. In modern times we may find it incredible that the close cooperation of two great Victorian physicists in one institution was prevented by differences in religious practice between two Protestant denominations, but Stokes was a person of strong principles and great integrity. We cannot help but ask what new results in mathematical physics might have emerged from a 50 year collaboration of Thomson and Stokes in the same university. Had he been in Glasgow, would his research efforts have been dragged down by the heavy weight of administration that Stokes carried in the second part of his life?

But Stokes did not have long to wait for his chair at Cambridge. Joshua King retired on grounds of ill health and on 23 October 1849 the College Masters elected Stokes unopposed to the Lucasian Chair of Mathematics. Two days later we find Thomson, who had seen the appointment in *The Times*, writing to Stokes for the first time as

My Dear Professor,

I am very glad for the sake of mathematics as well as for your own that you have got the chair as you will now have every inducement to go on as you have been doing, and we shall feel much surer of you than when you only had your fellowship to connect you with Cambridge. No wonder you have just discovered a theorem!

Set up by a statute of King Charles II in 1664, the achievements of many of the 17 holders of the Lucasian Professorship[9] have made it one of the most famous chairs of mathematics in the world. It has been held largely by applied mathematicians and theoretical physicists. The second professor was Sir Isaac Newton (1669–1702) and the eleventh was the originator of modern computing Charles Babbage (1828–1839). Since Stokes (1849–1903) there have been only four incumbents: the last major champion of the 'luminiferous ether' as a basis for the theory of light, Sir Joseph Larmor, (1903–1932); the Nobel Laureate in Physics Paul Dirac (1932–1969); the fluid dynamicist and admirer of Stokes, Sir James Lighthill (1969–1980); and the cosmologist Stephen Hawking (1980). But the chair was relatively poorly endowed from the fluctuating agricultural income of estates in Bedfordshire, and to augment his earnings Stokes also taught at the School of Mines in London throughout the 1850s.

Thus in keeping with the traditions of the Lucasian Chair, Stokes was far from being a pure mathematician. His mathematical results arose mainly from the needs of the physical problems which he and others studied. His paper on periodic series concerned conditions for the expansion of a given function in what we now know

Table 5.1. Table of Lucasian professors

Table of Lucasian Professors of Mathematics	
Isaac Barrow	1663–1669
Isaac Newton	1669–1702
William Whiston	1702–1710
Nicholas Saunderson	1711–1739
John Colson	1739–1760
Edward Waring	1760–1798
Isaac Milner	1798–1820
Robert Woodhouse	1820–1822
Thomas Turton	1822–1826
George Airy	1826–1828
Charles Babbage	1828–1839
Joshua King	1839–1849
George Gabriel Stokes	1849–1903
Joseph Larmor	1903–1932
Paul Dirac	1932–1969
James Lighthill	1969–1980
Stephen Hawking	1980–

as a Fourier series. In the course of this work he made use of what we now know as the Riemann-Lebesgue lemma some seven years before Riemann. Stokes is also credited with having had the idea of uniform convergence of a series, albeit in a restricted case. His major work on the asymptotic expansion of integrals and solutions of differential equations arose from the optical research of G. B. Airy on caustics behind rainbows, where he was the first to recognize what we know today as Stokes' phenomenon, that is fundamental to the modern subject of hyperasymptotics. Stokes employed the saddle point method for integrals in the complex plane a full decade before Riemann, to whom it is usually attributed. He also realized the link between his asymptotic methods for integrals and Kelvin's method of stationary phase. The well-known theorem in vector calculus which bears his name, and is fundamental to modern differential geometry, is however not due to Stokes, the result having been communicated to him in the postscript of a letter from Thomson in July 1850. The proof was set by Stokes as a problem in the Smith's Prize examination at Cambridge in 1854. It is said that James Clerk Maxwell was the only student to successfully attempt the question.

Although appointed to the Lucasian Chair for his outstanding research, Stokes showed a concern in advance of his time for the welfare of his students, stating that he was 'prepared privately to be consulted by and to assist any of the mathematical students of the university'. Although required to do so by the statutes, Charles Babbage never once addressed classes. Stokes immediately advertized that 'the present professor intends to commence a lecture course in Hydrostatics', which he was still delivering 53 years later, in the last year of his life. Stokes' manuscript notes still exist in the University Library in Cambridge, although he eventually became one of the first people in Britain to make regular use of a typewriter.

THE MECHANICS OF THE CORRESPONDENCE

This is perhaps the place to digress on the subject of Stokes's handwriting. Stokes was a prolific correspondent, using the penny post as a modern scientist might use e-mail. He communicated endlessly with the leading scientific figures of his day, his exchanges with Thomson alone comprise 656 letters. Sometimes he wrote to the same person twice in the same day, although this may have arisen from the need to catch the post, which departed at a fixed time each day. Often we find a brief letter announcing his result sent in the morning, with the detailed reasoning following in the evening. Many of his correspondents, Thomson included, found the writing of Stokes very difficult to read. In a letter sent to Stokes in March 1875 enclosing a recommendation for election to the Royal Society to be forwarded to C. V. Walker, Thomson jokes '... if you (who *ought* to be able to read any writing) can read his address from his note enclosed, send it direct to him'. Stokes acquired his first typewriter, which used only upper case letters, in 1878. The second, used from 1886, also had all capital letters and only the third, used from 1890, possessed both upper and lower cases. But these early machines were not without their technical problems and we find a letter on Rontgen rays on 17 March 1896 beginning:

"My dcar Kclvin,
Thc pull wirc of the lcttcr that lics bctwccn d & f got brokcn, and I havc scnt it to London to bc rcpaircd."

Apart from typewriting and postal difficulties, another problem faced by our two correspondents, especially by Stokes, was where to send the letter. Stokes's movements were predictable: the majority of his letters originate from Pembroke College, or later his home, Lensfield Cottage, in Cambridge. During the university vacations he could be found with his brother William in Denver Rectory in Norfolk or in Ireland at the homes of his eldest brother John Whitley in Aughnacloy Rectory, his sister in Malahide, County Dublin, or his father-in-law at Armagh Observatory. There is no evidence that Stokes ever took a holiday abroad and when in Ireland he generally tended to stay with relatives. But Thomson was much more adventurous and travelled a great deal, not only on academic or professional business (mainly in connection with the transatlantic cable or the sea trials of the nautical equipment that he designed), but also for pleasure. As early as October 1847 he writes:

> I have had a very pleasant tour in Switzerland ... and when we meet we shall have something to say about the Mer de Glace and the Faulhorn, as well as about hydrodynamics.

In another early letter he is about to depart on a walking tour of the Pyrenees. Summer vacation letters come from addresses on the Isle of Arran. Later we find him departing with his wife to tour Malta, Sicily, and Italy and again to pass holidays in Nice and Aix-les-Bains. After he bought his 126-ton yacht, the *Lalla Rookh*, in 1870, Thomson's letters could bear the postmark of any port on the western seaboard of Europe. Occasionally Stokes was requested to send his reply to the next anticipated port of call!

The correspondence also sheds light on the working practices of the two men. In our generation, Thomson would have been an exponent of time management. Some letters are dashed off in moments that he finds are unexpectedly free, or in boats, trains, and even horse-drawn carriages. There is even a substantial letter on wave motion in fluids written in September 1880 while Thomson is waiting in Largs with a friend who is to have a picture taken at Fergus's Photographic Studio, which is the address at the head of the letter. But even then Thomson breaks off with:

> Obliged to stop short to go out for an afternoon sail and use a good breeze to test relative merits of globular and elongated sinkers (16lbs). Which do you think best?

The opinion of Stokes on this latter topic has not survived, but he replied conscientiously to the question of wave discontinuities the next day.

This small incident shows well the difference in temperament between the two friends. Thomson as an undergraduate at Cambridge had typically set aside three hours each afternoon for exercise. He was a much more rounded character than Stokes and it is possible that he managed his working time efficiently in order to enjoy his recreational interests. This is not to say that he did not continue to ruminate on the great truths of physics during these outdoor activities and his frequent social distractions, for he was a warm and genial person who enjoyed the company of others. Stokes was a shy and self-effacing character, known for his reticence, and although not anti-social, he certainly had no reputation for small talk. He was interested only in mathematics, physics, and religion. There is a story in Cambridge about a dinner party[10] where Stokes once found himself seated next to a young American woman. The guests were amazed to find Stokes engaged in animated conversation and even smiling. When asked later about this rare event, the young woman replied that she had merely asked Sir George whether he preferred algebra or geometry.

Stokes had the custom of working long into the night at Lensfield Cottage, fortified only by strong tea from his trusty 'Brown Jenny' teapot. His careful and diligent approach is typified by the two letters to Kelvin of 13 March 1896.

I have a good deal to write you, and have not time to catch this post with it. I write just to say that I do not by any means think that Lord Blythswood's experiment proves the reflexibility of the X-rays.

He continues in a longer letter, started the same day, to give detailed reasoning for his assertion, but breaks off midway to restart the letter with:

March 14. So far last night, or rather the commencement of the small hours of the morning. I resume before breakfast.

Some letters were written in snatches over a longer period. We find in December 1898 a single letter on the discontinuity in the motion of a liquid contained by two pistons in a cylinder written over a seven day period, although it is true that the Stokes family were visiting Lensfield Cottage for Christmas. After one page of this letter, started on the 20th we find:

But now the clock has struck XII, and it is time to go to bed.
Dec 21. I resume ...

The letter was eventually completed and sent off on 26 December, but the amazing thing is that Stokes had sent four other letters to Kelvin in the interim! This controversy continued by return of post throughout the festive season until 7 January 1899. Their correspondence was not always of such frequency and in the years 1853 and 1891 no letters seem to have been exchanged. It could be that these letters have failed to survive, but this seems unlikely in the first case since Thomson in his letter of 20 February 1854 begins:

It is a long time since I have either seen you or heard from you, and I want you to write to me about yourself & what you have been doing since ever so long. Have you made any more revolutions in science?

and proceeds to describe what he himself has been doing the previous summer.

THE MIDDLE YEARS: THE LUMINIFEROUS ETHER AND THE ATLANTIC CABLE

Stokes's major advance in the 1850s was in the wave theory of light, by then well established at Cambridge, examining mathematically the properties of the 'luminiferous ether' which he treated as a sensibly incompressible elastic medium. 'Ether' was the old-fashioned name given to the medium filling all space which was thought to carry light waves as vibrations analogous to sound waves. It predated the theory of light as an electro-magnetic phenomenon introduced by the Scottish physicist James Clerk Maxwell who was appointed to the Cavendish Chair of Experimental Physics at Cambridge in 1871. After Hertz's 1887 experiment showed that electromagnetic waves could be generated by an electric circuit, the concept of the ether was attacked by another Irish physicist (and correspondent of Stokes) George Francis Fitzgerald, the leading follower of Maxwell.

Thomson was also an enthusiast of the ether. To begin with he used the model of the ether as an elastic solid, developed by George Green at Cambridge in the 1840s, but switched to the compromise of a very viscous liquid, likened to 'glue water' or 'jelly', developed by Stokes. While Stokes in his later years had begun to doubt the validity of the ether model, Kelvin was still elaborating the ether model at the turn of the century. His experiments with wax and pitch, both at the extreme of viscous liquids, are still to be seen in the laboratories in Glasgow.

But the concept of the ether had enabled Stokes to obtain major results on the mathematical theory of diffraction, which he confirmed by experiment, on polarization of light and on fluorescence, which led him into the field of spectrum analysis. When light is shone on to or through a material, its spectrum is modified because the material is absorbing light of certain preferred wavelengths. In some cases it may even happen that a part of the absorbed light is radiated again at a different wavelength. This is called fluorescence, whose discovery is attributed to Stokes. His last major paper on light was his study of the dynamical theory of double refraction, presented in 1862, although late in life, from 1896 onwards, he was involved in the early investigation of Rontgen rays, now known as x-rays, corresponding intensively with Thomson as we saw above.

Prompted by a question from Stokes in a letter early in 1854, Thomson turned his attention to the problem of sending signals through long underwater cables. Telegraphs had made their first appearance in Britain in the 1830s, mainly running along the developing railways, and shorter stretches underwater to Ireland and France were completed around 1850. But the link to America was of a different order of magnitude, with severe problems such as the attenuation of signals to be overcome. The usual 'try it and see' approach would not yield results and a physical theory was needed. There followed several letters between Thomson and Stokes on the solution of the partial differential equation governing the transmission of telegraph signals. By the beginning of December the theory had advanced to such a point that Thomson wrote to caution Stokes

I should be much obliged if you would not mention to any one what I wrote to you regarding the remedy for the anticipated difficulty in telegraphic communication to America, at present, as Rankine has suggested that I should join him in applying for a patent for a way of putting it in practice & improving conductors in ordinary circumstances, to which I have agreed . . . I would be greatly obliged if your letting me have my letters until I get a copy of them made.

W. J. M. Rankine was the Professor of Civil Engineering and Mechanics in Glasgow from 1855 to 1872. Stokes sent back the letters as requested, while expressing some doubt as to the neglect of the resistance of the water. The patents were successfully obtained and the Atlantic cable was completed, after several attempts, in 1866 with Thomson's close involvement. The success of the project made Thomson not only famous, but substantially rich. Along with others involved in the project he was knighted in 1867. Proceeds from his patents and from his directorship in the Atlantic Telegraph Company allowed him to purchase in 1870 the large sailing yacht *Lalla Rookh* (mentioned above under correspondence) and in 1874 the land in the coastal resort of Largs, not far from Glasgow, on which he built his splendid mansion, Netherhall.

Stokes also enjoyed a certain involvement in industrial applications, but not at the same level. Besides his links with the School of Mines, he acted, over a period of many years, as consultant to the lensmaker Howard Grubb who ran a successful and internationally-known optical works in Rathmines, County Dublin. He also acted as advisor on lighthouse illuminants to Trinity House. Stokes' collected works include a paper on a differential equation relating to the breaking of railway bridges and, following the Tay Bridge disaster, he served on a Board of Trade committee to report on wind pressure on railway structures. But his interest in industrial applications of physics came nowhere near that of Thomson and there is no evidence that Stokes ever made any money from his advice and consultancy so freely given. The Stokes family as a whole were not interested in material things and were somewhat unworldly.

LEADERS OF THE VICTORIAN SCIENTIFIC ESTABLISHMENT

The second half of Stokes' life was increasingly taken up with scientific and academic administration. A major reason for this change was that in 1851 he, along with Thomson, had been elected a Fellow of the Royal Society and shortly afterwards, in 1854, Stokes became a Secretary of the Society, where for 31 years he performed an important role in advising authors of research papers of possible improvements and related work. Besides editing their publications, Stokes also administered the distribution of government funds to support individual research. Much of the later correspondence between Thomson and Stokes concerns the refereeing of papers submitted for publication (Thomson could be a harsh critic, although Stokes did not hesitate to chide him when he was late in submission of his own work) and of research grant applications, from which Thomson's laboratories also benefited. In March 1877 we find Stokes writing to Thomson on 'the feeling in the Royal Society of Edinburgh and the general public on the matter', the matter presumably being that a scientist as rich as Sir William Thomson had no longer need of government grants. Stokes was elected President of the Royal Society in 1885, to be followed by Thomson from 1890 to 1895. Thomson was also President of the Royal Society of Edinburgh for 20 years and many of his best papers appear in their *Proceedings*.

Many famous scientists tried out their half-formed ideas on Stokes who, like Thomson, was extremely active in the British Association for the Advancement of Science. His close colleagues regretted his taking on these administrative duties and P. G. Tait (co-author with Thomson of the celebrated textbook *Treatise on Natural Philosophy*) even went so far as to write a letter to Nature protesting at 'the spectacle of a genius like that of Stokes' wasted on drudgery [and] exhausting labour'. Thomson wrote to him in 1859 of 'the importance to science of getting you out of London and Cambridge, those great juggernauts under which so much potential energy for original investigation is crushed', while attempting to persuade Stokes to apply for the vacant Professorship of Astronomy in Glasgow. But even though the religious tests, which had impeded him in 1849, had been removed, Stokes did not apply. It was perhaps unrealistic of Thomson to expect Stokes to leave such a prestigious chair in Cambridge, even although his administrative activities severely limited his time for personal research. Thomson of course also carried his share of government committees, university administration, and learned societies, not to mention the time spent in developing the industrial applications of his discoveries, but his boundless energy enabled him to maintain his research efforts in physics. Thomson was to make one more attempt in November 1884 to lighten the load on Stokes by urging him to apply for the Cavendish Professorship in Cambridge on the grounds that

> ...the income of the experimental physics chair is decidedly more than you have in the Lucasian and I thought possibly the difference might amount to even a money compensation for giving up the Royal Society work. Thus I thought of the whole thing rather as freeing you from fatiguing or possibly irksome work.

The reply of Stokes was typically altruistic:

> I feel that those who would be under me would be knowing more about the subjects than I do myself. Also it is hardly fair to block the way of promotion to younger men who might reasonably be expected to rise in their profession.

THREE MARRIAGES

In 1859 Stokes vacated his Fellowship at Pembroke, as he was compelled to do by the regulations at that time, on his marriage to Mary Susannah, daughter of Dr. Thomas Romney Robinson, astronomer at Armagh. Following a change in regulations, he was subsequently able to resume his Fellowship and for the last year of his life served as Master of Pembroke. Shortly after their marriage the couple moved to Lensfield Cottage, a happy and charming home, in which Stokes had a 'simple study' and conducted experiments 'in a narrow passage behind the pantry, with simple and homely apparatus'. Prior to their marriage Stokes, who, as we have already remarked, was a tireless writer of letters, had carried on an extensive (one letter ran to 55 pages) and frank correspondence with his fiancee. In one letter, the theme of which will be familiar to all spouses of research physicists, he states that he has been up until 3 a.m. wrestling with a mathematical problem and fears that she will not permit this after their marriage! Based on other remarks in this highly personal correspondence, David Wilson[11] suggests that: 'Stokes himself may have welcomed what others regretted—his abandonment of the lonely rigours of mathematical physics for domestic life and the collegiality of scientific administration'.

Thomson in 1852 had married Margaret Crum, the daughter of a prominent Glasgow industrialist, but, shortly after, she became an invalid, remaining in poor health until her early death in 1870. The purchase of the *Lalla Rookh* later that year must have provided a diversion from his loss. But in 1873 further cable-laying expedition took Thomson to the island of Madeira where he was entertained by the landowner Mr. Blandy. He returned to Madeira in the *Lalla Rookh* in 1874 and proposed marriage to Blandy's second daughter, Fanny, who became the second Lady Thomson. It was at this stage that the Thomsons moved into their Scottish Baronial style mansion, Netherhall in Largs. Thomson had no children by either marriage, but Stokes had two sons and a daughter who survived infancy.

Another result of Stokes's marriage was a rekindling of interest in water waves. He visited his father-in-law Dr. Robinson every summer and took regular trips with Mary and their children to resorts on the north coast, most frequently to Portstewart, from whence they made excursions to the Giant's Causeway. Although primarily a theoretician, Stokes was not afraid to experiment: he measured the waves breaking in the Land Cave there, and also on the sloping sandy beaches at Portstewart. Typical of his letters to Thomson about this time is one written from the Observatory, Armagh on 15 September 1880. Stokes had made a mathematical conjecture that the angle at the crest of the wave of greatest height should be 120 degrees:

You ask if I have done anything more about the greatest possible wave. I cannot say that I have, at least anything to mention mathematically. For it is not a very mathematical process taking off my shoes and stockings, tucking my trousers as high as I could, and wading out into the sea to get in a line with the crests of some small waves that were breaking on a sandy beach .... I feel pretty well satisfied that the limiting form is one presenting an edge of 120 degrees.

He writes on the same topic a week later and Kelvin replies from the *Lalla Rookh* at sea in the Clyde. In another letter Kelvin writes:

Will you not come and have a sail with us and see and *feel* waves? We would take you away out to the west of Scilly for a day or two if that would suit best.

Stokes replies somewhat pathetically:

It is not easy to say where to find a man who owns a yacht, but I write on spec, and at any rate you will soon I suppose be back in Glasgow.

Despite his seaside paddling, Stokes' interest in ocean waves was a serious one undertaken in consequence of his membership of the Meteorological Council. Stokes was aware that long waves radiating from distant storms travelled faster than short waves from the same source. In situations where unusually high seas were observed in the absence of a local wind, Stokes was able to analyse records of the direction and period of the waves to predict the location and direction of travel of the storm which had given birth to them. Larmor[12] describes his fascinating correspondence with the Admiralty Experimental station at Torquay and various sea captains, most notably Captain William Watson of *S. S. Algeria*, on observations in ships' logs. Returning with Fanny from the meeting of the British Association for the Advancement of Science in Toronto in 1897, Kelvin added to the database on the wave of greatest height, writing on 26 October:

We had waves on our homeward voyage on the Campania which cannot have been less than 60 feet high from hollow to crest ... We were on the Banks of Newfoundland, depth perhaps 40 fathoms at the time. But now!! I was told by several officers of the Campania that in both her and her sister ship the Lucania they had seen unbroken wave-crests right ahead of them and quite near, in line with the crows nest on the foremast. This is 90 feet above the sea ....

The estimate given by Kelvin is not far short of the theoretical height eventually established almost a century later. Stokes also advised the Council on meteorological instruments, and a sunlight recorder designed by him was in use in the station at Valentia until recent times.

CREATION AND EVOLUTION

Following the publication of *The Origin of Species* by Charles Darwin in 1859, the latter part of the nineteenth century was a time of conflict in British science between the supporters of Creation as set out in the Bible and followers of the Theory of Evolution. It is hard for us today to comprehend the importance of religious issues to the Victorian scientists and to the man in the street who, while excited by new scientific inventions and discoveries, saw a certain conflict with the role of God as creator. Stokes and Thomson were firmly in the creation camp, although for different reasons, and were certainly not as fundamentalist in their beliefs as the physicists Faraday and Maxwell. Neither Stokes nor Thomson went so far as to take literally the time periods in the Genesis account of creation. Thomson's scientific work on the age of the Sun and the Solar System showed that there had been insufficient geological time for evolution to have taken place at the pace suggested by Darwin. In response to Thomson's results Darwin subsequently modified his views on the rate of evolution, but the irony was that Thomson's estimate of the age of the Earth, based on its rate of cooling, did not take account of the contribution of the heat of radioactivity, which was yet to be discovered. Stokes believed that the Bible account was true in the sense of an ongoing creation of organic life through a process controlled by God.

On the other side, the physicist John Tyndall and Thomas Henry Huxley were leading supporters of evolution within the Royal Society and the British Association. Both these strong agnostics took more extreme views than Darwin. After Tyndall's provocative address to the Association in 1868, Stokes was moved to strongly refute it in his Presidential Address the following year. Matters were made worse by Tyndall's own Presidential Address to the Association in Belfast in 1874. The disagreement was still rumbling on in 1887 when Stokes was elected MP for Cambridge University. Huxley perceived a conflict of interest and wrote anonymously in *Nature* that, as President of the Royal Society, Stokes should not simultaneously be a Member of Parliament. But there was a precedent in the form of his predecessor in the Lucasian Professorship, Isaac Newton, who had successfully combined the holding of both offices. Thomson wrote quickly to support Stokes.

We were *very much* displeased with that article in Nature. I think on the contrary that your agreeing to be a member was most patriotic and public-spirited.

A profoundly religious man, Stokes had always been interested in the relationship between science and religion. From 1886 to 1903 he was President of the Victoria Institute, whose aims were:

To examine, from the point of view of science, such questions as may have arisen from an apparent conflict between scientific results and religious truths; to enquire whether the scientific results are or are not well founded.

In 1891 and 1893 Stokes delivered the Gifford Lectures in Natural Theology in the University of Edinburgh.

TWO SCIENTISTS IN PARLIAMENT

In the 1880s Thomson began to emerge as an activist for the Liberal Party[13]. At the behest of the Parliamentary whips, his brother-in-law, Alexander Crum, Liberal MP for Renfrewshire, urged Thomson to stand in 1884 as a University candidate, but he declined. When the Liberal party split over Gladstone's Home Rule measures for Ireland, Thomson campaigned vigorously during the 1886 general election for the anti-Home Rule Liberals. Seventy-eight Liberal Unionists were returned to Westminster to become a minority party in the Conservative led government of Lord Salisbury. Thomson was a strong supporter of this coalition, which contained among other smaller parties, his friend Stokes, who had been returned as an Independent for the Cambridge University seat at a by-election in 1887. We find him writing to Stokes in March 1888:

> I hope you have been enjoying Parliament. It must be very satisfactory, and pleasant, to see all going so well . . . I am now feeling quite hopeful that I might live to see the last of government by party.

It is not certain that Stokes did enjoy Parliament, as he spoke only thrice in five years and did not stand again at the 1892 election, finding the long hours uncongenial. In the meantime Thomson had become president of the West of Scotland Liberal Unionist Association whose honorary president was Lord Hartington (later eighth Duke of Devonshire), leader of the Liberal Unionists in Parliament. In a letter to the Prime Minister Salisbury, Hartington proposed Thomson for a peerage, and he became Baron Kelvin of Largs in the New Year honours list of 1892 in recognition of his 'most valued service to science and progress in this country'. Unlike Stokes in the Commons, Kelvin enjoyed attending sittings of the House of Lords and spoke on fourteen occasions, six times on shipping matters.

Stokes had been made a baronet by Queen Victoria in 1889. In contrast to Thomson, whose wealth and ability to support the lifestyle expected of a peer had been factors in Hartington's decision to recommend him, Stokes anguished long on this financial question[14] before accepting this honour, which in those days was hereditary. It was difficult enough for him in a university professorship to maintain appearances, but how would his elder son, Arthur, a schoolmaster in Shrewsbury, be able to shoulder the expenses?

The many honours given to Kelvin in later life have already been mentioned in other chapters. Stokes was awarded the Copley Medal of the Royal Society in 1893, and in 1899 given a Professorial Jubilee (50 years as Lucasian Professor) by the University of Cambridge. Kelvin's jubilee at Glasgow had already been celebrated in 1896 and he retired in 1899. Stokes died at Lensfield Cottage at 1am on Sunday, 1 February 1903. Kelvin died at his home Netherhall on 17 December 1907.

Although obsessed with his scientific work, Stokes had excited feelings of warmth and admiration among his contemporaries. Some colleagues felt that he could have done more in the field of physics in later life, but Stokes himself seemed to find fulfilment in his role as a gatekeeper and arbiter of the Victorian scientific establishment. Rayleigh and Kelvin both published obituaries of Stokes, Kelvin observing at the time that his heart was in the grave with Stokes. Sadly, the gravestone of Stokes in Cambridge's Mill Road cemetery has now vanished from view, while Lord Kelvin lies interred near to Sir Isaac Newton in Westminster Abbey, commemorated by a plaque at the hub of the Empire he did so much to bring closer together. The applications of Kelvin's physics, often by Kelvin himself, had changed the face of society through improving the efficiency of industry and the speed of communications. But we should not overlook the contributions of the relatively unsung Stokes.

6

Kelvin and Fitzgerald: Great Irish Physicists

Denis Weaire

COMPARE AND CONTRAST

The heroes of Victorian science were imposing figures, full of the vigour and energy of that age. George Francis Fitzgerald (1851–1901) and Lord Kelvin (1824–1907), as we shall call William Thomson throughout, were two such grand characters, inspiring awe and devotion in their followers and the respect of the public at large. They were both Irishmen and had much in common, including mutual admiration. We shall try to trace their similarities, and also the points at which they diverged. Some of these differences derived from their separation in age.

If Fitzgerald had lived longer we could talk of his reaction to quantum physics and relativity, on which he would surely have been a prime commentator, but he remains a classical physicist, expiring as did Kelvin on the threshold of the new era.

IRISH ROOTS

Both were Irishmen indeed, and both Protestants—but the country's history still leaves room for some divergence within that community at the outset. They were the products of very different classes and cultures.

Fitzgerald's books carry a heraldic crest and motto: Shanet Aboo. He belonged to the Anglo-Irish gentry, Anglican in religion, the class depicted by Somerville and Ross in their accounts of *The Irish R.M.* However eccentric, the Anglo-Irish ascendancy were 'no mean people', as Yeats insisted. Fitzgerald belonged to their intellectual wing, inhabiting the Georgian squares around Trinity College. It was an inbred coterie of

Fig. 6.1. Fitzgerald attempting to fly in the College Park of Trinity College, Dublin in the 1890s. (Courtesy of School of Physics, Trinity College, Dublin.)

thinkers and talkers, out of which Oscar Wilde was to emerge to dominate the dining tables of London. Michael Purser[1] has described the particular tangle of distinguished family roots (Purser, Jellett, Stokes ...) which Fitzgerald sprang from and married into.

Today those squares are silent by night, after the solicitors have returned to the suburbs, although the name plate of William Jellett (Fitzgerald's brother-in-law) is still polished by his descendants in a last hold-out of their beleaguered class. In the nineteenth century it steadily lost its self-confidence, after the Act of Union between Great Britian and Ireland had moved the centre of gravity of public affairs back to London. In Fitzgerald's later years, the tide turned again, towards Home Rule. As a Unionist, he was strongly opposed to it.

Kelvin was born in Belfast, but the masterful account of his life by Smith and Wise[2] rightly returns to his father's birthplace, nestling in the drumlins of County Down, near Ballynahinch. Emigrant Scots farmers populated this part of the country. Instead of the Anglican Church, they adhered to a wide variety of flavours of Presbyterianism and other forms of Nonconformist religion. Although many Ulster communities retain this character, and some are persistently in the news, the remarkable qualities of their ancestors have been largely forgotten. Liberal, industrious, and ambitious, they turned a sleepy agricultural province into one of the great productive centres of the empire. They saw to it that they were provided with educational establishments to match their aspirations: three great grammar schools of Belfast (Belfast Royal Academy, Royal Belfast Academical Institution, Methodist College) came out of that impulse. Kelvin's father taught at the Academical Institution (essentially a small university college at that time).

A PRIVILEGED EDUCATION

Both men received a privileged education, although in a different sense. Fitzgerald had the benefit of private tutoring at home (the tutors including the sister of George Boole). He took the conventional route to Trinity, where his father had been a professor, and excelled in mathematics and science. He never left.

In something of a contrast, Kelvin's father would round off a full day of instruction and scholarship in Belfast by teaching his young sons in the evening. After precocious success in Glasgow (where his father had become professor), and without taking the degree, Kelvin went to Cambridge. There he famously failed to become First Wrangler, but emerged triumphant, nevertheless. At which point, he returned to Glasgow.

In this way both men received an exceptional grounding in mathematics, and in particular an exposure to the works of Laplace, Lagrange, and the whole French school. In the early nineteenth century, the cutting edge of mathematics was to be found in France. As a free-thinking Ulsterman, Kelvin's father had no great allegiance to conservative British traditions in mathematics. Similarly, Trinity benefited from the inspired leadership of Bartholomew Lloyd, provost from 1831 to 1837, who founded its mathematical teaching on the works of the French school at a time when the English universities fell behind. It is no coincidence that William Rowan Hamilton belongs to that period of Trinity's history.

Nevertheless, both men had a characteristically blunt Anglo-Saxon attitude towards mathematical physics, distrusting the mere manipulation of symbols and emphasizing the reality of the objects of theory. They both liked to get their hands on instruments, and build models. The museum now devoted to Kelvin shows may examples. Unfortunately, Dublin can boast no counterpart of this, and it seems that shortly after Fitzgerald's model of the ether was last exhibited (at the Royal Dublin Society) it disappeared.

IN CORPORE SANO

Sporting endeavour, or at least intensive exercise, had a high moral value for the Victorians, including the academic community. Fitzgerald was a natural athlete and he features repeatedly in Trevor West's history of Trinity sports clubs[3].

Andrew Warwick[4] has described how the would-be Cambridge Wranglers were not exempt from the athletic imperative. In Kelvin's case, he was torn between the attractions of the river (which were as much social as sporting) and his father's strictures. The river won, the father relented.

SECOND CITIES

Among their admirers, Glasgow and Dublin each enjoyed the title of the Second City of the Empire. Glasgow had risen in importance, Dublin had declined, but both were mere provincial centres, from a London or Oxbridge perspective.

Kelvin remained in Glasgow for reasons of strategy as well as affection. He saw dangers in flying too close to the flame of the centres of power and influence. They could drain a man's energy, as he tried to persuade Stokes. His extraordinary productivity on so many fronts proves his point well. Fitzgerald probably set out with a similar attitude, but he became frustrated with the pace of reform of the educational curriculum, and the meagre support for applied research. He might have been happier in Belfast among the linen factories and shipyard cranes of the Lagan, looking across to the Clyde. In Dublin science and technology were at best a sideshow, university development a religious issue. His pleas fell upon deaf ears. Preposterously, yet apparently seriously, he proposed to move Trinity College, Dublin to England.

So it is that when they were considered for the Cavendish Chair in Cambridge, Kelvin declined it three times, and Fitzgerald (who actually applied) was passed over in favour of J. J. Thomson. Finding no such escape route, he seems to have worked himself to death.

OUTPUT

There is a wide disparity in the manifest output of these two scholars. Kelvin's publications, whose number varies smoothly from an early age to an advanced one with a maximum in his middle years, totalled about 650 papers and many books, as well as patents and reports. While one may point to the excellent technical support that was to be had in Glasgow, this was very much his personal achievement. He lost no time, working in carriages, in trains, and in bed, as many boxes of notebooks in the Cambridge University Library testify. Peter Tait may have had to cajole him persistently to complete their joint work, the *Treatise on Natural Philosophy* (always referred to as T and T'), but for most of the time he needed no such prompting.

In comparison Fitzgerald's publications are paltry, and he authored no books. Joseph Larmor gathered up his *Scientific Writing*[5], numbering just over 100 items. While he missed quite a few papers (including the letter on the now famous Fitzgerald Contraction), it is a fair reflection of what was committed to print. Much of it hardly deserves to be called 'scientific writings', being speeches and letters on science policy, short notes, commentaries on the work of others, and biographical essays. The actual science is often sketchy and incomplete. Fitzgerald recognized this fault in himself, saying that he often rushed out with his ideas or gave them to others, rather than working them out patiently himself. It was a fault of generosity: his legacy was to lie in his extraordinary influence through correspondence. The insights that he shared were always provocative and often highly imaginative.

EDUCATIONAL CONCERNS

The late nineteenth century was the time of the professionalization of physics. Until then natural philosophy had been the preserve of isolated professors, making experiments in their college rooms and demonstrating physical effects in the lecture theatre. Both Kelvin and Fitzgerald saw the need for organized departments in dedicated

Fig. 6.2. Fitzgerald, like Kelvin, saw the need for teaching laboratories and he strove to achieve something similar to Kelvin's Glasgow laboratory in Dublin. He died before it was accomplished, but his colleague John Joly carried his campaign forward, overseeing the erection of the elegant Physical Laboratory, now called the Fitzgerald building. (Courtesy of School of Physics, Trinity College, Dublin.)

facilities, to include teaching laboratories. Kelvin was at the forefront in the latter regard, and Fitzgerald strove to achieve something similar in Dublin. He died before it was accomplished, but John Joly carried his campaign forward, overseeing the erection of an elegant Physical Laboratory, now called the Fitzgerald Building.

Kelvin and Fitzgerald shared, in teaching, the common disability of great minds confronted by ordinary undergraduates, despite the respect they commanded. Kelvin's difficulties are well and humorously documented, Fitzgerald's less so—they were described to the author by Ernest Walton, as a folk memory of the Department.

Fitzgerald's advocacy of educational reform was vehemently directed at authorities at every level, and he deliberately expended much of his energy in that direction. He did achieve with others the foundation of a mechanics institute, now part of the Dublin Institute of Technology, but his speeches and letters reflect a growing disillusionment. Meanwhile Kelvin enjoyed a prospering, modernizing university setting.

ATTITUDES TO MAXWELL

In a scientific pantheon, our two subjects might occupy pedestals a little below that of James Clerk Maxwell. Even Kelvin, for all the breathtaking breadth and scope of his attainments, might not quite match the Scottish master's supreme achievement in unifying electricity, magnetism, and optics in four short equations. (So runs the convenient myth; in reality they were distilled from his work by others.)

For Kelvin, Maxwell's theory was too abstract, and it did not provide a fundamental physical basis for electromagnetic phenomena. For him 'the ether is a real thing', and he spent a lifetime in search of that real thing.

Fitzgerald shared his distaste for abstraction, but became so immersed in the Maxwell theory that for him it became real, and fulfilled his early hope that it would free us from the 'thraldom of a material ether'. More and more convinced of the validity and sufficiency of the Maxwellian doctrine, he attracted an invisible college of correspondents, the Maxwellians[6]. He was their St Paul, a constant source of encouragement, ideas, and interpretation. This crusade inevitably brought him into conflict with Kelvin.

LONGITUDINAL LIGHT

In two letters to *Nature* in 1885, Fitzgerald took issue with Kelvin over his telegraph theory, which was at odds with the Maxwellian method[5]. The argument found a distinct focus in the question of the existence or otherwise of a form of longitudinal light in which the oscillating electric field is in the direction of the wave, rather than transverse to it. Maxwell's theory permits only the latter. If light is conceived as a vibration in an elastic medium, as Kelvin and others saw it, then both kinds of wave should exist. Kelvin clung stubbornly to this notion until his death.

The debate over longitudinal light is spread over many decades and Kelvin was not alone in his beliefs. Its proponents were continually on the look-out for it, in unexplained optical effects. Roentgen (or x-) rays were a prime candidate when they were discovered.

One episode in Kelvin's career (not recounted by Smith and Wise) shows how far he was prepared to go to save the elastic solid model of light. George Green (whose work he had earlier championed) had recognized this problem, and with the detachment of a mathematician, declared that what was required was an elastic medium in which the elastic stiffness corresponding to compression in one direction only was either zero or infinite, and left it at that. In 1887, in bed at Netherhall, Kelvin woke with the idea that an ether that was a kind of foam would suffice,

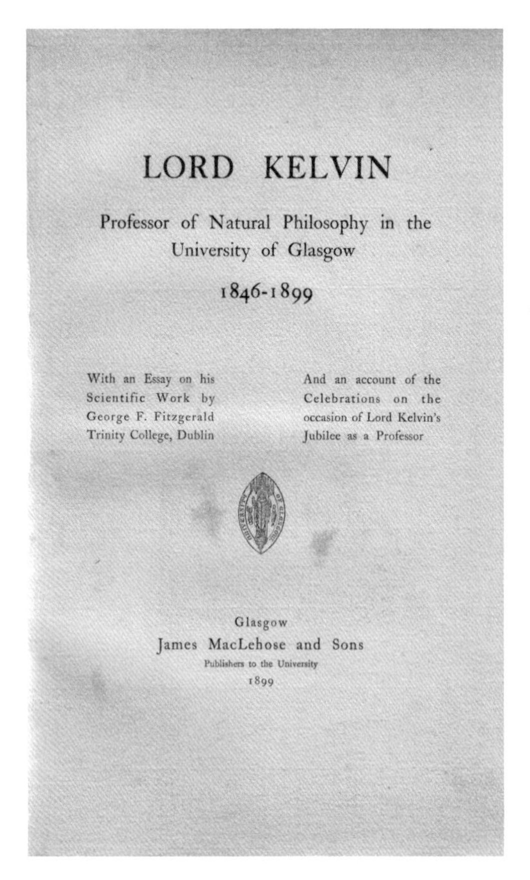
LORD KELVIN

Professor of Natural Philosophy in the University of Glasgow

1846-1899

With an Essay on his Scientific Work by George F. Fitzgerald Trinity College, Dublin

And an account of the Celebrations on the occasion of Lord Kelvin's Jubilee as a Professor

Glasgow
James MacLehose and Sons
Publishers to the University
1899

Fig. 6.3. An essay by Fitzgerald assessing Kelvin's work made up a major part of the book published by Glasgow University which formed part of the Jubilee celebrations for their most famous professor.

having zero stiffness. His new model, rushed into print, failed to impress anyone, but founded the 'Kelvin Problem', addressed to the ideal structure of a foam, which has occupied many minds ever since.

FINAL WORDS

In his old age Kelvin's eminence attracted many honours (though his peerage is attributed by Crosbie and Wise to his political friends in Unionism). One of the grandest celebrations of the great man was the Jubilee of his Professorship, at which Fitzgerald was a principal speaker. He was unstinting in his praise, and his conclusion illustrates his faith in science as a force for good in the world.

> He has advanced civilisation by making the all-pervading ether available for our use, by enabling us to measure its properties, and by teaching us how to lay the nerves of civilisation in the depths of the oceans. He has helped to unify humanity, to modify competition by cooperation, to push forward the federation of the world[5].

Fitzgerald's *Scientific Writings* are prefaced by remarkable obituaries and other anguished regrets at the passing of Fitzgerald. Kelvin expressed his admiration for a mathematician who, just like himself, had a keen interest in practical matters of trade and industry[5]. He said that his 'scientific sympathy and alliance' with his Dublin counterpart had greatly ripened in recent years; evidently there had been some real friction in their earlier disagreements.

As a final adieu he said:

> As for his whole life, it seems to me that no one ever attained more nearly than Fitz Gerald to the shorter catechism of the Church of Scotland, 'to glorify God and enjoy him for ever'.

7

Concepts and Models of the Magnetic Field

John Roche

THOMSON'S SCIENTIFIC STYLE

Thomson was so versatile, across the range and depth of his physics and in its applications, that it is very difficult to do justice to his achievement. Thomson was perpetually simmering with impatient creativity. Indeed, he had little time to polish ideas ready for publishing. He wished to rush off to the next concept or experiment or instrument or patent. As result, his works are bibliographical nightmares: the manuscripts sometimes seem to have been sent to the publisher during a rush of creativity, he flies off at a tangent, and inserts easy to miss paragraphs decades later than when they were first penned or printed. Apparently, a very long lifetime was too short to bring to maturity the richness of his insights. His students and his colleagues were infected by his creative drive. But they also found his speed of thought, and range of interest, difficult to absorb[1].

In later years Thomson became mildly eccentric, viewing vector notation as 'unmixed evil' and 'nihilism'[2], or choosing odd technical terms such as 'forcive'[3], for a force acting in a circle. It is difficult to understand the apparent contradiction between his fevered speech and impatience with publication on the one hand, and the clarity, logic and balance of his scientific thought, and the careful precision of his experimental work, on the other. Fortunately, the exegesis of Thomson is greatly facilitated by his own effort to publish almost every significant science paper and lecture that he wrote[4]. Also, Thomson's pious but critical friends, and later historians, have helped us to make sense of the context and multiple layers in his thought in two superior biographies; S. P. Thompson *The Life of William Thomson* (1910), and C. W. Smith and M. N. Wise *Energy and Empire* (1989)[5]. Between them they also include a wealth of letters, diary, and notebook entries.

Experimental philosophy flourished without interruption in eighteenth century Britain, and in the nineteenth century it's greatest representative was Michael Faraday (1791–1867)[6]. It continued the humanistic tradition, descending from the ancient Greeks. At its best, it incorporated a rigorous verbal logic and coherence in physics; great care in coining technical terms and definitions; a philology which often included Greek, Latin, French, and German; a learned style; an honest rhetoric; and a sensitivity to history[7]. These logical and philological skills were also embedded in analytical physics during the first half of the nineteenth century, when Thomson was educated in Scotland[8].

After the death of Newton, British analytic physics went into a decline, and did not recover until the 1820s[9]. Thomson represented the second generation of this recovery, and entered a well-established culture of mathematical physics, in various centres in Britain, especially in Dublin and Cambridge. But he became equally adept in laboratory physics, instrumentation, and metrology, giving him a depth and breath of understanding of physics, a clarity, and a practicality which was unequalled in nineteenth century Britain[10].

William Thomson was equally at home in both physics and in the humanities. He studied logic, Greek, Latin, French, and German. He worked through Goethe's *Faust*. He was a founder of the Cambridge Musical Society and played the french horn. He studied scientific methodology and rhetoric, especially in Francis Bacon. This deeply influenced the style of his physics. Thomson had the gift to recognize original ideas and promoted and developed them enthusiastically, especially in George Green (1793–1841), James Joule (1818–89), and Hermann Helmholtz[11]. He was generous in giving others credit, and he was always firm but courteous in criticizing others in print[12].

Like his equally learned friend, Hermann Helmholtz (1821–94), Thomson's physics was deeply intuitive, but he also had an extraordinary capacity to articulate and unify concepts with a deceptive ease and even earthiness[13]. He disliked metaphysics[14]. He was deeply committed to understanding the physical meaning of the formalism of mathematical physics. He referred to the *aphasia* of mathematics, its inarticulateness to express physical ideas, and insisted that all his formulas must have a physical meaning[15]. He dismissed concepts, hypotheses, or speculations not supported by diagrams, numerical calculation, physical models, experiment, and 'regular analysis'[16]. 'I never satisfy myself until I can make a mechanical model of a thing'[17]. Thomson was intensely aware of the importance of coining appropriate technical terms, and of framing rigorous physical definitions[18]. 'Nothing in science is more difficult than definitions'[19]. For example magnetic 'induction' was the *action* of inducing, for Thomson as for Faraday. But for Maxwell it was the resulting field intensity B, and Thomson saw this, with some justice, as a category error[20].

Thomson coined a vast array of technical terms and definitions in physics, such as the *capacity* of a conductor[21], *energy, kinetic* energy[22], *absolute temperature*[23], *simple harmonic motion*[24], *magnetic permeability*[25], *magnetic susceptibility*[26], *stress over strain*[27], *bulk modulus*[28], *circulation*[29], *vorticity*[30] and *vortex-sheet*[31].

Thomson, like Isaac Newton (1642–1727) was committed, throughout his career, to natural theology: 'I am a firm believer in design'. But he rejected the 'frivolities of teleology'. He also had a whole-hearted detestation of Spiritualism[32]. Thomson saw no conflict between his mechanical philosophy of nature and his theology.

In his early career, Thomson saw natural philosophy from Bacon's viewpoint—as generalizations induced from evidence and controlled by experiment leading to a larger theory. Natural Philosophy in this sense had very little role for hypotheses and speculations. It was 'positive'[33]. Throughout his life 'natural philosophy' in this sense was his ideal in physics[34]. Thomson, by analysing the concepts of physics, and through experiment, recognized new properties and structures and named and defined them[35]. In particular, through well-defined analogies and models, drawn from established principles in heat, elastic solids, hydrodynamics, and optics, he expanded electrostatics and electromagnetism.

Thomson had a golden age of success in expanding the foundations of electricity and magnetism, in energetics throughout physics, in thermodynamics, and in telegraphy, from 1840–60[36]. In the early phases of Thomson's trial by analogy—of mechanical and thermal formalism fit for electromagnetism—he advanced electromagnetism to an extraordinary degree. Thomson greatly advanced the theory of electrostatic force and potentials, macroscopic dielectrics and magnetization, the theory of the telegraph, the analytical theory of electric and magnetic fields, electrostatic energy, and the energy of electric currents and magnets. Methodologically, he expanded the cross-fertilization of fields by linking the formal properties of heat, elasticity, and hydrodynamics with electromagnetism. He also created the method of images in electrostatics and magnetism (see below). He had an intense commitment to metrology, especially in electromagnetism and electrometallurgy[37]. He also introduced many instrumental patents, the most famous of which was his marine compass, 10000 of which were manufactured between 1876 and 1907[38].

However, his obsession, after 1851 and up to his death, was in establishing the supposed mechanical foundations of electromagnetism and optics. In the later nineteenth century it slowly became clear that electromagnetism and optics cannot easily be reduced to mechanics. Thomson was highly reluctant to take this step. The younger men who took over the theory of electromagnetism, especially James Clerk Maxwell (1831–79) and Hendrik A. Lorentz (1853–1928), viewed the ether and mechanics as heuristic tools to expand, deepen, and unify the 'matter of fact' laws of electromagnetism. For Maxwell and Lorentz, the ethereal foundations of electromagnetism slipped further and further away from orthodox mechanics.

Thomson remained orthodox in interpreting the ether and electromagnetism and optics mechanically. However, Thomson never confused his 'positive' published work in physics with his underlying mechanical speculations.

It is important, therefore, to distinguish Thomson's 'positive' theory of electricity, magnetism, and optics, based on evidence and inductive generalization (which he calls 'Natural Philosophy'[39], or 'the matter of fact laws'[40]), from his analogies and models, but also from his hypotheses and speculative theories. For example, for his mechanical model of the 'Faraday effect' (the rotation of the plane of polarization in a magnetized medium) he declares that it is 'impossible to conceive any other [explanation] than this dynamical explanation'—for Thomson 'dynamical explanation' meant an underlying mechanical explanation. Nevertheless, Thomson also declares that it is 'impossible to decide, and perhaps in vain to speculate, in the present state of science'[41]. His positive science and his models were published in mathematical and physical journals and textbooks. His speculations were usually recorded in his letters, diaries, and notebooks, or in his *Baltimore Lectures* (1884) and in his *Popular Lectures and Addresses* (1889–94).

ANALOGIES AND MODELS IN THOMSON

Thomson worked on almost every area of mid-nineteenth century physics, both fundamental and applied. In certain areas he was an opportunist, in that he recognized the cutting edge of a variety of fields and was eager to contribute to each of them. A lesser mind, so driven, might have been an unproductive generalist, but the touch of genius—and unbounded energy—was active in each of his fields.

It was well recognized in the 1840s that electromagnetism, then only 20 years old, was undergoing a rapidly growing foundational period in Europe and America. To expand this new science new methodologies were needed, experimental, theoretical, technological, and metrological[42]. Thomson was a leader in each of these aspects of electromagnetism.

His breadth in physics recognized many parallels and analogies, perhaps because physics then was developing in many new directions. In examining recent electricity and magnetism, he added a boost derived from the cross-fertilization with formally analogous fields. His approach is well illustrated by the following passages from 1847 and 1848:

> There may be a problem in the theory of elastic solids corresponding to every problem connected with the distribution of electricity on conductors, or with the forces of attraction and repulsion by electrified bodies[43].

The application of [a particular] solution ... in the mathematical theory of elastic solids is analogous in some degree to a method of treating certain question in the theory of heat ...[44]

If the solid were incompressible, there would be a close connexion with the mathematical theory of electro-magnetism[45].

It quickly became part of Thomson's methodology to search for relevant analogies, thereby discovering unrecognized concepts and structures, as well as powerful mathematical developments and physical models.

Thomson viewed 'analogies' as comparisons between two well-defined physical systems. In his 'positive' theory of electromagnetism, Thomson frequently uses formal analogies between heat, optics, elasticity, and hydrodynamics to explore aspects of electricity and magnetism, which, without these analogies, might not have been recognized. However, Thomson never *directly* incorporates analogical concepts from one field into another. He would have regarded this as methodologically illegitimate. When writing about electrostatics, for example, he used the heat diffusion analogy, but he did not reduce electrostatics to the laws of thermal diffusion. He recognized that in moving from one physical system to an analogous system, the latter might require substantial modification[46]. Thomson used formal analogy as an heuristic device only.

Thomson's 'models' were usually artificially contrived and were almost always mechanical. They attempted to tailor a mechanical system to represent a given electromagnetic, or optical or thermal system, or other physical system. He writes[47]

My object is to show how to make a mechanical model which will fulfil the conditions required in the physical phenomena that we are considering, whatever that may be.

THOMSON'S EARLY HEURISTIC ANALOGIES IN ELECTROSTATICS

The generation which revived mathematical physics in Britain, that of John Herschel (1792–1871), Charles Babbage (1792–1871), George Peacock (1791–1858), George Green (1793–1841), William Whewell (1794–1866), George Airey (1801–1892), and Augustus de Morgan (1806–71)[48] were tutored by the French School of analytical physics. This tradition, by the 1810s had differentiated into an older, more abstract theory, which based observational physics on an underlying hypothetical molecular theory. It was led by Pierre Simon de Laplace (1749–1827) and Simeon Poisson (1781–1840). The newer tradition, derived from Joseph Fourier (1768–1830), Augustin Fresnel (1788–1827), and Augustin Cauchy (1789–1857) was committed to a positive analysis of nature, avoiding hypotheses. Thomson absorbed his theoretical and experimental physics from these sources, especially from the latter[49]. The practical orientation of Scottish philosophy and science, and his father's insistence that he be trained in advanced laboratory science and by the more 'positive' theoretical orientation of Cambridge physics, all shaped Thomson's physics. Fourier's *Théorie Analytique de la Chaleur* (1822), which he described as a 'great mathematical poem'[50], was the foundational influence in Thomson's physical theory[51].

In April 1841, at the beginning of his studies at Cambridge, stimulated by his enthusiasm for Fourier, Thomson recognized a formal analogy between the electrostatic force generated by a small charge, and the heat propagated by a small source in a uniform medium. Since Newton's day physics has made use of formal analogies between different areas of physics to stimulate theoretical exploration—in Newton's case between the swing of a pendulum and oscillation of a liquid in a U-tube. However, Poisson regarded such analogies as potentially misleading[52].

By a process of cross-fertilization Thomson discovered properties intuitive in one field which applied, *mutatis mutandis*, in the other, but far less intuitively. For example, using his electrostatic analogy he was able to prove that a uniform distribution of heat sources over a sphere sent no resultant heat into the inner space of the sphere, and has a uniform temperature throughout its volume[53]. Furthermore, following reflection on his analogy, he began to wonder whether the electrostatic force might also be propagated through an invisible material medium[54]. Gradually, Thomson erected formal analogy into a new heuristic methodology for physics.

In February 1845, after completing his undergraduate work at Cambridge, his father, James Thomson (1786–1849), informally arranged postgraduate work for his son in Paris, with the great Parisian physicists and physics laboratories. It was a seminal experience for Thomson, working at the white heat of his creativity. In Paris, stimulated by his recent reading of George Green and by his analogies between electrostatics and heat diffusion, he hit on the method of images in electrostatics. It was already well known that a small positive charge near a grounded sphere induced a negative charge distribution on the sheet, which broadcast an electric force distribution in the space between the original charge and the sphere. Thomson's insight was to recognize that the induced charge was equivalent to a smaller, imaginary, negative point charge inside the sphere. Thomson was able to calculate the magnitude and position of the image charge. The two charges generated zero potential on the spherical surface, and the correct electric potential and force distribution in the external space. Thomson developed the method of images into a powerful method of solving problems and expanded it to heat diffusion and later to magnetism. George Stokes (1819–1903) expanded it into hydrodynamics[55].

Again in Paris in 1845, Joseph Liouville (1809–82) asked Thomson to attempt to reconcile Faraday's theory of electrostatic induction with Coulomb's electrostatics[56]. Liouville published Thomson's analysis in 1845[57]. Faraday had argued, in a publication of 1839, that electrostatic induction is an 'action of contiguous particles' in a material medium. He also argued that induction sometimes occurred in 'curved' lines and therefore is 'utterly incompatible with action at a distance'[58]. As Thomson put it much later, this 'offended mathematical physics', which was based on Coulomb electrostatic law—according to which electric force acts at a distance along straight lines[59]. At first Thomson was rather dismissive of Faraday's mathematical naivété. He immediately saw, from his earlier paper, that Faraday's result followed deductively from Coulomb's laws. Pairs of equal and opposite sources and sinks, either electrical or thermal, have the same resultant curved pattern of action in the intervening region[60].

However, Thomson quickly learned to respect Faraday's experimental style and his physical reasoning. Indeed, suggesting an approach by Poisson, and Faraday himself, electrostatic induction could be interpreted as a state of electrically polarized conducting spherules[61]. However, for Faraday the air continued to conduct the electric force, even in the absence of a solid or liquid dielectric. For Thomson, the air was insignificant, and the theory then returned to ordinary action at a distance[62]. The net result for Thomson is that he was able to 'translate Faraday's results into the language of mathematical theory'[63]. In August 1845 Thomson sent a letter to Faraday explaining his manner of reconciling Coulomb and action at a distance with Faraday's theory of curved lines of force[64].

Thomson was not the first to use Faraday to construct a mathematical theory of dielectrics. From 1840, Ottaviani Mossotti (1791–1863) of Pisa, stimulated by Faraday's experiments, based his theory of dielectrics on a close analogy with Poisson's theory of magnetized media. He adopted Poisson's approach of beginning at an hypothetical discrete microscopic level, but ending with a macroscopic continuum[65]. Thomson's main innovations in the theory of dielectrics were to construct a 'positive' continuum approach from the beginning[66]. He also attempted to give equal weight to classical action at a distance and Faraday's propagation theory[67]. However, his most original innovation was his theory of electrostatic induction across the boundary between two dielectrics.

In 1846 Thomson published an analytical theory of dielectrics[68]. Basically, he used Faraday's experimental results, an analogy with Poisson's theory of the magnetic medium, Fourier's 'positive' theoretical style, and also an analogy with Fourier's theory of heat diffusion[69]. His analytical interpretation of Faraday involved the following main steps.

Faraday had created a concept of an extended or non-local 'intensity' or 'tension' in the dielectric medium[70]. Thomson translated this into two well known analytical concepts: the concept of the potential function and the electric 'force' or intensity in the medium. He considered a solid dielectric, and he had to deal with the transition from one dielectric medium to another[71]. For the latter he used the analogy with Fourier's heat diffusion. When the flux of heat moved from one medium to another, the flow of heat is continuous and the product of the thermal conductivity and the temperature gradient is also equal on both sides. Here Thomson recognized a major obstacle. For Thomson, heat flux was analogous the electric intensity[72], and temperature was analogous to potential. While exploring the boundary between two dielectrics, Thomson clearly recognized that the heat flux is continuous across a boundary in the absence of heat sources while the electric intensity cannot be continuous across the boundary. Indeed, Faraday's experiments showed him that the potential gradient inside the boundary of the dielectric is less than that in empty space on the outside. In established electrostatics the electric intensity has the same magnitude as the potential gradient. This meant for Thomson that the electric intensity suffers a jump in value across a dielectric boundary, analogous to the corresponding jump in temperature gradient. Thomson, at that time, was unable to measure the electric intensity within a solid dielectric. However, the above indirect evidence led him strongly to believe that the electric intensity is discontinuous across a dielectric boundary. He was also forced to regard the temperature gradient as analogous of the electric intensity, and ignored the original analogy with heat flux. Thomson now claimed that an electric intensity existed in a material dielectric and is equal to the potential gradient there[73]. Clearly, Thomson saw that the heat diffusion analogy, when applied to dielectrics, was helpful but inadequate.

Thomson then defined the ratio of the normal electric intensities just inside and just outside the boundary, as inversely equal to dielectric 'penetrability or conducting power', that is, to Faraday's 'specific inductive capacity'.

$$\frac{R'}{R} = \frac{1}{k}$$ [74]

He also argued from Coulomb and Poisson that the subtractive difference between R' and R is $4\pi\rho$ where ρ is an imaginary monopole layer of charge on the dielectric boundary, which generates the same external electric intensity as that of the whole polarized body. Also, he states that the components in the tangent plane are equal on both sides of the boundary[75]. For Thomson, the electric intensity within the dielectric was not simply an externally applied intensity as in Poisson's corresponding magnetic case[76]. Clearly, Thomson's electric intensity was influenced by the medium itself. The concept of an internal electric and magnetic intensity within material bodies became of fundamental importance for electromagnetism. At this time Thomson did not fully justify his assumptions. Nevertheless, his theory was very fruitful and, as we shall see, he subsequently discovered valid physical and operational interpretations for the field intensities in magnetized media and, by association, in dielectrics (see below).

Thomson insisted that the traditional action at a distance approach, and that of Faraday's propagation theory in a material medium, are equally valid. However, Thomson disagreed with Faraday for believing that induction was carried on by contiguous air particles, however low the pressure. But Thomson was willing to believe that induction propagated by an electric ether, or action at a distance across the vacuum, were equally acceptable hypotheses[77]. In 1893 Thomson states 'that before Faraday's death, in 1867, the notion that electric force is propagated by a medium was generally accepted by the rising generation of scientific men'[78].

THOMSON'S GUIDING MECHANICAL MODEL FOR ELECTROMAGNETISM

In 1847 Thomson attempted to unify electrostatics, magnetism, and electromagnetism by means of a mechanical analogy with various kinds of bulk and shearing deformation in an elastic solid[79]. The electric intensity of a charged sphere was made formally analogous to a sphere under pressure, and bodily displacing the solid medium outwards. The action of a bar magnet analogously dragged around the medium through a certain shearing angle about the axis of the magnet. An introduced current element analogously dragged the medium longitudinally in a shearing rotation[80].

Thomson primarily designed the analogy to account for Faraday's rotation of plane polarized light in a magnetized medium, itself partly stimulated by Thomson[81]. The magnetic forces were, therefore, represented as functions of rotational displacements but directed along their axes. Thomson, therefore, anticipated the 'axial' structure of the magnetic field[82]. He also identified mechanical structures in the medium which were formally analogous to what Maxwell later called the 'vector potential'. The curl algorithm applied to Thomson's 'vector potential' represented the magnetic intensity[83].

These three representations always remained the presiding analogies for his mechanical vision of electromagnetism. Thomson said to Faraday in 1847 that he 'did not venture even to hint at the possibility of making it a physical theory'. Nevertheless, he hoped that 'if such a theory could be discovered, it would also, taken with the undulatory theory of light, in all probability explain the effect of magnetism on polarized light'[84]. These mechanical analogies turned out to be fruitful for his 'positive' theory of electromagnetism. But more significantly for Thomson, they were the starting point for his 50 year long search for a comprehensive mechanical model of the ether, electromagnetism, and light[85].

INTRODUCING THOMSON'S 'POSITIVE' THEORY OF MAGNETISM

In 1833 Carl Gauss (1777–1855) transformed Poisson's rather abstract theory of magnetization into a metrological science[86]. For Gauss and Wilhelm Weber (1804–91) magnetic poles were the fictional foundations which led into various real quantities in magnetization theory. For example, the 'magnetic moment' of the body and the 'magnetic axis' were defined observationally, for the first time, by Gauss and Weber in 1836[87].

In the later 1840s, early 1850s, and again in the early 1870s, Thomson worked on the foundations of magnetism[88]. Here he sought to improve and develop the work of Poisson, André Ampère (1775–1836), Green, Gauss, and Weber in magnetism and electromagnetism. He introduced a 'positive' research programme in magnetism—'on the sole foundation of facts generally known .... With this object, I have endeavoured to detach the hypothesis of magnetic fluids from Poisson's theory ...'. According to Thomson 'the positive parts of this theory agree with [the positive parts] of Poisson's theory'[89]. Thomson did not accept Ampère's molecular currents in his theory as valid sources of magnetism[90]. But he did accept, for conventional reasons, that electric currents are 'the transference of matter along the conductor'. Given the state of contemporary magnetic theory, he regarded macroscopic currents and magnets as qualitatively different sources of magnetic fields[91]. Thomson always treated the magnetized medium as a macroscopic or 'molar' continuum.

Thomson developed a 'positive' analytical theory of magnetic forces and couples on a magnetized body expressed in terms of their elementary magnetic moments, ignoring monopoles[92]. Influenced by Gauss's magnetic metrology, he redefined Poisson's 'intensity of magnetization' observationally as the 'magnetic moment per unit volume'[93].

Thomson was strongly influenced by Faraday's experiments and concepts in magnetism and electromagnetism[94]. His attempt to be non-hypothetical in his work in magnetism is well illustrated by his particular definition of Faraday's 'magnetic field', a term which he used for the first time in print in 1851[95]:

> Any space at every point of which there is a finite magnetic force is called a 'field of magnetic force'.

This definition was crafted with the greatest care, and is still widely used in physics textbooks[96]. Thomson is attempting to formulate a definition of the magnetic field which would be acceptable to Faraday, to ether theory, to the positive tradition of Fourier, and even, to some extent, to the action at a distance tradition. Indeed, Thomson's definition is almost a common denominator of all four traditions. In old action at a distance theory a 'magnetic intensity' is, indeed, *applied* to the test body by a remote source body, but it does not also exist in nearby 'space' away from the test body. Later, Thomson transformed the old contact magnetic intensity into a 'strain' in space surrounding the magnet or current, even at points where there is no test body[97]. Otherwise, Thomson suspends judgement on the nature of the field, and he leaves it to others to add a richer concept of 'field'. Nevertheless, the helpful ambiguity of Thomson's phrasing made the concept of a magnetic field widely acceptable to analytical physics. Indeed, many physicists to this day add intuitive experience, rather than explicit meaning, to their minimalist 'Thomsonian' textbook definitions of the field.

THE 'POLAR' AND 'ELECTROMAGNETIC' INTENSITY INSIDE A MAGNETIZED MEDIUM

Poisson had discovered that a magnetized body can be mathematically transformed into a surface distribution of fictional poles, added to a volume distribution. The latter is needed only when the medium is nonuniform [98]. In 1849 Thomson developed Poisson's transformation for mensurational and interpretative reasons[99], but referred to it as 'conventional', or 'artificial', or 'imaginary magnetic matter'[100]. This is only to be expected given the abstract character of Poisson's transformation, and Ampère's respected hypothesis of molecular currents as the source of magnetism[101]. Nevertheless, Thomson used imaginary poles extensively in his magnetization theory[102]. Interestingly, he later calls this reduction of the magnetized body to monopoles an 'analogy'![103]

He followed Poisson and Green in basing his advanced mathematical investigations on the 'magnetic force' or intensity (later interpreted as the 'magnetic field intensity') on the magnetic potential function[104].

In his 1846 article on electrostatics Thomson had not specified the electrostatic intensity observationally inside a material dielectric[105]. But, from 1849, he began a long series of investigations, lasting more than 20 years, to define the macroscopic *magnetic* intensity within magnetized bodies[106]. This had followed earlier unsatisfactory definitions by both Poisson and Green[107]. Thomson's extraordinary research profoundly influenced all subsequent magnetization and dielectric theory.

In his June 1849 paper Thomson attempted to construct a measuring definition of the field intensity within a magnetized medium. The magnetic intensity or 'force' acting 'upon any small portion of an inductively magnetized substance' is equivalent to:[108]

> the actual resultant force which would exist within the hollow space that would be left if the portion considered were removed and the magnetism of the remainder constrained to remain unaltered.

The magnetic intensity in the cavity was further specified in terms of the force on a 'very small bar magnet . . . placed in a definite position in this space'[109]. The macroscopic magnetic intensity inside a magnetized body primarily meant for Thomson, therefore, the intensity experienced by a 'free body' element within the medium; but he argued that it was equivalent to that in a virtual cavity in the medium[110]. How did Thomson defend the latter definition?

Thomson recognized that the shape of the cavity itself would influence the intensity on the tiny test magnet (or on the actual free-body element itself). He therefore chose an 'infinitely small' *longitudinal* cavity tangential to the 'line of magnetization'[111]. He chose the latter because he recognized that Poisson's transformation now involved three imaginary monopole distributions: a surface distribution on the outer boundary of the magnetized body, one on the inner boundary of the cavity, and also a volume distribution of monopoles. The field intensity in the cavity due to the outer boundary, and also due to the volume distribution, were independent of the shape of the cavity. But the intensity due to the surface distribution on the cavity itself was insignificant in a longitudinal cavity. Any other cavity was shape-dependent. He also recognized that the negative gradient of Poisson's continuum magnetic potential function gives exactly the same result[112]. This made him confident that his definition of the magnetic intensity was 'the most direct definition . . . that could have been given'[113]. Nevertheless, Thomson was not entirely satisfied with his cavity definition, calling it 'conventional'[114].

On 7 November 1849 it occurred to Thomson that a magnetized body may also be represented by a distribution of imaginary electric currents, rather than by imaginary magnetic poles. This new analogy was based partly on Ampère's recognition that a uniform sheet of normal magnetic dipoles generates the same external magnetic intensity as a peripheral current with the same magnetic moment[115]. Thomson saw that a magnetized body was equivalent, in terms of its external magnetic intensity, to an imaginary distribution of currents circulating on its surface around the magnetic axis, added to a volume distribution of currents in its interior[116].

In 1850, combining his two imaginary analogies, Thomson introduced a far more sophisticated concept of the interior magnetic intensity[117]. He partitioned the magnetized body in two ways, in terms of long filaments or 'solenoids' (i.e. pipes), and also in terms of magnetic shells or 'lamella' (i.e. small sheets)[118]. The theory of the solenoidal distribution was a development of his monopole analysis of June 1849. The lamellar theory was a development of his new 'galvanic' analogy. Given the 'positive' nature of his theory of magnetization it was a considerable irony that his theory fell apart into two distinct theories, one derived from fictional magnetic poles and the other derived from fictional macroscopic currents.

In the course of his analysis of the magnetic medium in terms of magnetic shells or 'lamella', he discovered a second magnetic intensity in magnetized bodies. This was equivalent to the magnetic intensity acting on an 'infinitely small crevasse perpendicular to the lines of magnetization'[119]. By 1850, therefore, Thomson removed the notional cavities and replaced them by two infinitesimal defining 'crevasses'.

In 1872 Thomson came to regard both definitions as equally valid. Because of their origins he distinguished them as the 'polar' and 'electromagnetic' definitions of the field[120]. Indeed an electric current, regarded as a source of magnetic field, he always calls an *electromagnet*[121]. Outside the magnetic medium both definitions are reduced to one[122].

In his study of the magnetic intensity generated by electric currents, he relied heavily on the concepts and formalism of hydrodynamics[123]. He found that the 'electromagnetic' intensity definition applies for volume distributions of macroscopic currents, as well as to common magnetized materials. While the magnetic scalar potential did not easily apply to situations containing currents, he discovered that the 'electromagnetic' intensity can be generated in all cases by applying the curl algorithm to the 'vector potential'[124].

Thomson also discovered the analytical properties of the dual intensities. He demonstrated that the 'electromagnetic' intensity was greater than the 'polar' intensity by the added term $4\pi I$, where I is the intensity of magnetization[125]. He found that the divergence of 'polar' intensity was equal to the negative divergence of the magnetization, and that the divergence of the 'electromagnetic' intensity was always zero, as was the curl of the 'polar' intensity in the absence of macroscopic currents[126].

Thomson's dual definitions of the magnetic field intensity within a magnetized body were a remarkable achievement. Nevertheless, Thomson remained unsatisfied with his definitions, referring to them as 'complicated'[127].

Thomson's dual field magnetic intensities are mathematically, but not interpretatively, the same as the modern field intensities, $\boldsymbol{H}$ and $\boldsymbol{B}$. Faraday also introduced a dual theory of magnetic intensities in a magnetized body, with an interpretation very different to that of Thomson. Maxwell ambiguously merged the theory of Faraday and Thomson. Lorenz, in 1902, introduced yet a third tradition, in which Thomson's 'electromagnetic' intensity is real while the 'polar' intensity is an artefact[128]. This meant that, at the end of the foundational period of electromagnetism (*c.*1930), three incommensurable traditions of interpretation of the field intensities had been established for magnetized bodies.

In 1872, Thomson presented a 'positive' analytical approach to Faraday's theory of magnetized bodies. He drew all of the relevant concepts, except Faraday's 'conducting power of a magnetic medium for lines of force'[129], from the tradition of

Poisson, Fourier, and Augustin Fresnel (1788–1827). First he developed the theory of Poisson and Franz E. Neumann (1798–1895) of the relationship between the applied magnetic intensity and the resulting macroscopic intensity of magnetization as follows[130]:

the magnetic susceptibility of an isotropic substance is the intensity of magnetisation acquired by an infinitely thin bar of it placed in lengthwise in a uniform field of unit magnetic force'

Thomson's magnetic *susceptibility* μ, therefore, controls the ratio of the intensity of magnetization (I or α, β, y) to his 'polar' magnetic intensity X, Y, Z,

$$\alpha = \mu X, \quad \beta = \mu Y, \quad y = \mu Z,$$

where Thomson uses *Saxon* font for the 'polar' magnetic intensity. Thomson also applied this relationship as a correlation between every longitudinal 'free-body' element of the medium, and the component of 'polar' intensity applied to it[131]. He then extended his analysis to anisotropic materials[132].

Thomson's analytical version of Faraday's magnetic 'conducting power' then proceeded as follows. He examined the medium in terms of Poisson's monopole reduction of the medium. This involved the usual imaginary monopole layer on the boundary of magnitude I. He then recognized that the normal magnetic intensity outside and inside the boundary are related as[133]

$$N = N' + 4\pi I = (1 + 4\pi\mu)N'$$

Guided by this, by his earlier heat diffusion analogy, his earlier theory of dielectrics, with his 'hydrokinetic permeability', he interpreted Faraday's 'conducting power of a magnetic medium' as the 'magnetic *permeability*', expressing it by the ratio

$$\text{Magnetic permeability} = \frac{N}{N'}[= (1 + 4\pi\mu)]$$

Earlier, in 1855, he wrote. 'I am still strongly disposed to believe in the magnetic character of the medium occupying space'[134]. Thomson, therefore, effectively chose the ether as the unit of permeability[135].

In 1872 Thomson also gives the 'polar' and 'electromagnetic' intensities at the boundary between two media[136]. It was a development of the corresponding condition for dielectrics[137]. Indeed, a general theory of dielectrics, analogous to that of Thomson's theory of magnetization, was advanced three years later by Lorentz in his doctoral thesis of 1875[138].

ELECTROMAGNETIC ENERGY

Thomson was one the founders of the physical theory of 'energy', indeed he effectively coined the latter term[139]. Before 1851, rather than 'energy', he generally used the term 'mechanical effect'. However, 'mechanical effect' was an ambiguous term meaning both a property of a body and an action. But between his publications of November and December 1851 Thomson began to distinguish very clearly between the 'mechanical energy of a body in a given state' (a property) and 'doing work upon matter' (an action)[140]. From 1845 his theory of energy slowly began to pervade the whole of his physics, both interpretatively and mensurationally.

The formal development of his theory of field energy was originally built on Gauss's formal theory of potential[141]:

$$\int V \frac{dV}{dp} ds = -\int q^2 dT$$

where V is the potential, dV/dp the potential gradient perpendicular to the surface of the attracting body, q the intensity of the attracting force, and ds and dT represent surface and volume elements respectively. Thomson showed in 1843, by transforming some of Gauss's integrals, that a volume integral of the electric intensity (R) squared, over all of space outside the charge distribution, is equal to a function of the local charge (M) and its local potential (v)[142]:

$$\iiint R^2 dxdydz = 4\pi M(v)$$

In 1843 Thomson apparently saw this as a formal expression only.

From 1845 onwards Thomson gradually built his theory of electrical energy, as witnessed in his letters and diaries[143]. In April 1845, in analogy with the theory of heat engines and using the analytical theory of potential:[144]

> I got the idea, which gives the mechanical effect necessary to produce a given amount of free electricity, on a conducting or non-conducting body ... If m is any electrical element, V the potential of the whole system upon it, the mechanical effect necessary to produce the distribution m is ΣmV.

This is the work done to assemble the elementary charges from infinity to the surface of the conductor. The sum ΣmV involves both the 'test' element m, and the implicit source element m', which generates the potential V applied to m. It, therefore, involved a double sum or integration[145]. On 8 April 1845 Thomson writes in his diary[146]

Also the theorem of Gauss that ΣmV is a minimum when V is constant, shows that the double integral which occurs when we wish to express the action directly, may be transformed into a differential coefficient of a simple integral, taken with reference to the distance between the two spheres.

This transformation expresses the mechanical effect [the energy] in terms of the square of the electric intensity integrated over all space between the spheres, rather than in terms of the surface charges on the sphere. Furthermore, the energy minimization requirement applies, of course, to the space integral[147]. At this time, however, Thomson did not give it explicit physical meaning[148].

In January 1849 Thomson applied an identical formalism, but now with full physical significance, to the *vis-viva* [kinetic energy] Q of liquids[149].

$$Q = \rho\iiint R^2 dxdydz = \rho\iiint (u^2 + v^2 + w^2)dxdydz$$

Where R is the velocity of the liquid and ρ is the density of the liquid. He then shows that that the total kinetic energy for a frictionless, incompressible, fluid has the least possible value, given an arbitrary alteration in the boundary conditions.

He also transforms Q into the surface integral $= \rho\iint \phi H ds$ where ϕ was later called the 'velocity potential' of the fluid, and H 'expresses the given normal velocity at different points of the containing surface'[150]. He immediately draws attention to the formal identity of Gauss's and his own theorems in electrical and magnetic attraction to these hydrodynamics results. Indeed, Thomson's work on magnetic energy was carried out in parallel with his work on hydrodynamic energy, and he recognized in 1850 that:[151]

the distribution of 'electro-magnetic force' through the substance of the magnet, as well as through external space, corresponded to a possible distribution of motion in a continuous incompressible fluid filling all space . . .

Clearly Thomson was prepared, both mathematically and analogically, to give physical assent to the concept of a distributed energy in the space between charges or magnets. But a 'correspondence' was not an identity and he continued to suspend judgement, within his 'positive' physics, about the location of electromagnetic energy.

Systematically, Thomson expanded the role of work in electricity and magnetism, particularly influenced by Helmholtz's great essay of 1847 on the conservation of *kraft* [force][152]. In 1848 he derived Joule's proportion relating work done and the square of the electric current, and also the relation between the inducing current and the rate of change of what is now called the 'magnetic flux' function[153]. In 1851 he applied the concept of mechanical work to electromagnetic induction, and to electrolysis[154]. In 1853 he demonstrated that if two insulated charges spheres[155]:

be pushed towards one another ... the quantity of work which will be spent will be $F.dv$, since F denotes the repulsive force ... But the mechanical value of the distribution ... must be increased by an amount equal to the work spent in producing no other effect but this alteration. Hence $F.dv = -dW$.

This was a powerful principle for solving dynamical problems in physics.

Also in 1853, using the conservation of the energy of an electrical circuit, he introduced the theory of the discharge of a capacitor, controlled by both inductance and resistance. He examined a variety of cases using complex numbers. It is difficult to exaggerate the importance of this theory for electrical circuits.[156]

In seminal papers of 1853 and 1860 Thomson gives his first general summary of electric, magnetic, and current energy. He argues that the energy released when induced magnetization is reduced to zero, is composed of two terms. The first represents the arithmetical sum of the internal energies of infinitesimal bars partitioned along the lines of magnetization. The 'second term ... expresses the amount of work that would have to be done to put these parts together, were they given separately, each with the exact magnetization that it is to have when in its place in the whole'. His corresponding mathematical expressions are[157]:

$$\frac{1}{2}\iiint dxdydz\frac{q^2}{2\mu} + \frac{1}{8\pi}\iiint R^2 dxdydz\,,$$

where q represents the intensity of magnetization and μ the isotropic susceptibility[158]. R represents the magnetic intensity according to the 'polar' [H] definition. The total reduces to:

$$\frac{1}{8\pi}\iiint dxdydz(4\pi\mu+1)R^2$$

This is equivalent to Maxwell's expression[159]:

$$\frac{1}{8\pi}\iiint \boldsymbol{B}\boldsymbol{\cdot}\boldsymbol{H}dxdydz\,.$$

Thomson also transforms this in terms of his magnetic potential V as:

$$\frac{1}{2}\iiint dxdydzV\rho$$

where ρ is the imaginary pole density in the material (equal to div I, where I is the intensity of magnetization). V is valid because electric currents are not involved[160]. Thomson, therefore, has two expressions for the energy of a magnetized system, one applying to all space, and the second applying only to the magnetized body.

Turning to the energy of electric currents Thomson argues that,

If an electric current be excited in a conductor, and left without electro-motive force, it retains energy to produce heat, light and other kinds of mechanical effect. However, ordinary *inertia* ... [such] as that of a current of water ... does not exist for electricity in motion, or are but small compared with the [Faraday's] 'induction of the current upon itself'.

He then summarizes his theory of the energy of an electric current. It is the work done by the 'electro-dynamic force' [the induced electric intensity] on the electric current while it is being established[161].

In 1853 Thomson introduced the mechanical value or the 'actual energy' of a linear current as $1/2A\gamma^2$, where γ is the current[162]. Thomson calls A the 'electro-dynamic capacity' (today, the 'coefficient of self-induction')[163]. Thomson's 'actual energy' of a current, $1/2A\gamma^2$, which he later described as the 'kinetic energy' of the current, Maxwell describes as the 'electrokinetic energy' of the current[164]. This was *not* the mechanical kinetic of the charged particles, which Thomson demonstrated, from Faraday's experiments, as 'insensible'[165].

In 1860 he writes:[166]

that the mechanical value of a current in a closed circuit ... may be calculated by means of the following simple formula not previously published: $1/8\pi\int\int R^2 dxdydz$, where R denotes the resultant electro-magnetic force.

He now uses the 'electromagnetic definition' (Maxwell's '**B**' definition) of the field intensity, since his mathematical analysis of galvanic currents required the latter.

He says, modestly, that 'this expression may be useful in the dynamical theory of electromagnetic engines ...'[167]. Thomson, therefore, has two definitions of the energy of a current system, one suggesting that the energy is located in the conductors, the other in the surrounding space.

Although Thomson recognized the convenience and power of the spatial expression for energy, he never settled on either as the true physical location of the energy. Nor did he ever state that $(1/8\pi)R^2$ is the actual field energy in free space per unit volume.

Maxwell, in 1973, states that the energy expression involving 'electric currents ... is the natural expression of the theory which supposes the currents to act [at a distance] on each directly ...'. Maxwell, therefore, chose the magnetic field expression for the energy of electric currents. He states that $(1/8\pi)$ ***BH*** is the electrokinetic energy per unit volume in space, and represents 'the most significant form of kinetic energy'[168].

A KINETIC DREAM OF VORTEX ATOMS

1851 was a fundamental turning point in Thomson's view of the ultimate structure of matter[169]. The work of Julius Mayer (1814–78) and James Joule (1818–89), 'demonstrated the immateriality of heat'[170], and argued that heat is not a substance, it is a 'state of motion', 'a dynamical form of mechanical effect [energy]'[171]. In 1867 Thompson interprets elasticity as 'kinetic elasticity'[172]. In 1872 he goes further and argues that:

> motion is the very essence of what has hitherto be called matter. . . . I learned from Joule the dynamical theory of heat, and was forced to abandon at once many . . . statical conceptions regarding the ultimate causes of apparently static phenomena . . .[173]

Beginning in 1847 Thomson, along with Stokes, began to give increasing importance to the foundational status of hydrodynamics in physics[174]. He wrote in a letter of 1857, 'Now I think hydrodynamics is the root of all physical science'[175].

In 1856, he went further in his mechanical representations, and attempted to model the 'Faraday effect'—the rotation of plane polarized light. He notionally decomposed a ray of plane polarized light, parallel to the magnetic intensity in magnetized glass, into two oppositely rotating circular polarizations. He argued that these components are 'propagated at different rates'. The larger rotation occurs because Thomson's magnetic 'twist' is added to one of these circular polarizations. Its sense of rotation is 'the same as the nominal direction of the [equivalent] galvanic current in a magnetizing coil'. The result, therefore, is a rotation of the plane of polarization. He then develops this mathematically[176]. He also hesitantly interpreted the magnetized medium in terms of the molecular vortices introduced by William Rankine (1820–72). In 1873 Maxwell developed this interpretation further, and the modern theory of the 'Faraday effect' is partly derived from Thomson[177].

Helmholtz's seminal paper of 1858 on vortex motion[178] gave an enormous boost to Thomson's hope that all physical forces, especially elasticity, might be reduced to motional structures in fluids. In 1867, he was delighted by Helmholtz's demonstration of the stability of the vortex rings. According to Thomson a 'perfect fluid', has an 'absolutely unalterable quality', and are 'the only true atoms'[179]. He went on to declare that:

> . . . space is continuously occupied by an incompressible frictionless liquid acted on by no force, and that material phenomena of every kind depend solely on motions created in this liquid[180].

He also hoped to model the frequencies of light emitted by gases on vibrations of vortex rings[181]. He was particularly struck by the strong analogy between vortex rings with the magnetic field intensity due to a circulating current[182].

Previously, 'every ... property of matter has ... required an assumption of specific forces pertaining to the atom'. This reduction of specific forces to kinetics was a whole new programme of research for Thomson 'of an exciting character'[183]. However, speaking of his kinetic theory of forces in 1881 Thomson reacted against his own speculative theories[184]:

> But this kinetic theory is a dream, and can be nothing else, unless it can explain chemical affinity, electricity, magnetism, gravitation, and the inertia of masses (that is crowds) of vortices.

Between 1867–87 Thomson engaged in vortex research, hoping this research would model, in particular, elasticity and magnetic forces. To achieve this, Thomson developed the classical theory of vortices and solid and liquid gyroscopes.

In 1869–70, and also in 1887, he modeled the forces between magnets, electromagnets, and bodies with magnetic permeability, as solid bodies with apertures of irrotational circulation—that is, with coreless vortices[185]. He was also ready to replace the solid by a vortex, thus reducing the system to motion in liquid. In 1879 he modelled the forces between 'floating magnets' with corresponding 'groups of columnar vortices'[186].

Helmholtz also showed that closed vortex filaments in an infinite perfect fluid interact mechanically like electric circuits (although with attractions and repulsions reversed), and the velocity of the fluid corresponded with the magnetic intensity. This seemed a vindication of Thomson's 1847 analogy[187].

In 1875 Thomson gave a simple case of what he calls vortex statics, a 'circular vortex ring ... with an irrotational circulation outside all'[188]. He also considered a '*vortex-core*'[189] consisting of a '14-threaded toroidal helix' [Fig. 7.1.]. '[In] all of these cases ... the fluid-velocity at any point is equal to, and in the same direction as, the resultant magnetic [intensity] at the corresponding point in the neighbourhood

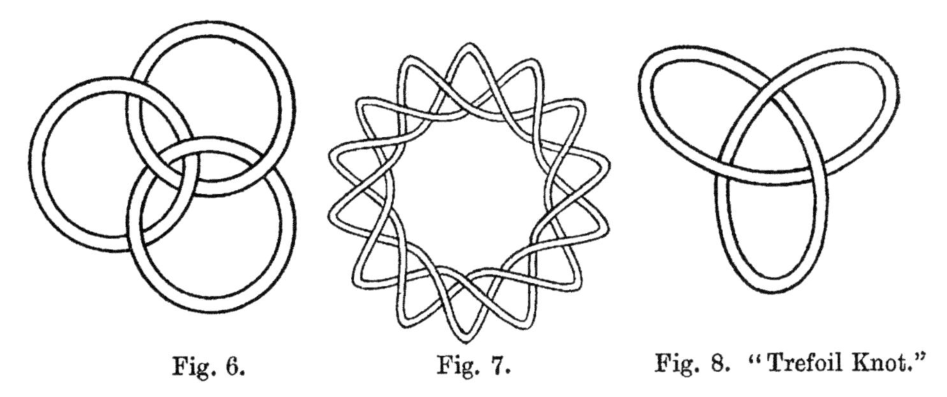

Fig. 7.1. Threaded Vortices, as illustrated in the case of 14 threads, all vortex threads run together into one.

of a closed galvanic circuit ...'. 'This analogy ... with the distribution of [the magnetic field] does much promote a clear understanding of [these] strange fluid motions'[190].

In 1884 he modelled a magnetic compass as a gyrostat with a horizontal axis, but with 'reference to "astronomical North" ... instead of "magnetic north"'[191].

In 1889 Thomson described his most sophisticated hydrodynamic analogy, which included alternating currents, electromagnetic induction, the magnetic field, and electrostatic field. He wrote that,[192]

> In respect to electromagnetic theory; we have a very fine analogy...with the diffusion of laminar motion into a viscous fluid ... [For example, in] the distribution of alternating electric current through a conductor such as copper, and the distribution of the motion of water in a viscous fluid disturbed by periodical tangential motions of its surface, follow identically the same law ...

He then gave 'an extension of that viscous fluid analogy which allows us to take into account all that goes in air and in metal, and in different metals'. Different metals represent different densities, but 'the viscosity must be the same in all'[193].

In this analogy the electric current is not the velocity of the fluid but the rate of change with distance of the viscous shearing rate. 'It also gives a motion of which the rotation is the magnetic [intensity]. This is what Maxwell calls "vector potential"'[194]. Wryly, he admitted that here 'we have a super-subtle mathematical definition of electric current which is not fluid velocity'[195]. Equally subtle, the representative of an insulator, including the air, in this analogy is a fluid without inertia[196].

The attempt to introduce the electrostatic intensity into any of his analogies is 'desperately difficult'. This is because electrostatic force is modelled by a pressure. But different normal electric intensities, on opposite sides of a dielectric boundary but with different dielectric constants, cannot be modelled in the 'interface between two fluids', since the normal pressures are equal. Thomson found no way of escaping from this problem in his model, describing the situation as one of 'despair'[197].

Thomson, therefore, was not happy with his 1889 analogy. '[It] cannot be considered as being in any respect a physical analogy'[198]. He states 'that the more we look at it the less we like it.' He also criticizes his analogy in terms of the distribution of energy[199].

But 'stability' was an even more difficult problem in his vortex theory. In 1880 he writes that 'I have not indeed succeeded in rigorously demonstrating the stability of the Helmholtz ring in any case'[200]. In 1887 'after many years of failure to prove that the motion in ordinary Helmholtz circular ring is stable, I came to the conclusion that it is essentially unstable...'[201]. Nor had Thomson succeeded in modelling purely transverse optical waves in an entirely liquid ether. Thomson then returned in 1889 to his earlier elastic-solid theory of matter and the ether.

THE MAGNETIC FIELD AND THE ELASTIC-SOLID ETHER

Thomson's vision was no less than a great comprehensive mechanical theory of the ether and ponderable matter, in all its modes. In a letter of 1862 he writes that:[202]

> The Sun may have ... prepared a luminiferous medium .. [or] ether ... for the propagation of light, and generally for the requisites of a 'world'.

Thomson never seems to have had any doubt about the ether. Indeed, the establishment of the wave theory of light seemed a necessary consequence in nineteenth century physics of the idea that light was carried by a universal material medium. Beside this general belief in the ether, Thomson created a large number of speculative theories about the constitution of the ether, driven by his own researches and also by contemporary experimental and theoretical research[203]. Here I will concentrate of the relation between magnetism and the ether.

Historically, the optical ether was postulated by Thomas Young (1773–1829), to explain the high frequency transverse vibrations of light. He suggested that the ether is composed of weightless matter, 'that it pervades of all material bodies with little or no resistance', that it was rigid with high elasticity, and yet the planets freely move through it[204]. This meant that the ether had to be visualized as a solid, since conventional fluids or gases did not sustain transverse vibrations. This created enormous difficulties of interpretation. Nevertheless, by 1830, the corpuscular theory of light had been marginalized[205].

Thomson was willing to accept that the ether, rather like wax, is brittle for light vibrations, but is perfectly mobile for planets, so that heavenly bodies can apparently sweep through it without being retarded[206]. His belief for the ether is expressed in his Baltimore lectures of 1884: 'A real matter between us and the remotest stars I believe there is [and it includes] motions in the way of transverse vibrations'[207]. In 1884 he stated that 'we have not the slightest reason to believe the luminiferous ether to be imponderable'. But in 1899 he wrote that 'I now see that we have the strongest possible reason to believe that the ether is imponderable'[208]. There is an ironic humour in Thomson's speculative writings, mixing exaggerated claims and an equally exaggerated scepticism about his claims.

In 1888 Thomson noted that George Green's 'jelly ether' had the required transverse waves to model light, since its longitudinal waves had infinite velocity. But such a property seemed unsatisfactory[209]. He then imagined an ether model as a:

> homogeneous air-less foam held from collapse by adhesion to a containing vessel, which may be infinitely distant all round, exactly fulfils the condition of zero velocity for the condensational-rarefactional wave, and a definite rigidity and elasticity of form, and definite velocity of distortional wave[210].

It thereby yielded transverse waves only. However, it was not yet rich enough to model electromagnetic waves.

With the emergence of Faraday's field theory, Thomson, with many other mid-nineteenth century physicists, began to wonder whether the luminous ether might have other properties, including the electric, magnetic, and gravitational field intensities[211]. In 1884 Thomson writes 'If I knew what the magnetic theory of light is, I might be able to think of it in relation to the fundamental principles of the wave theory of light'[212]. In 1889, returning from his long hydrodynamic excursus, he now views the ether as a solid, and seriously attempts to bring together electromagnetic waves and the waves of light. He writes that 'the propagation of electrical disturbances and light may be identical.'[213]. And that this propagation occurs 'by the influence of a medium, and [Maxwell] showed that that medium ... must be the ether'[214].

In 1889 he modelled a solid ether in terms of a 'jointed structure' of spherical atoms attached to bars, at the extremities of which were perfect frictionless, ball and socket joints. It is a 'skeleton framework, analogous to idealized plane netting consisting of stiff straight sides of hexagons perfectly jointed in their ends'[215]. To model 'luminous [transverse] waves, let us attach to each bar a gyrostatic pair composed of two Foucault gyroscopes' mounted 'perpendicularly to the line of the bar'. And both flywheels rotate in opposite direction. It could be also be represented by 'liquid gyroscopes'[216] [Fig. 7.2.]. This structure now has a kinetic 'quasi rigidity' against rotational deformation[217] [Fig. 7.3.]. That is, it allows free translation but contains rotational rigidity[218]. He then writes:[219]

> A homogeneous assemblage of points with gyrostataic quasi rigidity ... if constructed on a sufficiently small scale, transmit vibrations of light exactly as does the ether of nature.

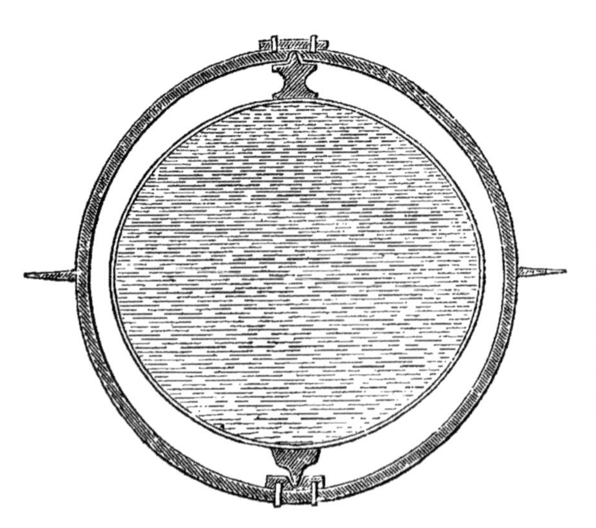

Fig. 7.2. The liquid gyrostat. 'In the liquid gyrostat the fly-wheel is replaced by an oblate spheroid made of thin sheet copper, and filled with water ... It is pivoted at the two ends of its polar axis in bearings fixed in a circular ring of brass surrounding the spheroid'.

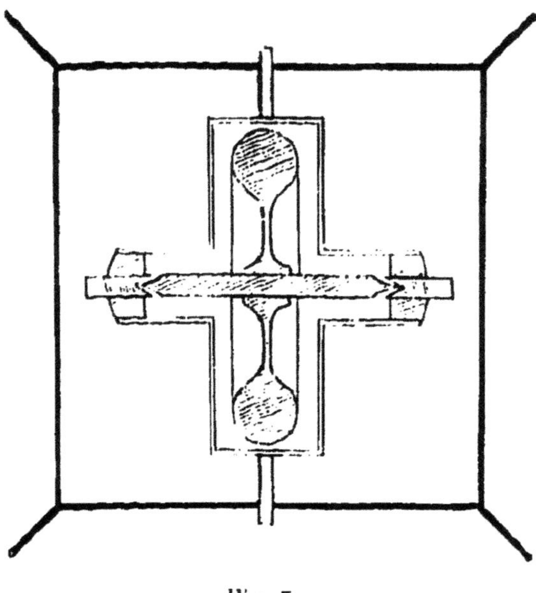

Fig. 7.3. Rotational rigidity. 'Thus we have a skeleton model of a special elastic solid with a structure essentially involving a gyrostatic contribution to rigidity'.

This ether does not support longitudinal vibrations because 'it would be incapable of transmitting condensational-rarefactional waves ...'[220].

Thomson's 1889 ether model was driven by the requirement to make it compatible with his mechanical model of magnetization[221]:

> The marvellous exigencies of an attempt to include inductive magnetisation in the mechanical representation, that compelled the assumption of a quasi-elastic force, depending on the absolute rotation ...

And his mechanical model of the magnetic field was, of course, driven by the requirement that it is compatible with the rotation of the plane of polarization of light in a magnetized body[222]:

> Whatever the current of electricity may be, I believe *this* is a reality: *it does pull the ether round* within the solenoid.

For Thomson, this twist of the ether *is* the magnetic field.

In more detail, he modelled electric currents generating a magnetic field within an elastic solid ether, of 'exceedingly small density', with an ether filled with porous tubes. These tubes are 'filled with a dense viscous fluid ... forced, by the aid of a piston .... to move through the tube'. This models various aspect of electromagnetism, including electromagnetic induction, 'electricity as the viscous fluid and ether an elastic solid'[223]. 'If it were not for the gross ponderable matter, I should be perfectly satisfied [with this model]'.[224]

Then, by comparing a copper core with an iron core inside a current bearing solenoid, he discovered that only an ether with gyrostatic rigidity has the correct magnetic boundary conditions[225]. This rotational rigidity must also be equal in copper and in other non-magnetic materials, in air, and in other non-conductors, but the rotational rigidity must be enormously less than in iron to allow for electromagnetic induction.

However, he ruefully conceded that this ether model fails to explain the forces acting on metals falling through magnetic fields: 'We do not get even get near the mutual attraction between the iron of an [current bearing solenoid]'[226]. Nor does it explain the 'molecules of air itself tearing through [the ether] in all directions'[227]. Furthermore, 'Our first love was electrostatics. That is absolutely left out in the cold'[228].

In 1890 he developed the gyrostatic ether further. Its work done per unit volume is $\frac{1}{2} nI^2$ where n is the rigidity and I is twice the absolute rotation[229]. In this model ether he does speak of the 'energy of the ether', modelling the energy of a bar electromagnet[230]. And it also 'represents perfectly the circumstances of the electromagnetic action on the space outside an infinitely thin circular cylindrical solenoid of electric current ...'[231]. Thomson with some gratification felt that he had completed the galvanic part of his 1847 analogy[232].

In 1890 Thomson appeared to accept Maxwell's theory that there is a 'magnetic stress' in a magnetic field, caused by a tangential drag on the surrounding medium derived from the current or magnet, and this is responsible for moving a free current-bearing wire[233]. In 1896 he also spoke of 'electrostatic stress' in the ether[234].

Thomson seemed to accept that the ether becomes liquid for planetary bodies moving through it. Then it becomes equivalent to a 'frictionless incompressible fluid'. It also obeys the minimum kinetic energy condition[235].

In 1890 he approached Maxwell's interpretation of light: '[If we are given an] electric current that is a periodic function of time. We have thus simply the *undulatory theory of light*'[236]. But he always interpreted Maxwell's electromagnetic waves as 'magnetic waves'[237] and refers to the 'magnetic theory of light,' and the 'so-called electromagnetic theory of light'[238]. But how does Thomson relate his magnetic model to Maxwell's electromagnetic theory?

THOMSON AND MAXWELL'S ELECTROMAGNETIC THEORY OF LIGHT

In his Baltimore lectures of 1884 Thomson writes that the:

> magnetic theory of light is a backward step from the definite mechanical motion put before us by Fresnel ... It introduces 'things we understand even less of'.

S. P. Thompson, in his biography of Thomson, expresses it less politely as an attempt to explain *ignotum per ignotius*[239]. Thomson also states that Maxwell's theory 'teaches us nothing about the actual motions of matter constituting a magnetic wave'[240]. Thomson was a generous spirit, and respected the positive achievement of Maxwell's theory—for example, that light is an electromagnetic vibration in the ether[241]. Also, some of Thomson's criticisms were undoubtedly eccentric[242]:

It is mere nihilism, having no part of or lot in Natural Philosophy, to be contented with two formulas for energy, electromagnetic and electrostatic, and to be happy with a vector and delighted by symmetrical formulas.

However, Thomson had very good reasons for disliking Maxwell's approach to electromagnetism. Maxwell's formal mathematical theory was not 'natural philosophy' or 'positive' in Thomson's sense. Maxwell, in his interpretations and his formalism, mixed together speculative analogies[243], speculative hypotheses[244], and deep ambiguities in the foundations of his theory[245]. For example, Thomson used the analogy of heat flux and the flux of a liquid (as a quantity of flow) even in his electrostatics and magnetostatics[246]. But in 1896 he objected 'to the damagingly misleading way in which the word "flux" as if it were a physical reality for electric and magnetic [intensities]', rather than a mathematical analogy[247]. But it was too late to withdraw the concept, and it remains to this day in modern electromagnetism.

Thomson also needed the physical interpretation of every line of formalism. However, he felt that much of Maxwell's formalism was almost impossible to understand[248]. Again, Thomson needed a mechanical model to understand a theory, and he felt he could not understand Maxwell's models[249].

Thomson particularly disliked Maxwell's 'displacement'. According to Thomson's model ether[250]:

Maxwell's 'electric displacement' [is] simply a to-and-fro motion of ether across the lines of propagation [of light],.. precisely the vibrations of ... Fresnel. But we have as yet absolutely no guidance ... of the relationship between this ... definite alternating motion, or any other motion ... of the ether, and the earliest known phenomena of electricity and magnetism...

Maxwell's 'displacement', and his electromagnetic theory of light, were incomprehensible to Thomson. This is partly because of the ambiguities in Maxwell, but equally because, in Thomson's mind, Maxwell's theory was read through the lens of Thomson's own mechanical theory of magnetism and the ether.

The methodological styles of Thomson and Maxwell were, therefore, very different. Nevertheless, despite its many flaws, Maxwell intuitively overcame the current dogma of a mechanical foundation for electromagnetism, discovered new 'matter of fact' laws, unified electromagnetism, and created the electromagnetic theory of light. However, there was a severe price to pay. To this day physics is still attempting to unravel the physical meaning and defects in Maxwell's electromagnetism.

THOMSON'S LATE THOUGHTS

In 1889–90 Thomson got as far as he could with his comprehensive programme of reducing elasticity, electricity, magnetism, and light to classical mechanics[251]. He recognized that the processes he had coherently modelled so far were far fewer than those required even to model the 'matter of fact laws' of electricity and magnetism.

In 1896 I have been trying almost incessantly since some time before 1846 for a mechanical-dynamical representation of electrostatic force, but hitherto in vain[252].

His greatest problem in his mechanical representation 'was the mobility of magnets and electrified bodies, showing the ponderomotive forces experienced by them...'[253]. While the mechanical model of a magnet or current does show an appropriate 'directional tendency', but it does not, in the 'slightest degree', explain the magnetic forces on iron[254]. This failure also applied to his hydrodynamic models:

no one has come with a million miles of explaining any one phenomenon of electrostatics or magnetism by hydrodynamic theory ... though we have really a wonderfully good knowledge of the matter of fact laws of electrostatics and magnetism ...[255].

Speaking to George Fitzgerald in 1896 of his various attempts to model the ether 'I now abandon everything I have ever thought of or written in respect to the constitution of ether'[256]. In 1904 he asks:

how much of the phenomena of light can be explained without going beyond the elastic-solid theory. We have now our answer: everything non-nonmagnetic; nothing magnetic[257].

In 1893 he writes:[258]

I have not had a moment's peace or happiness in respect to electromagnetic theory since Nov. 28, 1846. All this time I have been liable to fits of dipsomania, kept away at intervals only from rigorous abstention from thought on the subject.

And even more dramatically, during his Jubilee address of June 1896 to the Corporation and University of Glasgow, he writes:

One word characterises the most strenuous of the efforts ... during fifty-five years; that word is FAILURE[259].

There was a strong element of rhetoric and exaggeration—and even humour—in Thomson's 'reported' failure. He certainly failed in his underlying mechanical vision, but this was only one aspect of a career in physics of extraordinary distinction and success.

Furthermore, Thomson's presiding analogy was his view that magnetism is a twist in the ether. History has largely born this out when Arnold Sommerfeld

(1868–1951) in 1910 recognized that the magnetic field is a planar tensor, not a vector[260], and Hermann Weyl (1885–1955) pointed out that this tensor includes 'a unique direction of twist in . . . the magnetic field'[261].

From the 1890s until his death in 1907, Thomson continued to take great interest in the new developments in physics, especially in the discovery of the 'electrion' [*sic*], x-rays, and radioactivity[262]. He also broke the mould of his elastic-solid model by admitting *three* qualitatively different entities in physical nature:

> All of this essentially involves the consideration of ponderable matter . . . embedded in ether, and a *tertium quid*, which we may call electricity, a fluid go-between serving to transmit force between ponderable matter and ether[263].

Even in his 70s and 80s Thomson was flexible enough to recognize that his narrow mechanical view of nature was at an end.

8

'A Dynamical Form of Mechanical Effect': Thomson's Thermodynamics

Iwan Rhys Morus

INTRODUCTION

Late in the afternoon of Thursday, 24 June 1847, a young man sat at the back of a crowded hall at the British Association for the Advancement of Science's annual meeting in Oxford, listening to a rather rushed and abbreviated paper on something called the mechanical equivalent of heat.[1] It was the tail end of the proceedings of Section A—the section devoted to papers in physics and mathematics. The author of the hurried presentation was the Mancunian natural philosopher James Prescott Joule and the young man listening was William Thomson, recently appointed Professor of Natural Philosophy at his old university of Glasgow. Thomson was intrigued by what he heard for a number of reasons. He was impressed by the speaker's obvious skill as an experimenter, the ingenuity of his apparatus, and the seeming accuracy of his measurements. Despite the haste with which he presented his findings—he had been asked by the Chairman to keep his presentation short—Joule clearly knew what he was talking about and seemed in full command of the facts and arguments. What intrigued (and puzzled) Thomson most, however, was that he was quite sure that there was not (and could not be) any such thing as a mechanical equivalent of heat in the way Joule appeared to mean it. The very idea made little or no sense.[2]

Thomson knew there was no such thing as the mechanical equivalent of heat because he had been fascinated by the relationship between heat and work for a long time. He and his engineer brother James had been sharing their speculations on the matter for years. As a result he was well versed in the latest theories about heat and its origins so he knew that heat was best understood as a kind of substance, or fluid, filling hot bodies and flowing between hot and cold. He knew as well, of course, that there was an alternative view. Humphry Davy, the Cornish chemist who would become Professor of Chemistry at the Royal Institution and President of the Royal Society, had argued almost half a century earlier that heat was a form of motion rather than a fluid. Davy pointed to the fact that heat could be produced by friction as strong evidence of the fact that it could not be a substance. But Thomson was sure that Davy was wrong. All the best evidence appeared to be with those who argued heat was a substance, called caloric, and that work was simply the result of this caloric flowing from a hot to a cold body.[3]

Looking back at events from the vantage point of half a lifetime later, Joule recalled that Thomson had stood up and plied him with questions at the end of his presentation. Thomson remembered accosting the Mancunian natural philosopher after the end of the session, eager to continue the discussion. Joule was certainly glad of the attention. He had been trying for several years with little success to persuade other men of science of the importance of his experiments on the mechanical equivalent of heat. In attempting to draw attention to his work, Joule's background was certainly against him. He had few or no contacts amongst the metropolitan scientific elite that dominated British natural philosophy and its

Fig. 8.1. James Prescott Joule as a young man. The apparatus he used in order to determine the mechanical equivalent of heat can be seen in the background.

institutions. Fellows of the Royal Society had no reason to suppose that a brewer's son from Manchester had anything interesting to say on the thorny question of the relationship between heat and work. Joule kept in contact with Thomson, sending him copies of his recent publications. It was not long before they bumped into each other again—in Switzerland of all places. Joule was there on his honeymoon, whilst Thomson was embarked on an extensive European grand tour. According to Thomson, Joule had brought one of his thermometers with him and when they met was about to try the experiment of measuring the temperature of a waterfall at its top and bottom. The story may be apocryphal, but it underlines Thomson's recognition that one of the things that made Joule's experiments worth taking seriously was Joule's skill in measuring small differences in temperature.[4]

Thomson was, in fact, in a unique position to appreciate the scale of Joule's achievement with the paddle wheel experiments with which he had determined the mechanical equivalent of heat. Only a few years previously, just before his appointment to the Chair of Natural Philosophy at Glasgow and following the completion of his studies at Cambridge, Thomson had spent time working at Victor Regnault's laboratory in Paris. It was his introduction to the art of precision measurement. In 1843, a few years before Thomson's visit, Regnault, Professor of Physics at the College de France, had been commissioned by the French government to carry out an extensive experimental programme investigating the properties of steam and steam engines with a view to providing data that could be used to improve the design of steam engines. It was a project that—in Regnault's view at least—required careful and accurate measurement above all else. The experimental skills that Thomson acquired in Paris were to stand him in good stead as he aimed for the recently vacated Chair of Natural Philosophy at Glasgow in 1846. They also meant that he had a clearer understanding than most of his British contemporaries of just what Joule had achieved with his paddle wheel experiments.

Thomson would spend much of his time over the subsequent several years as he settled into his position as Professor of Natural Philosophy at Glasgow, pondering over the problem of heat and the difficulty of reconciling Joule's experimental determination of the mechanical equivalent of heat with Carnot's theory that seemed to demand that there could be no such figure. The eventual result of his ruminations was a series of papers presented to the Royal Society of Edinburgh in 1851 and published in their *Transactions* a few years later on the dynamical theory of heat. In these papers Thomson outlined the basic principles of the new science of thermodynamics that had emerged from his attempts at making sense of Carnot and Joule's apparently contradictory positions. Over the next few decades, Thomson, Joule and others would succeed in placing this new thermodynamic science at the very heart of nineteenth-century physics.[5] To make sense of this achievement, though, we need to start by going back to the 1820s and Carnot's essay *Reflexions sur la Puissance Motrice du Feu*. This chapter will start, therefore, with a careful look at Carnot's essay and responses to it, followed by an account of Joule's experiments during the first half of the 1840s that led him to the mechanical equivalent of heat. This should leave the reader in the same position as Thomson as he worried away at the problem of heat in the years between 1847 and 1851, and better able, therefore, to appreciate the magnitude of his achievements.

A REPUBLICAN SCIENCE

If the name of Carnot were to be uttered on the streets of Paris during the first half of the nineteenth century, it would not have been the obscure army officer and engineer Sadi Carnot that would come immediately to the listener's mind. They would think instead of Lazare Carnot, hero of the Revolution, general of the Grande Armée and member of the notorious Committee of Public Safety during the early revolutionary years. This was the man who was Sadi Carnot's father. Lazare was himself something of a *savant*. He had published extensively in engineering and mathematics, including a number of texts on the origins of work in machinery. He was known in particular for his mathematical analysis of water mills. Along with fellow natural philosopher Gaspard Monge, he was responsible for establishing the École Polytechnique as an elite engineering school for aspiring army officers during the early years of the French republic. Sadi followed his father's footsteps into the army and duly enrolled as a cadet at the engineering school his father had been so instrumental in founding. Like his father too, he was a staunch republican and a firm believer in the importance of placing scientific knowledge at the service of the French republic. It was his misfortune that following Napoleon's defeat in 1815, sentiments such as these were no longer popular under the restored Bourbon monarchy.

By 1820, Sadi Carnot was already retired and on half pay. As his father's son it was clear that his army career was going nowhere under the new regime. He turned his attention instead to the relationship of heat and work, studying widely in both natural philosophy and political economy. He visited factories and workshops to see machinery in action. *Reflexions sur la Pussiance Motrice du Feu* was published in 1824, a year after Lazare Carnot's death in exile. We can think of it in some ways as a son's tribute to his dead father. It was clearly written in the hope of attracting a broad and popular audience, with mathematical language kept to a minimum. It was also, quite clearly, both a deeply patriotic and a staunchly republican work. Carnot wanted to put his science at the service of the state by putting it to work to improve French industry. Looking at the old enemy across the English Channel, he argued that England's military superiority boiled down to its industrial might and the efficiency of its steam engines. As he put it:

To take away today from England her steam-engines would be to take away at the same time her coal and iron. It would be to dry up all her sources of wealth, to ruin all on which her prosperity depends, in short to annihilate that colossal power. The destruction of her navy, which she considers her strongest defence, would perhaps be less fatal.[6]

This was fighting talk.

To understand the steam engine, argued Carnot, one needed to understand the general relationship between heat and work. The key process taking place in an operating engine, he suggested, was 'the re-establishing of equilibrium in the caloric; that is, its passage from a body in which the temperature is more or less elevated, to another in which it is lower'. In a working steam engine, the:

caloric developed in the furnace by the effect of the combustion traverses the walls of the boiler, produces steam, and in some way incorporates itself with it. The latter carrying it away, takes it first into the cylinder, where it performs some function, and from thence into the condenser, where it is liquefied by contact with the cold water which it encounters there. Then, as a final result, the cold water of the condenser takes possession of the caloric developed by the combustion. It is heated by the intervention of the steam as if it had been placed directly over the furnace. The steam is here only a means of transporting the caloric. It fills the same office as in the heating of baths by steam, except that in this case its motion is rendered useful.[7]

In short, what happened in a steam engine was a movement of caloric from one place to another in order to restore its equilibrium.

It was a short step from this to the establishment of what Carnot regarded as the fundamental principle underlying the operation of any heat engine:

The production of motive power is then due in steam engines not to an actual consumption of caloric, but *to its transportation from a warm body to a cold body*, that is, to its re-establishment of equilibrium—an equilibrium considered as destroyed by any cause whatever, by chemical action such as combustion, or by any other.[8]

The presence of caloric of itself was not enough to produce work. The caloric had to fall from a body at high temperature to a body at low temperature, in much the same way that water in the water mills studied by Sadi Carnot's father had to fall from a height in order to perform useful work. The bulk of the remainder of Carnot's essay consisted of detailed accounts and figures of the actual operation of different kinds of engines in an effort to establish just how efficient such devices could be made to be. The final note was one of pragmatism in the name of efficiency:

To know how to appreciate in each case, at their true value, the considerations of convenience and economy which may present themselves; to know how to discern the more important of those which are only secondary; to balance them properly against each other, in order to attain the best results by the simplest means: such should be the leading characteristics of the man called upon to direct, to coordinate the labours of his fellow men, to make them cooperate towards a useful end, whatsoever it may be.

Carnot's essay almost immediately vanished, virtually without trace. There were a number of reasons for this. One reason undoubtedly was Carnot's status—or rather lack of it. Throughout Napoleon's rule, French physics had been dominated by the clique of natural philosophers surrounding that éminence grise of republican science, Pierre-Simon Laplace. They were exactly the sort of people who might have been expected to lend Lazare Carnot's son a lending hand in establishing a scientific career. By the middle of the 1820s, however, Laplace's influence was in steep decline under the new royalist regime.[9] Carnot was not a name likely to attract patronage under the new dispensation. His populist approach and avoidance of complex mathematics did little to help Carnot's case either. His little essay simply did not look like serious science to the men who ran French physics. When Thomson, on his Parisian jaunt 20 years later, looked out for a copy he could not find a single one. The only Carnot Parisian booksellers had heard of was either the infamous Lazare or Sadi's younger brother, Hippolyte, who had already made something of a name for himself in literary circles and was on the verge of a career in politics. Thomson's knowledge of Carnot's work was therefore second hand and based on his reading of the man who rescued Sadi Carnot from perpetual obscurity, Émile Clapeyron.

Clapeyron, like Carnot, was a graduate of the prestigious École Polytechnique. He graduated there in 1818 before going on to the equally prestigious École des Mines to train as an engineer. He spent the 1820s in Russia before returning to France in 1830 where he played an important role in a number of railway engineering projects. In 1834 he published a little memoir 'On the motive power of heat' in the influential *Journal de l'École Polytechnique*. Here he took up the cudgels in defence of Sadi Carnot. Carnot was already dead when this happened. He had died of cholera only two years previously. Clapeyron was well suited for the task of resuscitating Carnot's reputation. Like him he straddled the boundary between practical engineering and natural philosophy. He had spent his time in Russia working as an engineer on the railways as well as lecturing in physics. He continued with a foot in both camps after his return to Paris. Even as he was writing his memoir on heat he was involved in a project to build a railway from Paris to St. Germain. A few years later he visited England in the hope of collaborating with George Stevenson on locomotive design.

Clapeyron succeeded where Carnot had failed in bringing his theory to public attention by translating it into the abstract mathematical language favoured by French physicists of the time. As Clapeyron put it, 'Carnot, avoiding the use of mathematical analysis, arrives by a chain of very difficult and elusive arguments at results which can be deduced easily from a more general law', and this was exactly what he set out to do in the bulk of the memoir. He succeeded admirably, helped along the way by his use of indicator diagrams to demonstrate the relationship between the volume and pressure of steam in a steam engine and the amount of work produced. These diagrams were a closely guarded trade secret of the steam engine manufacturers Boulton & Watt. Clapeyron had doubtless encountered the trick whilst in Russia, where Boulton & Watt engineers were working under license on the railways. Clapeyron's memoir was the conduit through which Carnot's work eventually reached its audience, particularly after it was translated into German and published in Poggendorf's *Annalen der Physik* in 1843. It was certainly through Clapeyron's memoir that the young William Thomson was alerted to Carnot's theories and became convinced that work was the result of the flow of caloric from a hotter to a cooler body.[10]

THE MECHANICAL EQUIVALENT OF HEAT

By the time Thomson encountered James Prescott Joule at the Oxford meeting of the British Association for the Advancement of Science in 1847, the Mancunian natural philosopher had been an active experimenter for the best part of a decade. A weakly child, Joule had been largely educated at home by a succession of tutors. At the end of 1832, however, his father, doubtless with an eye to having his son cultivate a science that might prove useful in the family brewing business, sent him to be tutored at the Manchester Literary and Philosophical Society by the eminent chemist John Dalton. Dalton encouraged the young Joule's burgeoning chemical and philosophical interests with the result that within only a few years the brewer's son was already embarking on his own independent experimental inquiries. A further boost to his ambition to establish himself as a natural philosopher took place at the end of the 1830s when the pugnacious electrical experimenter William Sturgeon arrived in Manchester to become the superintendent of the Royal Victorian Gallery of Practical Science. The Gallery, modelled on similar enterprises in London, was a commercial concern that aimed to profit from middle class interest in useful natural knowledge and inventions by putting the latest thing on show. Sturgeon took the young Joule under his wing and was instrumental in teaching him the skills of electrical experimentation.[11]

Joule was already acquainted with Sturgeon through correspondence even before the electrician arrived in Manchester. The young natural philosopher had submitted his first scientific paper to be published to the *Annals of Electricity*, edited by William

Sturgeon, in 1838. His communication, a 'Description of an electro-magnetic engine', was in many ways typical of the kind of material that Sturgeon published in his *Annals*.[12] Sturgeon was a leading figure in the London Electrical Society—the American natural philosopher Joseph Henry rather unkindly described him as being 'at the head of the second rate philosophers of London'[13]—a group of enthusiastic electrical experimenters of whom Joule was himself a corresponding member. London Electrical Society members had a particular interest in finding ways of making electricity useful.[14] Locomotion and producing useful work through electromagnetism were particular concerns. It was widely anticipated that before many more years had passed:

> Half a barrel of blue vitriol, and a hogshead or two of water, would send a ship from New York to Liverpool; and no accident could possibly happen, beyond the breaking of the machinery, which is so simple that any damage could be repaired in half a day.[15]

Even Joule, in a subsequent contribution to Sturgeon's *Annals,* was moved to prophesize that 'electro-magnetism will eventually be substituted for steam in propelling machinery'.[16]

In fact, despite this apparent optimism, Joule was inclined to be a little more circumspect regarding the immediate prospects for electromagnetic engines than at least some of his fellow members of the London Electrical Society. His caution was rooted in the careful measurements he was carrying out in the course of his own experiments on electromagnetism and work. Throughout his work on electromagnetic engines, Joule was overwhelmingly concerned with efficiency. In typically hard-headed, industrial Mancunian fashion, he wanted to know not just whether electromagnetic engines could work, but whether they would be cheaper to run than their rival steam engines. This was the point behind his careful measurements of the 'duty' (a practical engineer's term) of his engines. Duty, defined as the number of pounds his engine could raise per second to a foot's height, was a measure of the engine's power—the amount of work it could do. It was by comparing these numbers to those produced for steam engines that Joule hoped to determine whether or not electromagnetism had a serious future on Britain's railways and in its factories. The more he measured, the less sure he became about electromagnetism's economic prospects, and by the mid-1840s he had effectively abandoned his efforts in that direction. It was nevertheless his electromagnetic enthusiasms that first led him to consider the wider question of the origins of mechanical work.

Joule's famous paddle wheel experiments on the relationship between work and heat were therefore, in many ways, a direct offshoot of his earlier, electromagnetic experiments. They were certainly motivated by the same sort of concern with understanding and maximizing efficiency.[17] Throughout the first half of the 1840s, Joule carried out a range of experiments aimed at trying to understand the relationship between work and heat. In 1843 he published 'On the calorific effects of electricity, and on the mechanical equivalent of heat' in the *Philosophical Magazine*. This was the first time that he had employed this particular term—'the mechanical equivalent of heat'—to describe the quantity that he thought he was measuring in his experiments. In the experiments leading to this publication he had carefully measured the heat produced by an electromagnetic machine powered by a set of descending weights. It was the constant ratio between the work done by the weights in falling and the heat produced by the electric current induced in the electromagnetic machine that was to be characterized as the mechanical equivalent of heat. It was a loaded term, implying some kind of identity between work and heat as it did. In case any of his readers missed the point, Joule rammed it home, emphasizing that if:

we consider heat not as a *substance*, but as a *state of vibration*, there appears no reason why it should not be induced by an action of a simply mechanical character.[18]

To substantiate his suspicion that experiment could demonstrate the identity of heat and mechanical effect, Joule needed to find an arrangement that could prove the point. What he needed was a way of directly correlating an increase in the temperature of some substance to a set amount of mechanical work. Over the next few years he published a number of such experiments, but with very little impact on his scientific contemporaries. He was turned down for publication in the Royal Society's prestigious *Philosophical Transactions*. In 1845 he developed yet another method of measuring the mechanical equivalent of heat and establishing the interconvertability of heat and motion. In this method, a pair of weights was attached to a paddle wheel sealed in a cylinder full of water. When the weights were allowed to descend, they caused the paddle wheel to rotate, heating the water by friction in the process. If Joule could show that the same amount of work carried out by the falling weights constantly produced the same increase in the temperature of the water in the cylinder then he could argue that this demonstrated a constant relationship between the two quantities. A certain amount of work always produced the same amount of heat and vice versa.

Whilst the paddle wheel experiment appeared, on the surface at least, to be relatively straightforward, in fact it required extremely careful and highly skilled measurement. The temperature differences involved were such that highly sensitive thermometers and a great deal of experience were needed to make the necessary measurements successfully. This was where Joule's background in the brewing industry came to the fore. Brewers had a long tradition of carefully regulating the heat at which their brews fermented and knowing how to measure very small but crucial changes in temperature was part of that traditional practice. This was one of Joule's big advantages—and one of his problems. His thermometric measurements were at the edge of current technology and it was difficult to persuade others that such small changes in temperature had the huge significance that Joule accorded them.[19] One of the reasons that William Thomson took Joule's work seriously was that as a result of his own familiarity with Victor Regnault's work in Paris, he knew what kind of measurements were possible and what their significance might be. Another problem was the ambition of Joule's claims. By now he wanted to do far more than just establish the mechanical equivalent of heat. He was convinced that his experiment was crucial evidence for the interconvertability of all the forces of nature and for the conservation of force.

By the mid-1840s, Joule was convinced that more than the mechanical equivalent of heat was at stake in his experiments. He was increasingly explicit in his assertions that his experiments provided incontrovertible ammunition in favour of a dynamical theory of heat and against the existence of caloric. He was also starting to argue that 'the power to destroy belongs to the Creator alone', so that 'any theory which, when carried out, demands the annihilation of force, is necessarily erroneous'.[20] Just a few months before the fateful 1847 Oxford meeting of the BAAS where he encountered William Thomson for the first time, Joule gave a public lecture 'On Matter, Living Force and Heat' at St. Ann's Church in Manchester, during which he spelled out the ramifications of his natural philosophy. He argued that his experiments demonstrated how:

> the phenomena of nature, whether mechanical, chemical, or vital, consist almost entirely in a continual conversion of attraction through space, living force, and heat into one another.

This was part of a cosmic Divine plan through which 'order is maintained in the universe' and 'nothing is deranged, nothing ever lost, but the entire machinery, complicated as it is, works smoothly and harmoniously'. In this view, despite the fact that:

> every thing may appear complicated and involved in the apparent confusion and intricacy of an almost endless variety of causes, effects, conversions and arrangements, yet is the most perfect regularity preserved—the whole being governed by the sovereign will of God.[21]

This too was an aspect of Joule's work that the young William Thomson would find conducive.

THE DYNAMICAL THEORY

In many ways, William Thomson and James Prescott Joule—only six years his senior—had a great deal in common. Both were members of a new generation of British natural philosophers that was in many ways dissatisfied with the gentlemanly London, Cambridge, and Oxford club that had dominated early Victorian science.[22] Both were outsiders to that group, though Thomson, with his Cantabrigian connections, as we shall see, less so than was Joule. They came from similar backgrounds as well. Joule was a product of industrial Manchester. Thomson, though his father was an academic rather than a hard-nosed industrialist like Joule senior, had grown up in Belfast and then in industrial Glasgow. There he had imbued more than a little of the industrial ethos. Through his father, he had connections to the Glasgow Philosophical Society, founded in 1802 for the 'advancement of the arts and sciences' in the city. The Society was dominated by industrialists and business men who wanted to know how to turn science to profitable purposes. Thomson's first presentation to the Society after he had been elected a member in 1846 shows just how well he understood its members' concerns and how close they were to his own interests. He gave them an account of the theory underlying the working of the Stirling air engine, based, significantly enough, on Carnot's principles.[23]

As well as having imbibed what Crosbie Smith has characterized as a 'north British' industrial ethos in his physics, Thomson had studied mathematics at Cambridge.[24] By 1841, when William Thomson arrived there as an undergraduate after completing his studies in natural philosophy at Glasgow College, the Cambridge Mathematics Tripos probably offered the best mathematical training in the world. A new generation had completely revamped the curriculum, rescuing it from its eighteenth-century sterility and, in Charles Babbage's mischievous terms, replacing the old Newtonian 'dot-age' with the 'Principles of Pure D-ism'.[25] New French analytic techniques had been introduced along with written examinations and exhaustive training. Even Thomson found it tough going. 'Three years of Cambridge drilling is enough for anybody', he declared to his father in 1844.[26] The aim of the mathematics tripos was to produce, not mathematicians—and certainly not mathematical physicists—but highly trained minds ready to be sent out to govern the Empire. Mathematical physicists were one of the by-products, however. By the time he graduated in 1845 as Second Wrangler (the second place in his year) and first Smith's Prizeman he had the mathematical tools that would be needed to transform the science of the relationship between heat and work.[27]

Thomson clearly had an abiding interest in the relationship between heat and work that predated even his arrival at Cambridge. His first public activity as a man of science was in 1840 when the British Association for the Advancement of Science visited Glasgow. William along with his brother James was roped in to help organize an exhibition for the occasion. The final display gave pride of place to a model of James Watt's steam engine. Whilst at Cambridge, William kept an ongoing correspondence with his brother, apprenticed to William Fairbairn's famous engineering firm, on the relationship of heat and work. The correspondence demonstrates both brothers' familiarity with the latest literature on the problem and their ambition of finding a satisfactory solution. It was a topic that fitted well with their commitment to practical and useful natural philosophy. They shared information about the latest technology, with James regaling William with news about the newest engines and his younger brother responding with mathematical theories about their mode of operation. They were interested in more than just knowing how engines worked. They were interested in finding out how to make them work as efficiently as possible, with the minimum of waste in their operation.

Much of the Thomson brothers' correspondence was based around their common reading of Émile Clapeyron's mathematical rendering of Sadi Carnot's theory of the relationship of heat and work. William Thomson did not interpret the theory in quite the same way that Carnot himself had. Whereas Carnot quite explicitly portrayed heat as a fluid, Thomson, with his Cambridge mathematical training, characterized it as a state of matter. His brother characterized things in much the

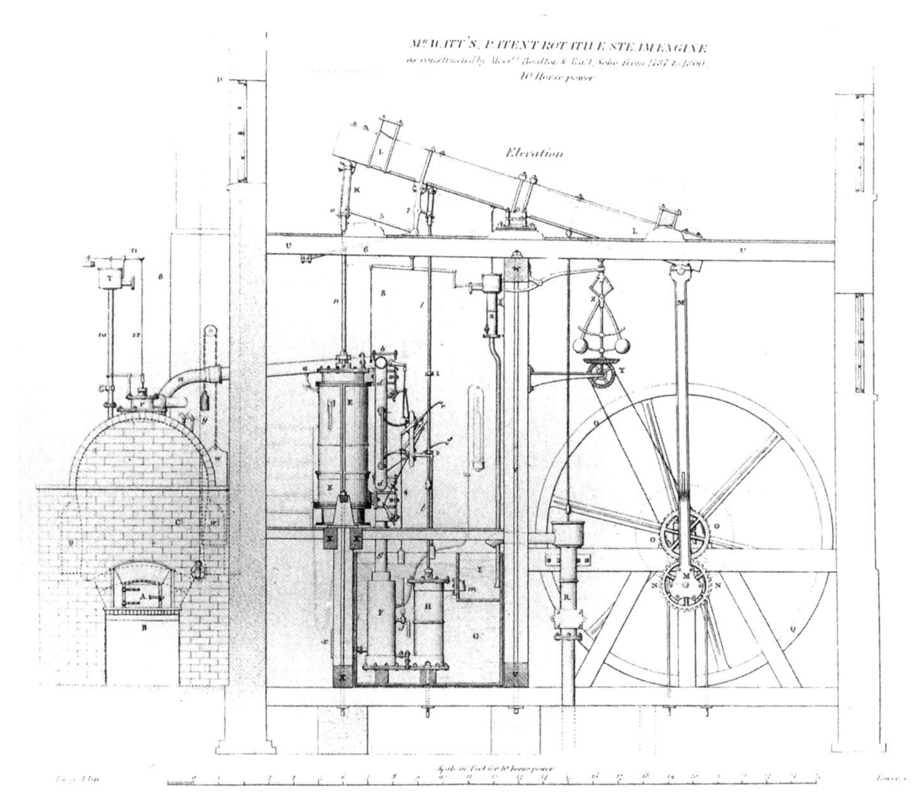

Fig. 8.2. James Watt's steam engine as illustrated in patent specifications. Trying to understand the relationship between heat and motion in engines like this played an important role in William Thomson's development of thermodynamics.

same way. As he expressed their mutual understanding of the theory in a letter of 1844:

> during the passage of heat from a given state of intensity to a given state of diffusion a certain quantity of mec[hanical] eff[ect] is given out whatever gaseous substances are acted on, and that no more can be given out when it acts on solids or liquids.[28]

Both of their attentions were focused on the problem of efficiency—of making engines in the real world as similar as possible to the ideal engines Carnot and Clapeyron's theory described—particularly as James was in the process of developing and patenting his own vortex turbine engine.[29] It was these kinds of considerations that underlay William Thomson's search for an absolute thermometric scale as well. For Thomson, it was Carnot's principle of the relationship of work produced to absolute quantities of heat and defined temperature differences that made an absolute, rather than a merely arbitrary, temperature scale possible at all.[30]

So when William Thomson encountered James Prescott Joule at the Oxford BAAS meeting in 1847, heat and its relationship to mechanical work had been at the forefront of his mind for several years. He was convinced that Carnot's theory as developed by Clapeyron was the best one available and was interested, along with his brother, in finding ways of developing that theory as a way of understanding how best to maximize the practical production of mechanical work from heat. This was what made Joule's claims concerning the mechanical equivalent of heat appear both disturbing and challenging. Thomson was committed to Carnot's theory that work was produced by the 'fall' of heat from one temperature level to a lower one and that no heat was lost during that process. Joule, on the other hand, argued that heat was quite literally lost in the process of producing work—or rather was converted into mechanical work. It was difficult to see how these two contradictory positions could possibly be reconciled. Yet despite his firm belief in the correctness of Carnot's theory, Thomson was equally convinced that Joule was on to something as well. The experimental evidence that he offered seemed extremely solid, after all. Thomson therefore found himself in the position of needing to find a way of reconciling the apparently irreconcilable if he wanted to make any progress in his search for a useful and comprehensive theory of the relationship between heat and work.

William Thomson worried away at this problem for the next three or four years. His first inclination was to find a halfway house that would accommodate both Joule and Carnot. In his *Philosophical Magazine* discussion of the possibility of establishing an absolute thermometric scale he conceded the possibility that Joule might be right in saying that work could be converted into heat—after all, strictly speaking, this was all that the paddle wheel experiment really demonstrated.[31] Whilst that and other experiments might be taken as demonstrating 'an actual conversion of mechanical effect into caloric', there appeared to be no 'experiment... in which the converse operation is exhibited'. In Thomson's view, it seemed only to be the 'converse operation' of turning heat into work that violated Carnot's principle.[32] But there was a problem with Carnot's principle as well. There seemed to be circumstances in which it violated the personal theological commitment that Thomson shared with Joule regarding conservation. If work took place simply be means of the flow of caloric from one temperature level to another, what happened in situations—like straightforward heat conduction—when no work seemed to be being done? It seemed like a violation of conservation. As Thomson put it:

> When 'thermal agency' is thus spent in conducting heat through a solid, what becomes of the mechanical effect which it might produce? Nothing can be lost in the operations of nature—no energy can be destroyed. What effect then is produced in place of the mechanical effect which is lost?[33]

The series of papers that William Thomson read before the Royal Society of Edinburgh on the 'dynamical theory of heat' during the early 1850s was an attempt to answer that question. The outcome was the making of a new science—thermodynamics. By February 1851, Thomson was sure he had cracked it. He wrote to Joule with the news and received an enthusiastic reply urging him to publish as quickly as possible. Thomson's dynamical theory rested on two propositions. In the first, based on Joule's mechanical equivalent of heat, he posited that:

> When equal quantities of mechanical effect are produced by any means whatever from purely thermal sources, or lost in purely thermal effects, equal quantities of heat are put out of existence or are generated.[34]

In the second proposition, based on Carnot, Thomson posited that:

> If an engine be such that, when it is worked backwards, the physical and mechanical agencies in every part of its motions are all reversed, it produces as much mechanical effect as can be produced by any thermo-dynamic engine, with the same temperatures of source and refrigerator, from a given quantity of heat.[35]

The two principles—now known as the first and second laws of thermodynamics—added up to the declaration that both Joule and Carnot were correct. According to the first principle, Joule was right in declaring the interconvertability of heat and work. According to the second principle, the production of work from heat was always accompanied by a dissipation of heat from a higher to a lower temperature and that dissipation was irreversible. In effect, Thomson had given the universe a sense of direction through time.

Thomson's new dynamical theory was taken up eagerly by British natural philosophers and engineers. The new science was given its name in 1859 by the Scottish engineer William John Macquorn Rankine in his *Manual of the Steam Engine and*

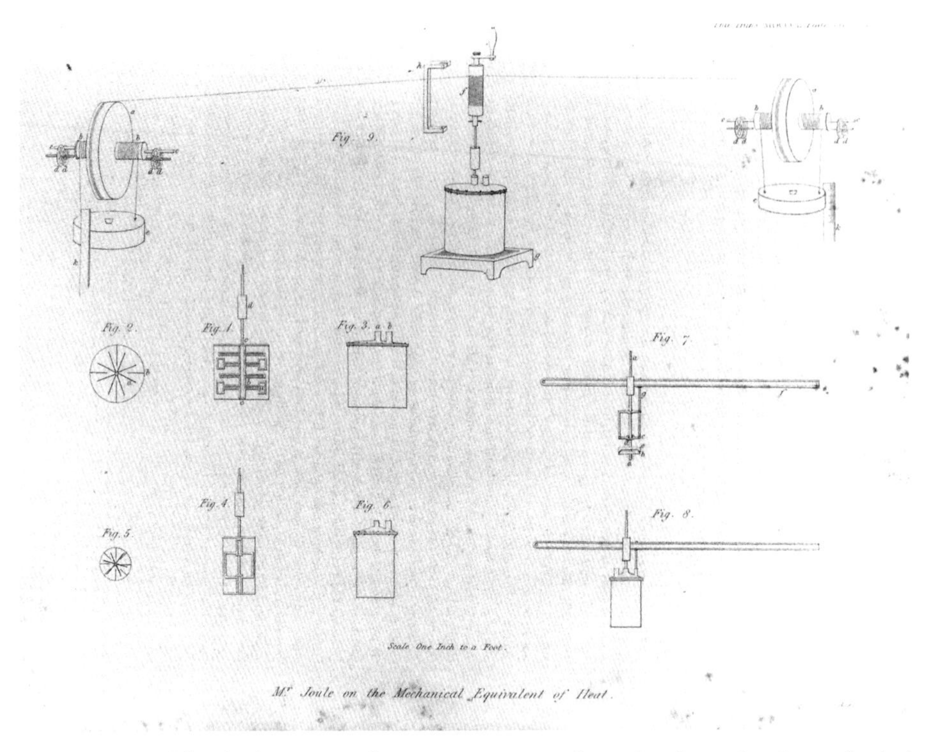

Fig. 8.3. The paddle wheel apparatus that James Prescott Joule used to determine the mechanical equivalent of heat as illustrated in the *Philosophical Transactions of the Royal Society.*

Other Prime Movers. Rankine entitled the extensive section of the book that dealt with steam engines, 'The Principles of Thermodynamics'.[36] Rankine's book was an attempt to straddle the boundary between engineering and natural philosophy—to show that physics mattered (or ought to matter) for engineers and that engineering problems could provide fruitful insights into the operations of nature for natural philosophers. Thomson himself was soon collaborating with another Scot—Peter Guthrie Tait, previously Professor of Mathematics at Queen's College, Belfast but appointed Professor of Natural Philosophy at Edinburgh in 1860—on an ambitious project to place thermodynamics at the centre of a powerful new physics. From his inaugural lecture at Edinburgh onwards, Tait had made his position as Thomson's thermodynamic ally clear. Thomson and Tait's *Treatise on Natural Philosophy*, eventually published in 1867, was no less than an attempt to re-write Newton, putting thermodynamics and the conservation of energy (as encapsulated in the first law of thermodynamics) at the core of the new science. It provided the blueprint for exporting the values and practices of thermodynamics throughout physics and beyond.

One thing in particular distinguished Thomson's thermodynamics and the universe it described from previous efforts. Thomson's universe had a sense of direction built into it by the second law of thermodynamics. Once heat had been dissipated, unless useful work had been produced in the process, there was no way of getting that work back. In effect, this meant that the universe was running down. Not only did it have a definite beginning with the original act of creation, but it had a certain and foreseeable end as well. Gone was the static universe of Descartes and Laplace, or even Newton's universe where stability was maintained through the exertion of God's will. Every time a transformation of energy took place, some of that energy became unavailable for further work. It was still there—after all it was central to Thomson's view of things that only God could create or destroy energy—but it could no longer be made useful. Eventually, there would come a point when the entire universe was at the same temperature. When that happened no further work could be extracted from any transformation since any conversion of heat to mechanical work had to be accompanied by a fall from one temperature level to another. This was what thermodynamics' popularizers called the 'heat death of the universe'.

During the 1860s, William Thomson became involved in one particular controversy that provides a good example of the way in which he regarded the importance of thermodynamics and the model of the universe it presented. In a series of papers, popular articles and lectures he used thermodynamics to attack the basic principles underlying the controversial new theory of evolution by means of natural selection put forward by Charles Darwin in the *Origin of Species* just a few years previously in 1859. Darwin's argument explicitly rested on the assumption that the Earth had existed for a practically indefinite period for natural selection to have taken place. Thomson used thermodynamic arguments in order to demonstrate that neither the Earth nor the Sun could have existed for more than the tiniest fraction of the time required for such an evolutionary process to take place. The Sun, for example, could not have been in existence:

> through an infinity of past time, since, as long as it has existed, it must have been suffering dissipation, and the finiteness of the sun precludes the supposition of an infinite primitive store of heat in his body.[37]

Thomson estimated that the Sun could not have been emitting light for more than 500 million years at the very most. Darwin, on the other hand, as Thomson gleefully pointed out, assumed at least 300 million years for such a comparatively recent geological event as the denudation of the Weald in southeast England.[38]

The cooling of the Sun and age of the Earth controversies show how Thomson wielded thermodynamics to attack what he saw as dangerously heterodox doctrines.[39] He was certainly not alone amongst thermodynamics' founders and promoters in this respect. James Prescott Joule, for one, was more than happy to egg Thomson on to 'expose some of the rubbish which has been thrust upon the public lately'.[40] But more than this, these controversies were symptomatic of Thomson's view that thermodynamics really was the solution to the riddle of the universe. As such, it provided the model not just for the rest of physics, but for all of natural science. Thermodynamics showed how science could straddle the abstract and the practical, just as Thomson himself was to do throughout his own career. The *Treatise on Natural Philosophy* he co-authored with Peter Guthrie Tait, was explicitly modelled on Newton's *Principia* precisely because its authors saw themselves as unifying the whole field of natural philosophy in just the same way. The preface to the book made clear their view that all they were doing was reformulating physics as Newton himself would have done if he knew what nineteenth-century experimental physics had taught them:

> One object we have constantly kept in view is the grand principle of the *Conservation of Energy* ... It is satisfactory to find that NEWTON anticipated, so far as the state of experimental science in his time permitted him, this magnificent modern generalization.[41]

Thermodynamics—as far as Thomson and his fellow workers were concerned—exemplified a new and powerful approach to understanding the natural world. One particularly striking feature of almost all early nineteenth-century contributors to the strand of thought that culminated with thermodynamics was their interest in straddling the boundary between natural philosophy and engineering, between the abstract and the practical. If Carnot, Clapeyron, Joule, and Thomson had nothing else in common, they certainly had that. Thermodynamics was the science of the steam engine. In Thomson's eyes, the search for a satisfactory physics of the relationship between heat and work was at the same time a search for economic efficiency. Thomson's thermodynamics straddled other boundaries as well. It brought together different kinds of skill. To make thermodynamics work, the skills of the practical engineer, the experimenter, and the mathematician needed to be harnessed and brought together. One of the things that made William Thomson stand out as a nineteenth-century physicist was the way in which he combined these skills and qualities. He shared with his brother, James Thomson, an abiding practical concern with engines. His experiences with Regnault had taught him the value of precision experiment and put him in a position to recognize Joule's achievement with the paddle wheel experiment. His Cambridge training had provided him with the mathematical tools needed to develop the new theory.

But thermodynamics was far more than just a physical theory as far as Thomson was concerned. It was an expression of the way in which he saw his world. Thermodynamics, or so it seemed to many of its promoters, embodied a particular set of values in its operations. It revealed an universe where economy, efficiency, thrift, and the avoidance of waste were built into the very fabric of things. It also revealed an universe ruled by God. The principle of conservation—for William

Thomson as for Joule—was a theological as much as a physical imperative. Matter and energy were conserved because, being created by God, they could not be destroyed by any other agency. It was also an universe that had a sense of direction, forever running down towards its ultimate dissolution. If the dominant metaphor of eighteenth-century natural philosophy was of a balance, with nature's forces forever working to restore equilibrium, the metaphor for thermodynamics was the steam engine continually propelling nature forwards. Like the industry it encapsulated, Thomson's thermodynamics was also ambitiously expansionist. Thomson and his fellows regarded it as the ultimate science and therefore as the ultimate arbiter over nature. Other sciences, like biology or geology, would have to pay due obeisance. Thomson's triumph was to produce a new physics that not only explained how to build better steam engines, but accounted for the age of the universe and foresaw its end at the same time.

9

Kelvin and Engineering

Bernard Crossland

INTRODUCTION

The nineteenth century was considered by many as the golden age of science, distinguished by a galaxy of great men of science, of whom many considered Lord Kelvin to be the greatest[1]. However the nineteenth century was also the golden age of engineering, and equally distinguished by a galaxy of great engineers, and many considered Kelvin to be one of them. They also recognized that he epitomized a new breed of engineer more based on science, as well as being innovative with manufacturing and marketing acumen[1, 2].

Perhaps the foundation of Kelvin's great contribution to science and engineering was the remarkably broad education he and his nearly equally famous elder brother, James Thomson, received from a very young age. It is recorded that when he was 10 years old and his brother 12, they informally attended their father's lectures on natural philosophy in the University of Glasgow. Subsequently at home they repeated some of their father's demonstrations, building for themselves electrical machines and Leyden jars, which they used to give electrical shocks to their friends. There is no doubt that this combination of theory and experiment proved to be Kelvin's great strength, not only in theory but also experimental verification and application.

Following his completion of the Mathematical Tripos at Cambridge in 1845 when he was turning 21, Kelvin spent some months in Paris learning about experimental methods, an experience which was not available at Cambridge at that time nor elsewhere in the UK. This gave him the foundation for creating the first teaching laboratory in science, when he was appointed Professor of Natural Philosophy in the University of Glasgow in October 1846.

For the first time undergraduates obtained hands-on experience in experimental methods, which no doubt greatly enhanced their physical understanding of science. It also gave them the opportunity to participate in Kelvin's experimental researches, at a time when there was little if any technician support and no research assistants or research students. It also provided facilities for research to back up the engineering works in which he became involved, and which brought great public recognition as well as great personal wealth.

During the course of his long life the engineering profession recognized his great engineering works. The Society of Telegraph Engineers made him President in 1874, and when it formed the foundation of the more broadly based Institution of Electrical Engineers in 1888 Kelvin became its first President, and was elected for a second term in 1907, the year of his death.

The Institute of Marine Engineers, a more practically based professional body, recognized his great contribution to marine engineering in, for instance, the development of depth sounding machines and the mariner's compass that were widely adopted, by electing him a Life Member in 1891 and President 1892–93. The first great engineering institution to be created, the Institution of Civil Engineers, made him a Life Member in 1891. Clearly the engineering profession recognized Kelvin as one of them.

Though he demonstrated the power of the scientific approach to solving practical engineering problems, Kelvin also clearly recognized that science had much to learn from practical applications. In his Presidential address to the Society of Telegraph Engineers in 1874, he reflected on the benefits which science gains from

Fig. 9.1. Kelvin's cable galvanometer as used on-board the US *Niagara*. (Courtesy of the Photographic Library, Hunterian Museum and Art Gallery, University of Glasgow.)

practical application, and quoted as an example the development of systems of measurements and units of electricity which grew out of the requirements of practical telegraphists.

TRANSATLANTIC TELEGRAPH CABLE

Up until the mid-1850s Kelvin was mainly immersed in thermodynamic studies and the properties of matter. It was not until 1854 that he became interested and involved in the proposal to lay a transatlantic telegraph cable and the considerable problems that had to be overcome. He recognized that the Fourier mathematics he had been interested in since 1840, in relation to the flow of heat in solids, was equally applicable to understanding the transmission of electrical pulses through a conducting cable. This led to his paper[3] 'On the theory of the electrical telegraph' in 1855. With William John Macquorn Rankine and John Thomson he patented[4] the idea of using a multi-strand instead of a single conducting wire, which gave much greater flexibility.

In 1800 Count Alessandro Volta constructed a 'voltaic pile', an early form of battery, which produced a steady flow of electrical current. This promoted the study of electricity by the early pioneers such as J. Henry and M. Faraday. In 1835 Henry recognized the basic concept of the telegraph, but did not pursue it. In 1837 C. Wheatstone, after visiting Henry, and in collaboration with W. F. Cooke constructed the first commercial telegraph, which was a great success. Thousands of miles of telegraph lines were constructed within the next decade. In America, S. F. B. Morse, after also meeting Henry, persuaded Congress to provide the financial support to construct a 40 mile telegraph line between Washington and Baltimore in 1844. Morse's first telegraph message, when the line was opened, was in 'Morse Code', a series of dots and dashes.

The overland telegraph was mainly carried by overhead lines supported on poles, but a few were placed underground. The question was then, if underground why not underwater? It was made possible by the opportune discovery of gutta-percha—the sap in the form of a milky latex produced by a few tropical trees—which Faraday recognized as having excellent insulating properties. This formed the ideal coating material for copper wire conductors. Morse was the first to lay an underwater cable across the East River in New York in 1842, although an anchor destroyed it shortly after it was inaugurated. In 1851, T. R. Crompton successfully laid a cable across the English Channel from Dover to Calais, using a copper conductor insulated with gutta-percha with an outer protective layer of galvanized iron wire. This design of cable was widely adopted for undersea cables laid subsequently.

After the successful laying of the cross-channel telegraph cable numerous other undersea cables were laid. England was connected with Ireland, Scotland, Holland, and other countries. In 1856 Newfoundland was connected to Cape Breton and thence overland to New York. The military value of telegraphy was recognized and in the Crimean War in 1854 a 270 nautical mile cable reaching a depth of 950 fathoms was laid from Varno in Bulgaria across the Black Sea to Balaklava in the Crimea, and then by land line to the battle front in front of Sebastopol.

It was not long before the idea of a transatlantic cable between America and Britain was mooted, but it is debatable who was the first to come up with the idea. But before the idea could be realized there were many practical problems to overcome. Importantly there was the question of which route to follow? Fortunately the shortest direct route from Newfoundland to Ireland, involving 1600 nautical miles, proved the best in terms of depth, profile, and composition of the seabed. There was the design of the cable to consider, but even more importantly the design of new machinery to manufacture the long length of cable required. The laying of a cable involving depths of up to three miles posed considerable problems including the design of cable laying gear. At the time there was no ship large enough to carry the entire cable, so that splicing of two lengths of cable carried in separate ships would have to be carried out in mid Atlantic. But these and other practical problems paled into insignificance with the fundamental problem of transmitting signals over the large distance involved at a speed which made economic sense, and this is where Kelvin became involved and made a major contribution to the ultimate success of the enterprise.

In 1854 an American, W. Cyrus Field, set up the New York, Newfoundland and London Electric Telegraph Company, with the ultimate objective of providing a direct telegraph link between New York, London, and Europe via a transatlantic cable between Newfoundland and Ireland. After successfully completing a telegraph between New York and St John's in Newfoundland, including an underground cable from Cape Breton Island to Newfoundland, Cyrus Field launched the Atlantic Telegraph Co. in London in 1856, with the intention of attracting capital and involving the British Government and cable manufacturers based in England. Of the 18 directors elected in December 1856, 2 were from Glasgow including Professor William Thomson, later Lord Kelvin, elected by the Scottish shareholders. The Board appointed C. T. Bright as Engineer-in-Chief, E. O. W Whitehouse as Electrician and Cyrus Field as General Manager. Kelvin held no technical appointment but he was to prove to be essential to the success of the enterprise.

The cable between England and Holland, which was 110 miles long, had shown a slight retardation of the signal transmitted. If a sharp, sudden impulse is transmitted, the effect of retardation on the impulse as it travels to greater distances is to smooth it out into a longer lasting impulse, which slowly rises to a maximum and then decays. The effect of this is to slow down the rate of transmission of message, which, if excessive, threatens the financial viability of the enterprise.

Whitehouse, the Electrician to the company, believed that retardation could be readily overcome by applying a higher voltage impulse, and he believed, wrongly as it happened, that he had experimental evidence to support his contention. Kelvin, however, adopted a fundamentally based analysis of the problem using Fourier's mathematics, which formed the basis of his 1855 Royal Society paper[3]. He showed that the retardation varied in direct proportion to the electrical capacity and also its resistance. For a cable with constant proportions along its length it followed that the retardation would be proportional to the square of the distance, which was known as 'The law of squares'. If a cable of 200 miles long showed a retardation of 1/10th second, then one of 2000 miles would have a retardation of 10 seconds. Kelvin clearly recognized that this could pose a serious economic problem for the transatlantic cable.

According to Kelvin's analysis the retardation for long lengths of cable could be kept constant by increasing the diameter of the central wire and the thickness of the gutta-percha insulation in proportion to the length of cable. However, this was clearly not practical as it would have led to an excessively heavy cable. Improvements in retardation could also be achieved by improving the electrical properties of the central conductor and of the insulation, but such improvements were limited to materials readily available.

The problem remained of how to reduce the retardation of the signal to ensure that the rate of transmission was such as to make the whole enterprise financially viable. Kelvin's theoretical understanding gave him an insight and ability to assess ideas he generated. This led to a paper[5] 'On rapid signalling by electric telegraph' published in 1856. The solution was to generate a rapidly rising and falling pulse, using some mechanical contrivance to control the pulse, followed immediately by the end of the cable being earthed.

At the receiving end he envisaged the use of a modified Helmholtz galvanometer with a copper damper to give critical damping. The deflection of the mirror was detected by an observer looking through the telescope at the reflection in the mirror of an illuminated scale. An electrical pulse deflected the mirror to a maximum position, and with critical damping it would fall back to the position of rest. The dynamic characteristics of the magnet/mirror and its suspension had to be sufficiently high to achieve the required rate of transmission.

He envisaged 13 positive and 13 negative strengths of pulse bringing 1 of 26 letters on the illuminated scale into view. But he also devised alternative strategies to improve the rate of transmission of cablegrams. Perhaps more surprisingly he proposed a modified galvanometer which would generate an electrical pulse recorded on electro-chemical paper driven by a clockwork motor. This suggestion obviated the very tedious recording of messages by a telegraphist looking through a telescope at an illuminated scale, but that was to be very much in the future.

Early on in the formation of the Atlantic Telegraph Co. the provisional committee decided on the specification and manufacture of the cable without wide consultation. The cable which was agreed to had a copper core weighing 107 lbs per nautical mile, insulated with 261 lbs of gutta-percha, and finally sheathed with stranded iron wire, giving a total weight of 1 ton per nautical mile. Too late Bright, the Engineer-in-chief, C. F. Varley, an eminent telegraph engineer, and Kelvin advocated a heavier cable. The manufacture of the cable was split three ways: the Gutta-Percha Company produced the copper conductor with its insulation, while the iron wire sheathing was split between Messrs Glass, Elliott and Co. of Greenwich, and Messrs Newell and Co. of Birkenhead. Lack of adequate specification led to the two lengths of cable being manufactured with opposite handed twist of the iron wire sheath. Inadequate thought had also been given to the continuous testing of the cable in manufacture to detect faults.

Kelvin at that stage took the opportunity to investigate the conductivity of copper from various suppliers, which he reported in a paper[6] 'On the electrical conductivity of commercial copper'. He discovered a wide variation in the conductivity of the various samples, related to the purity of the copper. But it was too late to influence the manufacture of the first cable used in the abortive 1857 attempt to lay the cable. However, high purity copper was used in the manufacture of a replacement length of cable for the second attempt, and in subsequent cables.

In 1857 there was no vessel afloat capable of carrying the entire length of cable, so two vessels had to be employed. One of these would lay the cable from Valencia Island in the furthermost corner of SW Ireland to the centre of the Atlantic, where its cable end would be spliced to the cable carried in the second vessel, which would continue on to Trinity Bay in Newfoundland. The British and American Governments made the *HMS Agamemnon* and the *US Niagara*, both steam powered and screw propelled, available.

The two vessels were fitted with hurriedly designed and constructed cable laying-out machinery, which in the event proved to be inadequate for the tasks. The brake-wheel needed to control the tension in the cable was governed by a hand operated clutch, which needed the continuous and close attention of the operator. This proved to be an extremely tedious task requiring considerable concentration to avoid excessive loads being placed on the cable. Kelvin turned his attention to the mechanics of cable laying and considered the settling curve of the cable between the cable laying vessel and the sea-bed up to 3½ miles deep, which formed the basis of a paper[7] 'On machinery for laying submarine telegraph cables'. It also led to his patent[8] 'Apparatus for applying and measuring resistance of rotating wheels, shafts, etc', in which he claims 'Thirdly the using of the arrangements herein —before claimed for applying and measuring the resistance to the motion of machinery for laying submarine cables'.

The two cable laying vessels met at Queenstown, now Cobh, in Southern Ireland on 21 July 1857. For the first time it was possible to splice the two lengths of cable, and to test the ability to transmit a message through 2200 miles of cable, which was successfully achieved. On 7 August the *Niagara* made a connection to the shore station at Valencia Island, and then started to lay its cable with the aim of reaching the mid-Atlantic. The *Agamemnon* with Kelvin on board to replace the Company's Chief Electrician, who was ill, was to take over the cable laying from mid-Atlantic to Newfoundland. The *Niagara* was able to maintain telegraphic contact with the telegraphic station on Valencia Island through the cable. On 11 August—with 338 miles of cable laid and a water depth of 2000 fathoms or 2¼ miles—the cable failed when the mechanic operating the brake failed to release it fast enough to control the tension in the cable. There was little hope of recovering the cable as grappling for a cable at a depth of 2¼ miles had never been attempted, so cable laying was abandoned for that year.

The inquest into the cable failure, to which Kelvin contributed, concluded that staff levels and competence of the mechanics on the cable playing-out machinery was inadequate. It was also agreed that too little attention had been paid to the

Fig. 9.2. HMS *Agamemnon* (left) and US *Niagara* taking the transatlantic cable on-board at Keyham, Plymouth. (*Illustrated London News*, 28 May 1858.)

design of the machinery. It is interesting to note that in his paper[7] in *The Engineer*, Kelvin had declared that:

> the mechanism should, like the actions of a fly-fisher in yielding to the sudden tug of the fish, afford sufficient play in the event of one sudden strain of the cable to avoid failure.

After his conductivity experiments on copper, Kelvin was insistent that when a replacement length of cable was ordered in late 1857, high conductivity copper should be specified for the core of the cable. To ensure compliance Kelvin set up testing equipment in the factory, which some consider to be the first example of a factory based testing laboratory.

During the time before the next attempt to lay the cable in 1858, Kelvin developed a more sensitive galvanometer by lightening the movement. Instead of using a telescope to detect the rotation of the mirror, he used a focused light beam from an oil lamp reflected by the mirror onto a fixed scale. An observer could read the letters being transmitted to a clerk who recorded them. The marine mirror galvanometer patented[9] in 1858—'Improvements in testing and working electric telegraphs'—proved to be the mainstay of submarine telegraphy until the 1870s. Kelvin also devised the suspension system for the use of the galvanometer at sea to isolate it from the movement of the ship.

On 24 May 1858 the *Agamemnon* and *Niagara* with their escorting vessels sailed from Plymouth to the Bay of Biscay to gain experience of laying and recovering cable at depths of 2500 fathoms, prior to making the second attempt to lay the transatlantic cable. Before their departure Kelvin received his latest instruments from Glasgow, and he also got reluctant agreement for his instruments to be inserted in the circuits. The management at that stage relied on the equipment devised by Whitehouse, their Chief Electrician. The cable laying gear on the two vessels had been modified and the brake patented by Kelvin had been fitted.

After the successful completion of trials in the Bay of Biscay the ships returned to Plymouth. For the second attempt it had been agreed that the cable would be laid in both directions from the centre of the Atlantic. This meant that the two cable laying ships would remain in telegraphic contact with one another, but not with the shore. Again Whitehouse, claiming ill health, was unable to go to sea so Kelvin sailed in the *Agamemnon*.

The vessels sailed from Plymouth on 10 June, at a time of year which promised good weather. In the event they experienced extremely severe gales, so severe that the *Agamemnon* was in great danger of sinking. However the two cable laying vessels and their escorts assembled on 20 June and on the 26th a splice was made. After paying out 6 miles of cable the cable on board the *Niagara* failed. A second splice was made but after 80 miles of cable had been laid the cable failed again. A third splice was made and after each ship had paid out 100 miles of cable, the cable failed at the stern of the *Agamemnon*. Being unable to locate one another both vessels returned as agreed to Queenstown, the *Niagara* on 5 July and the *Agamemnon* a week later.

There was considerable dismay amongst the directors, and even some suggestions that they should cut their losses and give up. Some of the directors, strongly supported by Kelvin, decided on a final attempt. On 17/18 July the two cable laying vessels and their escort sailed and met on 29 July, and again spliced their cables. On 5 August, the *Agamemnon* arrived at Valencia Island and on the same day the *Niagara* arrived at Trinity Bay. The first message was transmitted between the two continents, which was followed by great celebrations both sides of the Atlantic, though these proved to be rather premature.

It seemed a successful conclusion to the great enterprise, but after the cable was handed over to Whitehouse to organize its operation, the cable showed increasing deterioration of the insulation. Subsequently it appeared that the successful messages, supposedly using Whitehouse's equipment, had mostly been received using Kelvin's galvanometer. Probably the insulation of the cable as manufactured was far from perfect, as there was no adequate quality control in place. Added to this the insulation may have deteriorated in storage during the 1857/58 winter. Further damage was incurred on board the two cable laying vessels during the severe storms they experienced in June 1858. But the final coup de grâce was Whitehead's belief in very high transmission voltages to activate his receiving equipment. These voltages were produced by a large induction coil and massive battery installation, which probably produced voltage as high as 2000 volts.

Increasing acrimony between Whitehead and the Board of Directors resulted in his dismissal, and the Board's request to Kelvin to take over. He attempted to salvage the line, but by the end of August the last intelligible message was received at Valencia.

Subsequently there was much viperish communication in the press and elsewhere between Whitehouse and the directors trying to allocate blame. Nearly the only person to come out with his reputation unscathed was Kelvin. Despite the recriminations the possibility of laying a transatlantic cable and transmitting messages through it had been clearly demonstrated, and it was only a matter of time before another attempt would be made. Many lessons had been learned; Kelvin clearly saw the need of continuous inspection of the cable in manufacture. He had also clearly demonstrated the viability of his transmitting and receiving equipment. He had also been largely influential in understanding and improving the cable laying machinery.

By the end of 1858 the Atlantic Telegraph Company had used all its assets, and after the failure of the cable laid earlier in the year there was little appetite for investors to provide further capital. However some of the early promoters and scientists, in particular Kelvin, had faith in the ultimate success of a transatlantic cable and felt its day would come. During the next few years numerous attempts were made to put together a consortium to finance the next attempt. However it was not until 1863 that it appeared that a financially viable consortium was in place, which would allow the company to go ahead. Even then there were hiccups until the next attempt.

In the meantime the technology and the science were being actively pursued. Further experience was gained in the design and manufacture of under-sea cables and their laying. Not all the cables laid were successful but there were increasing successes, which gave some confidence. Units of electrical measurements were agreed, much due to Kelvin's effort, which made it easier for engineers and scientists to communicate with one another. Testing methods for cables during manufacture were greatly developed. Kelvin and Fleeming Jenkin, a telegraph engineer, developed an improved sending key for despatching messages, and also came up with an automatic signalling device, which eased the transmission of messages. Varley, another telegraph engineer, and one of Kelvin's collaborators, invented the signalling condenser to sharpen up the pulse, which improved the rate of transmission. So the technology and science to support a further attempt were steadily being developed.

In 1863 the Atlantic Telegraph Company set up a Scientific Committee, which included Kelvin, Galton, Wheatstone, and the great engineer Joseph Whitworth, to consider the specification for the cable, and examine samples of cable submitted by the cable manufacturer, Glass and Elliot Co. After exhaustive testing the cable design agreed had a seven strand copper core embedded in Chatterton's compound, which weighed 300 lb per nautical mile, a substantial increase over the previous cable as urged by Kelvin. This was covered with eight alternating layers of gutta-percha and Chatterton's compound, and finally it was armoured with cotton covered iron wire. The total weight was 1.8 ton per nautical mile. The cable was ordered in March 1864 from the Gutta-Percha Co. and Glass and Elliot Co., who subsequently amalgamated to form the Telegraph Construction and Maintenance Co., known as Telcom.

Cyrus Field, the early promoter of the cable, had in 1860 taken a short cruise from New York on Brunel's *Great Eastern*, by far the largest vessel to have been built in her day. No larger vessel was built until the *Oceanic* was launched in Belfast in 1899. Cyrus Field recognized the potential of the *Great Eastern* as a cable laying vessel, not only because of her great carrying capacity but also because she had both screws and paddles which made her remarkably manoeuvrable. However she had proved to be a financial disaster to a succession of owners because she was too large for the port facilities and the trade of that day and age. So in 1864 the vessel was purchased by the great engineer, Daniel Gooch, and refitted with three large water filled cable storage tanks, greatly improved cable laying machinery, and unproven grappling equipment to recover broken cable.

By July 1865 the *Great Eastern* had taken on board the entire transatlantic cable. Kelvin and Varley, using their new sending key and signalling equipment, found that they could nearly double the rate of transmission to 6 words per minute. They were confident that they could double it again to 12 words. Kelvin and Varley were on board as consulting experts along with William Howard Russell of *The Times*, famous for his reporting of the Crimean War. On 23 July the *Great Eastern*, having spliced its cable to the heavy cable which had been laid from Valencia Island, departed for Newfoundland. After detecting and repairing several faults in the cable, and while repairing yet another fault on the 2 August, the cable failed. They had laid 1250 miles and only had another 600 miles to go. They attempted to recover the cable, and though they located it and partly raised it three times, on each occasion a bolt on the grapple failed so they gave up and returned to Valencia Island, where it was decided to delay a repeat attempt until 1866.

Kelvin took the opportunity to further develop testing equipment to test the replacement cable during manufacture, and to further improve his mirror galvanometer. On 13 July 1866 the *Great Eastern* again picked up the new shore end and spliced the new cable to it. Kelvin was again on board with his improved galvanometer, and during the voyage they were in regular contact with Valencia and the rest of the world, to the extent that they produced a regular *Test Room Chronicle*. On the 27 July, 14 days out from Valencia, they arrived at Heart's Content Bay in Newfoundland. Within 24 hours the line to Europe was in full use. The *Great Eastern* then returned to try to locate and pick up the end of the cable which had failed in the previous year. After several weeks of grappling they managed to pick up the cable end on the 2 September and splice it to a new length of cable. They then completed the laying to Newfoundland, so the Atlantic Telegraph Company finished up with two working cables.

There was no doubt that the Atlantic Cable Company, the scientific community, and the public recognized the great contribution that Kelvin had made to the success of the enterprise. In October 1866 various Royal Honours to those associated with the successful laying of the transatlantic cable were announced, including a Knighthood to Kelvin, Professor William Thomson as he then was. Banquets were held in Liverpool and London to celebrate the opening of telegraphic communication between Europe and America. The city of Glasgow conferred the freedom of the city on Kelvin.

However this was not the end of Kelvin's contribution to telegraphy, and he was to remain closely associated with transocean telegraphy for many years. Kelvin, Varley, and Jenkin formed a partnership in 1865 to protect their patents and enforce the payment of royalties. Kelvin and Jenkin founded a partnership as consultant engineers, which resulted in their association with many telegraph companies laying transoceanic cables worldwide. One of these was the laying of a French

Fig. 9.3. Cable paying out machinery on *The Great Eastern* 1865. (*Illustrated London News*, 2 September 1865.)

transatlantic cable from Brest to St. Pierre, a French island off Newfoundland, a distance of 2580 miles, which was laid by the *Great Eastern* without a hitch. These developments not only brought great kudos to the UK, but also a considerable income for the cable manufacturers, equipment providers, and the service industry, as well as making Kelvin a very wealthy man.

In 1872 Kelvin was involved with the Hooper Telegraph manufacturing company in the design and construction of a purpose built cable laying vessel, being built by Mitchell and Co., Newcastle. To increase the manoeuvrability he proposed to use four transverse water jets, and he consulted with his brother James who was an authority on hydraulics and hydraulic machinery. Ultimately he opted for a transverse screw, which must be an early, or possibly the first, use of transverse propellers, which are now commonly used.

In 1873 when the cable-ship *Hooper* was completed it was first employed laying the transatlantic cable from Lisbon via Madeira and the Cape Verde Islands, to Recife in Brazil, with Kelvin on board. They were forced to halt in Funchal Bay in Madeira while they checked for a fault in the cable on board. During the stay over Kelvin was entertained by one of Madeira's leading citizens—Mr Charles R. Blandy and the Miss Blandys. In the following year he sailed to Madeira in his 126 ton

Fig. 9.4. The Great Eastern. (Illustrated London News, 8 February 1865.)

yacht—the *Lalla Rookh*–and married Frances Anna Blandy, his first wife having died some years earlier.

In his paper[5] of 1856 Kelvin envisaged a spark recorder for recording the electrical pulses transmitted through a telegraph cable from a distant sending station, and this forms part of his patent[10] of 1858. The basic concept was that the electrical signal activated an indicator, which was made to take a to and fro movement. The indicator was connected to a Ruhankorff coil, capable of producing a succession of sparks to a horizontal metal plate positioned below the indicator. A ribbon of specially prepared paper, capable of being acted on by the passage of a spark either by chemical action or perforation, is passed at a constant speed produced by a clockwork drive. Thus a permanent record of the signal is recorded on the tape and can be transcribed at one's leisure.

For its day and age this was a remarkably novel idea, but it was not developed into a practical recorder. However it forms the basis of an even more revolutionary idea, which is covered by his patent[11] of 1867. He refers to this instrument as the Siphon Recorder, which he developed into a fully commercial instrument widely used by the cable companies. In this patent the siphon tube is attached to the body moved by the current in the telegraphic cable. One end of the siphon tube dips into a reservoir of ink and the ink is spurted from the opposite end by an electric force. To quote the patent:

When the electric method is used the paper is drawn over a metal plate electrified say positively, and the siphon tube is electrified say negatively, and a powerful difference of potential is maintained between the tube and the metal plate such as would tend to cause a succession of sparks to pass between these, and, which in the circumstances produce a fine stream of ink or a succession of fine dots squirted from the tube onto the paper, leaving a record of the position of the tube at each instant, and drawing a sensibly continuous line on the paper without impeding by friction, the motion of the tube as directed by the receiving instrument. I prefer to let the paper move in a vertical plane and to use a small glass siphon with its short leg dipping in the ink reservoir and its long leg pointing obliquely downwards at the paper and close to it.

This describes in 1867 what must be the first jet ink printer, which nowadays are been extensively used in conjunction with the personal computer.

It required three years of incessant work to bring the development of the Siphon Recorder to a successful conclusion. In 1869 the recorder was first installed for trial on the French Atlantic cable. Later in 1870 it was installed on the Falmouth, Lisbon, Gibraltar, Malta, Alexandria, Suez, Aden, and Bombay cable, which required eight

Fig. 9.5. Kelvin's siphon recorder made by Muirheads in the 1880s being demonstrated in the 1940s. (Courtesy of Porthcurno Telegraph Museum and Archive.)

instruments. It was readily accepted by the telegraph operators and it largely replaced Kelvin's mirror galvanometer.

CONTRIBUTIONS TO MARINE TECHNOLOGY

Kelvin, by his own admission, was at heart a sailor, a passion which he indulged by purchasing the yacht—The *Lalla Rookh*—in 1869/70. After this he spent much of the six months between academic sessions sailing. At the time the academic world led a gentlemanly life, not the sweat shop of learning of the modern university. However these periods afloat, sometimes with his scientific friends, gave him the opportunity to recharge his batteries and to think and plan ahead. They were in fact periods of intense intellectual activity. He also loved these periods at sea in which he developed his marine compass, depth measuring machine, tide analyser, and tide predicting machine, and much else besides. He had gained much maritime experience of common problems during his involvement in many of the early deep sea cable laying operations.

When in the early 1870s he was asked to contribute an article to *Good Words*[12] he chose as his topic the mariner's compass, which had been in use for many centuries. This focused his mind on the many defects which he had observed. These included its sluggish response and occasional sticking. In the Navy, under conditions of heavy firing and the resulting concussions felt by the vessel, the compass proved to be useless. Under stormy weather the rolling caused the compass to oscillate violently misleading the steersman. Another considerable problem was the magnetism of iron vessels, which could produce significant errors in the compass. These problems aroused Kelvin's intellectual interest.

Kelvin had demonstrated in his design of the mirror galvanometer and the siphon recorder his innate ability as an instrument designer, which he now turned to the design of the mariner's compass. From his study of various mariners' compasses he reached some important conclusions, some much at variance with the thinking of that time. With iron ships the inherent magnetism of the ship structure greatly influenced the accuracy of the magnetic compass. To correct for this it was practice to position two soft iron globes, one each side of the binnacle, together with permanent magnets in the vicinity of the compass. Kelvin concluded that the magnetic needles placed symmetrically each side of the NS axis of the card, must be kept short to keep the size of the soft iron globe to an acceptable size. The compass card must be sufficiently large to achieve an acceptable reading accuracy, while at the same time it should be as light as possible to reduce the friction and sticking at the supporting pivot. To achieve steadiness in rough weather the moment of

Fig. 9.6. Kelvin's compass card showing the eight magnets slung under the card and disposed symmetrically about the North / South axis (From Silvanus P. Thompson, *The Life of William Thomson Baron Kelvin of Largs*, 2 volumes, Macmillan & Co., London, 1910, p. 715.)

inertia about a diameter of the disc should be made as large as possible by concentrating as much mass as possible in the rim of the compass card.

These considerations led to a succession of patents from 1876 through to 1907, the year of his death. In patent[13] he describes the basic construction of his compass:

The outer edge of the compass card is supported on a thin rim of aluminium, while the rest of the card rests on a series of 'spokes' or radial threads stretched from the rim to a small central boss of aluminium. The card itself is of strong paper with all the central part cut away, leaving only enough of it to show conveniently the points and degree division of the compass. The central boss consists of a thin disc of aluminium with a hole in the centre, which rests on the projecting lip of a small aluminium inverted cup mounted with a sapphire cup which rests on an indium point. In the 10 inch Admiralty Compass eight small needles of thin steel wire from 3½ to 2 inch long are fixed like the steps of a rope ladder, on parallel silk threads and slung from the light aluminium rim through eyes in the four ends of the outer pair of needles. These magnetised needles are symmetrically disposed about the NS axis of the card and parallel to it. The small size of the needles allows the magnetism of the ship to be completely compensated for by soft iron globes of an acceptable size.

In patent[14] he describes an improved instrument for taking azimuths.

Clearly the form of construction must have made it finicky to assemble, and it would have required a skilled and well trained instrument mechanic. For the 10 inch compass the total weight of the assembly was a mere 170 grains (≈11 g), a seventeenth of the weight of the 10 inch compass previously in common use. It also had double the period of oscillation if tilted a few degrees and released, and the frictional error was not more than one quarter of a degree. The small size of the steel needles allowed for the magnetism of the ship to be completely compensated by two soft iron globes of a reasonable and acceptable size.

The compasses and earlier instruments had been manufactured by James White, Optician, of Glasgow of which Kelvin was a leading partner. Ultimately the company became Kelvin and James White, and later in life Kelvin's patents were taken out jointly in his name and the name of his company. By 1878 the compass had been widely tested in 60 large steamers and was widely accepted. After it was tested in the German Ironclad *Deutschland*, the German Imperial Admiralty ordered further compasses. They were also widely supplied to other world navies. But the British navy at the time was ultra conservative and resistant to change, and it was not until 1889 that the compass was adopted, as its standard compass. The Kelvin compass was widely used throughout the world until well into the twentieth century.

During the early cable laying expedition one of the escorting vessels took soundings at regular intervals, which necessitated halting the vessel while lowering a weighted hempen rope overboard. This was a tedious operation, more especially at water depths of up to 3000 fathoms (3½ miles), experienced in mid-Atlantic. If the ship was underway the drag on the hempen rope made it impossible to take what were termed flying soundings.

In a paper[15] to the Society of Telegraph engineers, Kelvin reported on soundings carried out in 1872 on board his yacht—the *Lalla Rookh*—to depths of up to 2700 fathoms (3 miles), using a 30 lb sinker supported by a line made up of lengths of pianoforte steel wire spliced together. In 1873 he fitted a sounding machine on the cable laying vessel—the *Hooper*—and took soundings in deep water off Brazil. The sounding machine had an auxiliary hauling-in wheel, which took the chief strain off the light sounding wheel, based on his patent[8] and described in ref[16].

He claimed that it was a much more rapid operation than using a weighted hempen rope. The drag on the pianoforte wire when the ship was underway was much less than for a hempen rope, so that flying soundings could be taken, and he demonstrated this while travelling to America on the Cunarder, *Russia,* when he took soundings at a speed of 14 knots. He recognized the importance of a readily viable sounding machine as an aid to navigation, especially when approaching land in dirty weather.

Kelvin also developed several depth recorders to be attached to the sinker, such as the depth gauge described in ref[17]. The recorder has three independent gauges for measuring depths between 11 and 27½, 27½ and 60½, and 60½ and 126½ fathoms. Each gauge has an outer chamber in the form of a brass tube, open at the bottom, and connected by a short, very fine passage to the inner chamber, which is in the form of a graduated glass tube sealed at the bottom. This basically forms an inverted U tube. As the recorder descends the air in the first outer chamber is compressed until at 11 fathoms it is full of water and water begins to be forced into the inner chamber. At 27½ fathoms the inner chamber is half full, and at that stage water in the second gauge begins to spill over into the inner chamber of the second gauge.

In ref[1] a piston gauge developed by Kelvin is described in which air in an inner cylinder is compressed by a piston acted on by water pressure. The compression of the air is recorded by a marker pushed along the piston as it enters the cylinder.

He also made a contribution to the identification of lighthouses, by proposing that the flashes should be in Morse code with long and short flashes. He urged this on the Commissioners of Irish Lights, the Authority of Trinity Commissioners, and the Clyde Commissioners. A study of tides led to a simple tide-gauge, and a Mechanical Tidal Harmonic Analyser[18], which made use of the globe-cylinder integrator invented by his brother James. It determined the harmonic components of the tidal fluctuations. This in turn led to the Tide-Predictor[18], installed in the National Physical Laboratory to predict tides in any future epoch.

Kelvin's interest in nautical matters led to his being appointed, along with Professor W. J. M. Rankine and W. Froude, to the Admiralty Committee upon the Design of Ships of War in January 1871. This committee was set up as a result of a newly-built turret ship—the *HMS Captain*—going down with all hands in 1870 in a gale in Vigo Bay. As Chairman of the Scientific Committee, Kelvin reported on the stability of naval vessels and their structural strength. Interestingly they were also concerned with experiments on the Waterwitch which employed 'hydraulic' propulsion using a water jet, an early forerunner of the modern day high speed ferries. In Victorian days they worked such committees very hard, and from January 1871 they met in London fortnightly for two days and finally delivered their report in July 1871. Its conclusions laid down the rules governing the construction of future naval vessels.

Later in December 1904, when Kelvin was already 80, he was appointed by the Lords Commissioners of the Admiralty to a Committee to review the types of fighting ships which the Board of Admiralty proposed to adopt for the British Navy. The result of this committee was to greatly influence the outcome of the First World War, especially the war at sea. In particular the committee considered the outline design including the armament of a 21 knot battleship and a 25 knot armoured cruiser, including a recommendation to adopt steam turbine propulsion made by Sir Charles Parsons. Kelvin proposed the underwater form of hull to increase the carrying capacity without adding to the wave-resistance. He also advised on technical matters related to compasses and their protection, as well as the provision of instruments for fire control, ranging, and the communication of orders. The committee reported in 1905 and their recommendations were accepted and implemented.

THE HEAT PUMP

One of the less well known contributions made by Kelvin in 1852 was the invention of the heat pump, either for heating or cooling of buildings. This formed the basis of the Bell-Coleman refrigerator used later on in the century for cooled cargo carriers for the frozen meat trade, such as the 3239 ton *Star of Victoria* built in 1887 by Workman and Clark Shipyard in Belfast.

His papers[19, 20] relate to the heating/cooling of the Queen's College, Belfast, one of the three constituent colleges of the Queen's University of Ireland incorporated in 1849. He noted if an engine is employed to circulate air for heating and ventilation, that if all the heat in the fuel was used in circulating and heating the air, then it would be possible to utilize 100% of the heat content in the fuel. Though this limit might appear to be the absolute maximum, he goes on to suggest that it could be surpassed.

To quote Morley[21], the machine envisaged by Kelvin consisted of two cylinders with pistons and inlet and outlet valves driven by a steam engine. The cycle is as follows: air from the external atmosphere is admitted to the 'ingress' cylinder for part of the stroke, the inlet valve is then closed and for the rest of the stroke the air expands resulting in a fall of pressure. The resulting fall of temperature is reduced as much as possible by making the ingress cylinder of conducting material with a large surface area for heat transfer. On the return stroke the air is discharged through the outlet valve into a receiver, also designed to encourage heat transfer, and having its external surface exposed to the external atmosphere, or better still to a stream of water, the intention being to obtain as nearly as possible isothermal expansion and to obtain in the receiver air at a pressure below atmosphere and as close to the outside air temperature as possible. The receiver might take the form of a coil of pipe to increase the heat transfer to the air in the receiver.

The low pressure air is then passed into the 'egress' cylinder and compressed therein until the atmospheric pressure is reached, the operation being adiabatic. As a consequence, the air temperature at the end of the compression is a maximum and above the atmospheric temperature. The air is then finally delivered into the building through the outlet valve of the egress cylinder. The work spent on the air in the egress cylinder exceeds the work generated in the ingress cylinder, the difference being provided by the steam engine.

As an example Kelvin considered heating air from an external atmospheric temperature of 50°F to a final temperature of 80°F. To heat 1 lb of air per second requires 7 Btu/s, whereas using his ideal heating machine the heat equivalent of the power to drive it is only 0.2 Btu/s*. If the engine is assumed to have a thermal efficiency of 10% then the heat consumed to produce the power to drive the heating machine is 2 Btu/s, or the heat given to the air is 3.5 times the heat expended in the engine.

If the ideal engine is reversed and the air is compressed isothermally and then expanded back to atmospheric pressure, the air delivered is cooler than the atmospheric air, so buildings could be cooled in hot weather.

Kelvin considered it an idealized machine, and for many reasons the performance he calculates is unachievable in practice. Nevertheless his basic concept formed the basis of the Bell-Coleman refrigerator, which opened up the highly profitable international frozen meat trade, and cooled banana carriers. His concept has also led to heat pumps for heating and cooling buildings, which are now being extensively exploited.

CONCLUDING REMARKS

Design depends on creative and innovative ideas, but successful design depends on much else. Clearly there is a need to understand the underpinning science, and an ability for detailed design which can be profitably produced at a cost acceptable to the customer. If one studies the 70 patents filed by Kelvin from 1854 to 1907, the year of his death, he demonstrates his remarkable ability in all these aspects of design. It makes him one of the greatest engineers of the nineteenth century.

There is no doubt that his greatest contribution was in resolving the scientific and practical problem of high speed telegraphic communication over great distances. Not only did he resolve the electrical problems but also the practical problems in successfully laying a deep sea cable. Without his contribution there is little doubt that the successful exploitation of long distance telegraphic communication would have been greatly delayed. The world impact of the successful development of intercontinental telegraphic communication was to effectively shrink the world, and to encourage communication between the peoples of the world.

*1 Btu–British thermal unit or heat to raise the temperature of 1 lb of water 1°C.

10
William Thomson's Determinations of the Age of the Earth

Patrick N. Wyse Jackson

INTRODUCTION

For four decades in the nineteenth century William Thomson held the centre-stage in the debate on the age of the Earth. His domination of this subject began in the 1860s and lasted largely unchallenged until the close of that century, and he was pivotal in promoting the viewpoint of the physicists that often ran contrary to the opinions held by the geologists.

By the mid-1800s it was widely recognized that the Earth was much older than the earlier studies of both the biblical texts and pseudoscientific methodologies had suggested. Gone were the chronologies of Archbishop James Ussher (1581–1656) and countless others who had scanned the pages of the Bible for the answer to this question. Confined to the pages of fantasy were the thoughts of Edmond Halley (1656–1742), the astronomer who in 1715 had suggested a method that examined the saltiness of the ocean to date the Earth. Ignored too was the method of the Oxford academic and Keeper of the Ashmolean Museum, Edward Lhwyd (1660–1709) who had suggested that counting the number of boulders lying on the floor of the Vale of Llanberis could point to a similar conclusion.

Geology as a discipline was firmly established in the early 1800s. The Geological Society of London was founded in 1807, universities began to establish professorships in the subject, and in 1845 the Geological Survey of Great Britain and Ireland commenced work. By this time most practitioners of this emerging science realized that the Earth was very ancient, but few could say how old it was in absolute figures. The Scottish geologist-gentleman farmer James Hutton (1726–97) had in 1785 published an abstract of his theory of the Earth in which he argued that the Earth's history was cyclical and basically of infinite duration. Charles Lyell (1795–1875) who, through his popular geological books had brought the new subject to a huge audience, had championed the idea that the processes acting on the Earth at present were similar to the processes that had acted on it in the past. Change was slow, and uniform; this conclusion was in marked contrast to the earlier ideas in which events were sudden or catastrophic. By the 1840s most geologists had accepted Lyell's ideas on the uniformitarian nature of Earth's history, and adhered to the doctrine that the Earth's history was infinite.

Geologists had, by the mid-nineteenth century, also established a well-defined stratigraphical order of the rocks exposed at the Earth's surface. On the basis of geological mapping the distribution of various geological horizons was well-known and in Britain it had been pointed out by William Smith (1769–1839), the surveyor and author of a celebrated geological map of England and Wales, that each rock unit could be recognized and identified through examination of its fossil content, and that those found at the top of a sequence were younger than those found at its base (unless of course there was clear evidence that the whole sequence had been overturned). This allowed geologists such as the Cambridge professor Adam Sedgwick (1785–1873) and the soldier-turned-geologist Sir Roderick Impey Murchison (1792–1871), and others, to define various geological periods each of which contained rocks with particular physical characteristics and fossils. These periods make up what is termed the 'geological column' and by 1860 the relative positions of one period to another had been worked out. However, the absolute ages of the boundaries of the geological periods, and their duration, were not known and would remain unresolved until the 1920s.

Palaeontology, the study of fossils, played an important role in the development of the geological column. Today it is largely forgotten that Charles Darwin (1809–82) began life as a medical student and that it was when he was unable to continue his studies in Edinburgh that he turned his attention to geology. Prior to sailing on H.M.S. *Beagle* he undertook fieldwork in north Wales, and many of his observations made during his epic journey were geological in nature. He published on volcanoes and coral reef islands. On his return to England he turned his attention to biological evolution, but in *The Origin of Species,* published in 1859, he also wrote on geological time; he argued that the topography of the Weald in southeast England had taken 306 million years to form. His theory of evolution had many skeptics because it appeared to them that there was not enough time available for biological evolution to have taken place, yet through his Wealden example Darwin had provided enough time. His ideas brought him into conflict with many thinkers and men of science who fundamentally could not accept his ideas on biological evolution. One of these was John Phillips (1800–74), successively Professor of Geology in London, Dublin, and Oxford, who in 1861 examined the rate of deposition of sediments and argued that the pile of rocks that constituted the rock succession could only have taken 54 million years to form, and that Darwin's estimate based on the denudation of the Weald was very wide of the mark.

Darwin also came into conflict with William Thomson, the Professor of Natural Philosophy in the University of Glasgow. However, as Stephen Brush put it in 1982, the rationale that the argument between Thomson and the geologists and Darwin was about evolution was, and continues to be, overstated.[1] Thomson's main bone of contention with Darwin's logic was that it didn't follow the laws of physics. That lay at the root of the debate between the physicists and the geologists. He didn't take a stance against biological evolution: he even went so far as to suggest that it could proceed at a faster rate than imagined by the biologists, and given this his shorter timescale would not be problematic.

THOMSON ON THE AGE OF THE EARTH: A THREE-PRONGED ATTACK

Thomson was infuriated by Charles Darwin's dabbling in geochronology and was also not impressed by writings of the 'modern geologists' as he termed those who followed the uniformitarian scheme, and so he set out to prove the actual age of the Earth through the application of the laws of physics. Early on, while formulating his ideas, Thomson had the sense to seek out some geological advice and discussed the validity of Darwin's ideas with none other than John Phillips. Given Phillips' stance against Darwin, Thomson wittingly or unwittingly had acquired an ally within the ranks of the geologists. Nevertheless Thomson's findings did not, by and large, satisfy the geologists, and thus begun a feud that lasted until the death of Thomson over 40 years later. This stand-off between the physicists and the geologists continued until the mid-1920s when the latter were arguing that the Earth was older than the age suggested by the former for the universe.

Thomson applied himself to determining the age of the Earth in three ways: the first was in relation to the Sun, and he attempted to estimate how long it had been shining and used this as a corollary for the age of the Earth. Secondly, Thomson investigated the effect that friction caused by tides might have had on the shape of the Earth. The third method took the secular cooling rate of the Earth, and it is for this work that Thomson is chiefly remembered in the geochronological field. He revisited the research of George-Louis Leclerc, Comte de Buffon (1707–88) nearly a century earlier. Buffon, a French aristocrat had been Curator of the celebrated Jardin du Roi (now the Jardin des Plantes) in Paris, and had built a forge on his estate near Dijon, where he heated spheres of different materials and sizes and measured their cooling rates. From this he estimated that the Earth was nearly 75,000 years old. This figure was considerably higher than that thought correct by the bulk of Buffon's contemporaries, Thomson would arrive at a figure that was higher still.

Whether by design or coincidence these very different schemes worked-up by Thomson ultimately came to broadly the same conclusion in terms of age determination, and this led to his unwavering belief that his determination was correct. It led too to considerable support from colleagues, and was partially instrumental in his preeminence in British science for such a long time.

THE AGE OF THE SUN AND ORIGIN OF ITS HEAT

Thomson produced a large volume of work on the Sun and its heat, and first published on the subject in a number of papers published in the *Transactions of the Royal Society of Edinburgh* and in the *Philosophical Magazine* in 1854. Drawing on the work of James Prescott Joule (1818–89) published ten years earlier, who demonstrated the relationship between energy, heat, and force, Thomson discussed the origins of the mechanically-derived heat of the Sun. Energy, lost as radiant heat and light could not, he noted, ever be replenished. Noting some earlier ideas that accounted for the source of the Sun's heat, Thomson rejected out-of-hand the notion that the Sun was a heated body, one assumes with its own source of heat. He considered that the Sun formed due to the attraction meteorites which built up its mass. As they were pulled towards the Sun they burnt up and the frictions caused by the vaporized gases were released as heat. In this he was following the scheme suggested by John James Waterston (1811–83) the previous year. While the bulk of this paper was theoretical he published only one quantitative calculation. He argued that it was possible to determine how long it would have taken for the meteorite component of the Sun's mass to accumulate. This was only a proportion of the Sun's bulk. This he said took 32,000 years, and suggested that the Sun would not have shone on the Earth for many more times than 32,000 years. He remarked:

> We may be confident, then, that the gradual augmentation of the Sun's bulk required by the meteoric theory to account for this heat, may have been going on in time past during the whole of the human race.[2]

In terms of geological time, he pointed out that the Sun could have taken two million years to grow, which in his opinion was an adequate length of time required for the geological processes that had been described by leading scientists such as Charles Lyell.

Initially the surface of the Earth, however, would have remained in a molten state due to the gravitational energy produced when meteorites collided with it. Following this period the surface crust developed but it would have been too hot to allow vegetation to cover it.

Computing the amount of heat lost from the Sun to be 2,781 thermal units per second per square foot, which would equate to removing a layer 55 miles thick per year, Thomson rationalized that the Sun would burn out in 8,000 years time, and conversely that it must have been eight times its mass 8,000 years earlier. He went on to state:

> We may quite safely conclude then that the Sun does not get its heat by chemical action among particles of matter primitively belonging to his own mass, and we must therefore look to the meteoritic theory for fuel . . .[3]

Eight years later, Thomson returned to the question of the age of the Sun in a three-part paper published in *Macmillan's Magazine* in March 1862. His earlier premise that meteorites falling into the Sun and the heat generated by friction of the vaporized gases as they burnt up simply was untenable. Meteorite falls would not have generated the volume of heat that he had suggested they would. Following the meteoritic theory of Hermann Ludwig Ferdinand von Helmholtz (1821–1894) Thomson modified his ideas on the role of meteorites in generating the Sun's heat, and rather than vaporization being the source of the heat, it was the fact that as they moved the meteorites had a potential energy, which became kinetic energy as they moved towards each other, and which was converted into heat upon collision with themselves or the Sun. Consequently he had to alter his time estimates and concluded:

> It seems, therefore on the whole most probable that the sun has not illuminated the earth for 100,000,000 years, and almost certain that he has not done so for 500,000,000 years.[4]

However he was more comfortable with the conclusion that the Sun had been operational for between 20 and 60 million years. Towards the end of his life, following persuasion by his associate Peter Guthrie Tait (1831–1901), Thomson favoured the lower age limit of 20 million years. This by association gave a limit to the age of the Earth.

TIDAL FRICTION

Immanuel Kant (1724–1804), the German philosopher and Professor of Mathematics at the University of Königsberg, realized that the Earth was slowing down due to the effects of tidal friction. Running with this idea, Sir George Howard Darwin (1845–1912), son of Charles, wrote extensively on the orbit of the Moon and the Earth's retardation rates. Some of his contemporaries even went so far as to hypothesize that the Moon had been formed by metastasis from the Earth. Terrestrial material was lost from the surface due to rapid spinning of the parent body and consolidated into the lunar sphere—a theory now known to be incorrect. Darwin calculated the time taken for the Earth and Moon to settle down from this initial rupture to their present condition, and came up with a minimum of 56 million years.

In 1868 Thomson examined the Earth's history and chronology by focusing on changes in the Earth's shape over time, changes which were caused by a slowing of its rate of rotation over time. The rate he calculated was 3.6 seconds per annum. It was well known that due to friction on the surface waters as the Earth spins, tidal waters tend to become banked up and do not act in the predictable manner normally expected. The Earth, Thomson said, assumed its flattened spherical shape soon after its formation, while it was still molten. He realised that if one took the present rotation rate of the Earth, and used this to calculate what the shape of the globe would have been if this had been the primordial spinning rate, one would expect a spheroidially flattened globe of a particular shape. This expected shape he found was appreciably no different from the actual shape of the globe, and so he deduced that very little time had elapsed since the formation of our planet. He acknowledged that there were difficulties of actually determining accurately the parameters that fed into this methodology, but was confident that the Earth was no more than 1,000 million years old, and that 100 million years old was the more plausible figure.

SECULAR COOLING OF THE EARTH

In his research into the cooling rates of the Earth, Thomson was greatly influenced by the work of French physicist Jean Baptiste Joseph Fourier (1768–1830) which he had studied when aged only 16. Fourier had written that the source of the Earth's heat was threefold: from primitive internal heat, from heating by the Sun, and from heat in the universe, but had also suggested that the Earth was cooling down by conduction and provided mathematical models to quantify this.

Thomson believed that the primordial Earth was molten throughout, but that it developed an outer solid crust soon after its formation. He believed that it then developed a solid nucleus surrounded by molten rock, and a solid thin crust, and that it continued to solidify from its centre outwards as the internal heat migrated through the rocks by conduction and heat was lost from its surface. Once solid, but still hot, it lost heat by conduction rather than convection. Thomson drew on experimental work undertaken between 1837 to 1842 by James David Forbes (1809–68) which had revealed a temperature gradient at three stations near Edinburgh (the readings continued to be recorded at Calton Hill, Edinburgh for the next 13 years until 1854). Forbes had placed thermometers at three, six, twelve, and twenty-four feet below the surface at each station.

In a paper published in 1861, Thomson discussed the findings and implications of this work at length, and observed that the effects of diurnal changes in temperature become obsolete at depths of greater than three feet below the surface.[5] The earliest studies of temperatures beneath the Earth's surface had been carried out in Paris from 1671[6] and it was well known by the mid-1800s that temperatures increased as one went down into the Earth. To further research the British Association for the Advancement of Science, at its 1867 meeting in Dundee, appointed a committee to carry out more work in this area. It met between 1868 and 1877 and was awarded the considerable sum of 330 guineas for this work. Geothermal gradients had previously been investigated by taking temperature readings from various depths in mines or from boreholes; however, this work was not very accurate until the invention in the 1830s of specialized thermometers designed for the task. Even by the 1860s results were not conclusive, and it was believed that the temperature at the Earth's centre was approximately 3,600 degrees Fahrenheit which is about half of what we now consider the temperature to be.

In April 1862 Thomson read his first paper on the cooling of the Earth that tackled some of the problems raised in his mind by the geologists. This paper was published two years later in the *Transactions of the Royal Society of Edinburgh* (and summarized shortly afterwards in *The Times*) and began with a full-fronted attack on their uniformitarian philosophy and the need for an extended or indefinite Earth history:

> For eighteen years it has pressed on my mind, that essential principals of Thermo-dynamics have been overlooked by those geologists who uncompromisingly oppose all paroxysmal hypotheses, and maintain not only that we have examples now before us, that the earth, of all the different actions by which its crust has been modified in geological history, but that these actions have never, or have not on the whole, been more violent in past time than they are at present.[7]

Thomson stated that the temperature increased 1 °F every 50 feet downwards, and that this indicated that heat was being lost from the Earth's surface over a long period of time, and that the whole globe was cooling down. He rejected the hypothesis that the earth had an internal, or as he put it 'temporary dynamical', or chemical source of heat. He argued that there was less volcanic activity at the time of writing that had been 1000 years earlier, but that this fact had been ignored by most geologists.

Less volcanic activity pointed to a cooler Earth. However, in the 1890 reprint of the 1862 paper he added, no doubt with some satisfaction, a footnote in which he said that most geologists in 1889 now believed the statement to be true! One wonders what Thomson made of the major volcanic eruption at Krakatau of 1883.

He carried out experiments to determine the conductivity of various rock types, and took 7,000°F as the temperature of fusion of rocks, an estimate that had been determined a short time earlier. He argued, given this formula, that consolidation of the surface would have taken place 98 million years ago, or a maximum of 200 million years ago if one took 10,000°F (a high estimate, in his opinion) as the melting temperature of rock. Given the underground temperatures known, and the loss of heat due to conduction, he noted that if the Earth was only 40,000 years old then the temperature gradient would be 1 degree per foot. If 4 million years old it would be a tenth of a degree Fahrenheit per foot, and if 100 million years old it would be one-fiftieth of a degree Fahrenheit per foot. The latter figure is the temperature gradient figure that Thomson used, and gave a coincident age with that of the consolidation age of the crust of approximately 100 million years.

Thomson also noted that it was possible, however, to generate the observed temperatures if the whole of the Earth's surface had been heated up to 100°F at some time in the last 20000 years, but that this scenario was an impossibility as all

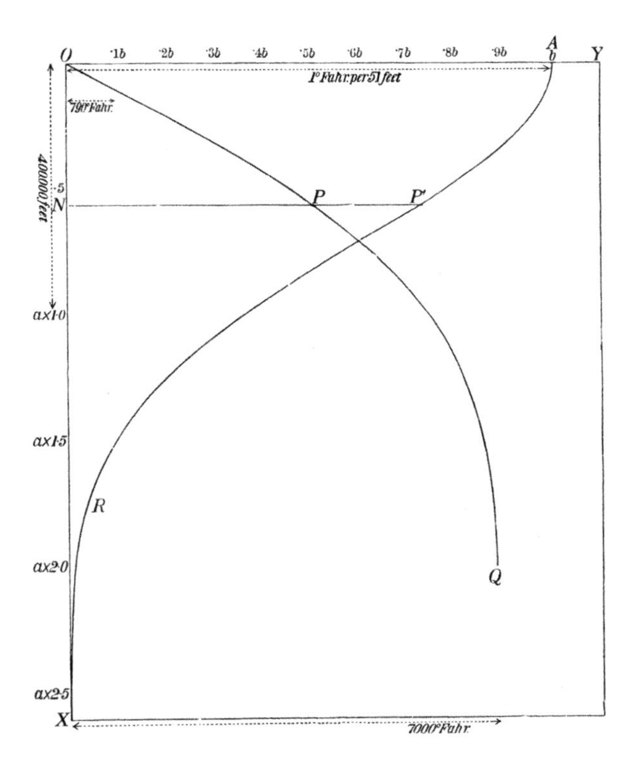

Fig. 10.1. Graph showing the temperature gradient (R–A), and the temperature curve (O–Q) for the Earth. The left-hand axis is depth below the surface; the top axis 1 degree Fahrenheit per 51 feet, and the bottom axis 7,000 degrees Fahrenheit. (From W Thomson, *Mathematical and Physical Papers* [MPP], Cambridge University Press, vol. 3 (1890) p. 303.)

traces of life would have been killed if the surface had attained this temperature. From examination of the cooling of the Earth, Thomson believed that was formed between 20 and 400 million years ago, with 98 million years being its likely age.

In arriving at this figure, Thomson made four assumptions about the conditions of the Earth. First, it was solid; secondly it began at the same temperature throughout; thirdly, that it must be homogeneous and have an identical conductivity throughout; and finally, no internal heat source was present. Subsequently all of these assumptions were challenged and found to be untenable and incorrect. We know through the work of Richard Dixon Oldham (1858–1936) that the outer part of the Earth's core is liquid. The Earth is not homogeneous, and crustal rocks are quite different in mineralogy and density to those found in the Earth's mantle. Following the discovery of radioactivity by Antoine Henri Becquerel (1852–1908) in 1896, it was soon recognized that disintegration of radioactive elements provided an internal heat source for the Earth.

On 27th February 1868 Thomson revisited the subject of the Earth's antiquity when he addressed the members of the Geological Society of Glasgow. He used the occasion to try to bring the geologists around to his viewpoint, and gave the assembled company a synopsis of the methods and results used in both the tidal friction and secular cooling schemes to age the Earth. He concluded that the Earth was no more than 100 million years old.

> ... we find ourselves driven to the conclusion in every way, that the existing state of things on the earth, life on the earth, all geological history showing continuity of life, must be limited within some such period of past time ...[8]

DETRACTORS AND SUPPORTERS

Many of Thomson's detractors questioned his figures regarding the internal temperature and gradient in the Earth, and also questioned his reliance on a theory which they considered to be without any geological foundation. In 1867 during the British Association for the Advancement of Science meeting, Thomson fell into conversation with the geologist Andrew Crombie Ramsay (1814–1891) and the question of timescales arose. Thomson asked Ramsay:[9]

> Do you think it [the Sun] has been shining on for a million million years?

to which Ramsay responded:

> I am as incapable of estimating and understanding the reasons which you physicists have for limiting geological time as you are incapable of understanding the geological reasons for our unlimited estimates.

Immediately Thomson retorted:

You can understand physicists' reasoning perfectly if you give your mind to it.

Thomson was not the only one prone to caustic remarks. Thomas Mellard Reade (1832–1909), a civil engineer and geologist from Liverpool, wrote in 1878 that 'Facts are safer than theories'[10] implying that Thomson relied too heavily on the latter.

Thomson's ideas on a cooling Earth were at variance with the views of many other scientists, but not all. In 1865 the Rev. Samuel Haughton, (1821–97) a Fellow of Trinity College, Dublin and its Professor of Geology, and Darwin critic, supported his fellow Irishman's work. Haughton reworked the calculations applied to the cooling method for age determination. In 1865 in his *Manual of Geology* Haughton published the results of these calculations and which showed that the Earth was 2,298 million years old—considerably higher than Thomson's findings. Haughton was no mathematical slouch and his calculations were accurate, but his problem was that the basic figures he used for the calculation were flawed.

Thomson's timescale was questioned in 1869 by none other than Thomas Henry Huxley (1825–95) who asked in his Presidential address to the Geological Society 'has it ever been denied that this period *may* be enough for the purposes of geology'.[11] In doing so Huxley laid the seeds of discontent between the views of physicists in one corner and the geologists in the other corner. However, he did imply that the for the biologists Thomson's timescale didn't necessarily cause them problems, as they, the biologists, simply used the geologists timescale in the absence of having one for themselves.

Fig. 10.2. Rev. Samuel Haughton (1821–1897), Professor of Geology, Trinity College, Dublin supported Thomson's views and calculated the cooling rate of the Earth for himself (Courtesy of Dr Patrick Wyse Jackson.)

Fig. 10.3. Thomas Henry Huxley (1825–1895), President of the Geological Society (of London) in 1869, was a major critic of Thomson's views on the age of the Earth. (Courtesy of Dr Patrick Wyse Jackson.)

In 1895 Thomson penned a short paper for *Nature* in which he discussed a recently published paper by Clarence King (1842–1901), who had been the first Director of the United States Geological Survey between 1879 and 1881. King's paper reexamined Thomson's work and using a new figure of 1,950°C for the fusion temperature of rocks, a figure which he had obtained from the physicist Carl Barus (1856–1935), arrived at twenty-four million years as the age of the Earth. This limit was accepted by Thomson who acknowledged that Barus' data would have rendered his 100 million year limit too large, and that while the method would point to 10 million years, the effects of pressure on the geological processes would push the age determination to that of King's. In a short letter to *Nature* in 1897 Thomson settled on a range of 20 to 40 million years old, but said that the 24 million year estimate reached by King in 1893 was probably correct.

The cleric and geophysicist Osmund Fisher (1817–1913) stated in a paper published in the *Geological Magazine* in 1895 that 'no reliable estimate of the age of the world, based on considerations of the present temperature gradient at the surface, has hitherto been made'.[12] James Geikie (1839–1915) Professor of Geology at Edinburgh wrote in the February 1900 issue of the *Scottish Geographical Magazine*, 'there are certain other considerations which increase one's doubts as to the adequacy of Lord

Kelvin's theory'. Thomas Chrowder Chamberlin (1843–1928), President of the University of Wisconsin, was probably Thomson's most outspoken critic, at least in America. He believed that the Earth had formed thanks to the accretion of cold material and that it had never been never fully molten. He remarked caustically in 1899, in a paper published in the journal *Science,* that 'the postulate of a white-hot liquid earth does not rest on any *conclusive* geological evidence'. Even the biologists were concerned that Thomson's age limits were too short for biological evolution. Edward Bagnall Poulton (1856–1943), Hope Professor of Zoology at Cambridge, weighed-in at the annual British Association meeting in Liverpool in 1896. As President of the Biological Section he could create quite a stir, and in his address he attacked the findings both of Thomson and those of some geologists.

The main and most serious objection to some of Thomson's conclusions came from another Ulster Protestant, John Perry (1850–1920), who coincidently spent 1874 working as a research assistant in Thomson's laboratory in Glasgow before moving to Japan. Following a stint as a Professor at the Imperial College of Engineering in Tokyo, Perry returned to Britain. He took a position at the Finsbury Technical College in London and in 1896 moved to the larger Royal College of Science in the city.

In 1894 Robert Arthur Talbot Gascoyne-Cecil, 3rd Marquess of Salisbury (1830–1903) was President of the British Association for the Advancement of Science meeting held that year in Oxford. Although not a scientist, he was a Fellow of the Royal Society, and he was influential and people listened to what he had to say. He had recently twice served as Prime Minister, when his Conservative party held power alternately with William Gladstone's Liberal Party, and he would hold this office again for a further eight years between 1895 and 1902. In his address Salisbury attacked the basis of Darwin's theory of natural selection saying that there was not enough time for natural selection to have taken place. To back his assertion Salisbury depended on the age estimates and reputation of Lord Kelvin.

Having read the printed paper, Perry wrote to Kelvin three times and outlined a number of objections to his work and gave some suggestions as to how the calculations could be improved, but he received no response. Not content to be brushed-off, Perry sought and received support for his objections from other scientists, and he then felt forced to commit himself to print. The following year his objections were outlined in two papers in *Nature*: in brief he said that convection, and not just conduction, played a role in heat loss from the Earth and that this invalidated Thomson's findings. Perry's objections were immediately countered by Kelvin in a note in the same journal, and this was followed by a third missive from the younger man. As Brian Shipley (2001) has observed, Perry showed that:

> the faster heat was conducted outwards from the Earth's core, the *longer* it would take to obtain the present observed temperature gradient at the surface.[13]

This was because Kelvin had based his calculations on the conductive properties of surface crustal rocks and he didn't include the denser sub-crustal rocks known to exist. *Longer* was counter to Kelvin's conclusion.

However, there was some support at this time, and it came from the American, George Ferdinand Becker, who in *Science* in February 1908 reexamined Thomson's methods and concluded:

> Notwithstanding the inadequacy of the data, I can not but believe that the 60-million year earth here discussed is a fair approximation to the truth and that with better data this age will not be changed by more than perhaps 5 million years.[14]

Becker pointed out that this age was broadly concurrent with those derived from the sedimentation accumulation and oceanic sodium methods. Even Mark Twain (1835–1910), who wrote a short essay on geological matters, said of Thomson that 'I think we must yield to him and accept his view'.[15]

Thomson's initial reasoning behind taking up research on the age of the Earth was, as we have seen, his difficulty with the Darwinian timescale, and Darwin himself acknowledged that it caused him trouble. In terms of his influence on the biologists who adhered to the longer time frame required by natural selection, it would appear to have been slight if we believe Archibald Geikie (1835–1924), Director of the Geological Survey, and later (like Thomson) President of the Royal Society. In fact many biologists didn't require the longer timescale suggested by Darwin as many didn't accept natural selection. They preferred the Lamarckian means of biological change which required little time. Natural selection and thus the need for greater time was generally accepted by most biologists by the 1930s. In his Presidential address to the geological community at the British Association for the Advancement of Science meeting held in Dover in 1899, Geikie firstly acknowledged the debt geologists owed Thomson:

> Geologists have been led by his criticisms to revise their chronology. They gratefully acknowledge that to him they owe the introduction of important new lines of investigation, which link the solutions of the problems of geology with those of physics.[16]

but he then delivered a sharp reprimand to the ennobled physicist and a reminder of the leanings of palaeontologists towards a chronology longer than that proposed by him:

> It is difficult satisfactorily to carry on a discussion in which your opponent entirely ignores your arguments, while you have given the fullest attention to his ... [Geologists] have been willing to accept Lord Kelvin's original estimate of 100 millions of years as the period within which the history of life upon the planet must be comprised ... yet there is undoubtedly a prevalent misgiving, whether in thus seeking to reconcile their requirements with the demands of the physicist they are not tying themselves down within the limits of time which on any theory of evolution would have been insufficient for the development of the animal and vegetable kingdoms.[17]

Although critical of Thomson's work in the geological sphere, Geikie was rather friendly with him on a personal level and certainly respected his genius. In April 1899 he shared a railway compartment with Thomson and his wife on a journey to Genoa and the following day they explored the hills and scenery together. Geikie attended Thomson's funeral and in his autobiography remarked:

> ... we laid to rest all that was mortal of one who was not more reverenced for his towering genius and his manifold achievements in science, than he was beloved for the rare modesty, simplicity, and goodness of his character.[18]

THOMSON'S CHRONOLOGICAL CONCLUSIONS AND THE DEMISE OF HIS TIMESCALE

Thomson's final contribution to the chronological debate was a paper entitled 'The age of the earth as an abode fitted for life'. This was published in several journals in England and America in 1898 and 1899 and is basically a summation of his ideas. He remarked that his determinations were based on science (i.e. physics) and haughtily concluded that if his methods proved incorrect then geologists and biologists were essentially free to plough their own furrows:

> If science did not allow us to give any estimate whatever as to whether 10,000,000 or 10,000,000,000 years is the age of the earth, then I think that Professor Huxley would have been perfectly right in saying that geologists should not trouble themselves about it, and biologists should go on in their own way, not inquiring into things beyond the power of human understanding and scientific investigation.[19]

He held on to his view that the Earth had consolidated no less than 20 million years ago and no more than 40 million years previously (but favoured the precise date of 24 million years suggested by Clarence King), and that the Sun was anything between 20 and 60 million years old (he favoured an age trending towards the lower time limit). One can see that Thomson's coeval age of both Sun and Earth coincided in and around 20 to 30 million years.

Thanks to his continual tinkering with his age determinations Thomson more than likely reduced the credibility of his work, and his research in the area of geochronology was overtaken by the emergence of the study of radioactivity in geology. He failed to fully acknowledge this new science and see the full implications that it heralded, although interestingly he had in 1862 anticipated an internal heat source in the Sun:

> As for the future, we may say, with equal certainty, that inhabitants of the earth cannot continue to enjoy the light and heat essential to their life, for many million years longer, unless sources now unknown to us are prepared in the great storehouse of creation.[20]

Shortly before Thomson (by then Baron Kelvin of Largs) died, Ernest Rutherford (1871–1937), who was Professor of Physics at McGill University, Montreal, was waiting to give a lecture on his work to an assembled audience at the Royal Institution in London in 1904. Looking out from the lectern Rutherford was horrified to see the 79 year old sitting in the front row, and he realized that the final section of his lecture would discredit a great deal of the older man's geological research on the age of the Earth. He had no option but to begin, and was highly relieved shortly afterwards to see Thomson begin to slumber. Throughout the lecture Thomson remained lost to the world but awoke at the critical point when Rutherford was about to deal with geochronology. He recalled later:

> I saw the old bird sit up, open an eye and cock a baleful glance at me! Then a sudden inspiration came to me, and I said Lord Kelvin had limited the age of the earth, *provided no new source of heat was discovered*. That prophetic utterance refers to what we are now considering tonight, Radium! Behold! the old boy beamed upon me![21]

Initially, while Thomson recognized that the radioactive material Radium produced heat, he refused to believe that it produced it itself, and rather argued that the Radium must have gained the heat that it subsequently emitted from an external source. However, soon afterwards he privately accepted that the discovery of Radium had made some of his conclusions regarding secular cooling of the Earth difficult, though he never made this view publicly known. Perhaps it would have been best if he had slept through Rutherford's lecture, but there is little doubt that he saw the dawn of radioactivity herald the demise of his own geochronology.

In public Thomson remained defiant to the end of his life. None of the criticisms of his methods thrown at him by geologists or the biologists made him moderate or alter his views substantially, and he continued to hold that the Earth was rather young.

Acknowledgement: I thank Brian Shipley for his kindness in providing me with copies of John Perry's publications.

11

Thomson and Tait: The *Treatise on Natural Philosophy*

Raymond Flood

INTRODUCTION

On 29 October, 1902, Lord Kelvin unveiled a portrait in Peterhouse College, Cambridge of his friend and collaborator Peter Guthrie Tait who had died in July of the previous year, aged 70. Although they had both been students at Peterhouse their times there did not overlap, Thomson being the elder of the two by some six years. The *Cambridge Chronicle* reported Kelvin, in his address at the unveiling, describing their friendship as beginning about 1860, when Tait came to Scotland from Queen's College, Belfast to succeed Forbes as Professor of Natural Philosophy at Edinburgh. The report said that Kelvin:

... remembered Tait once remarking that nothing but science was worth living for. It was sincerely said then, but Tait himself proved it to be not true later. Tait was a great reader. He would get Shakespeare, Dickens, and Thackeray off by heart. His memory was wonderful. What he once read sympathetically he ever after remembered.[1]

And Kelvin went on to describe their collaboration:

Thus he was always ready with delightful quotations, and these brightened their hours of work. For they did heavy mathematical work, stone breaking was not in it. A propos, perhaps, of the agonies (he did not mean pains, he meant struggles) of the mathematical problems which thcy had always with them.[2]

Probably the most influential legacy of their collaboration on 'heavy mathematical work' was the production in 1867 of The *Treatise on Natural Philosophy* by Thomson and Tait—still so called even after Kelvin was ennobled in 1892. They started collaborating on it in 1861 shortly after they met and although it fell far short in scope of the original intentions of its authors, it was to be highly influential in identifying and placing conservation of energy at the heart of its approach.

Fig.11.1. Reid portrait of P.G. Tait (1831–1901) at Peterhouse College.

The two men were of very different natures. Thomson frequently went travelling while Tait did not leave Scotland after 1875. Tait could also be argumentative with bitter disputes, for example, with Heaviside and Gibbs over the relative merits of the vector approach as compared to using quaternions. However it was Tait who drove the collaboration towards publication—in turn cajoling, coaxing, and berating Thomson to try to get him to keep to deadlines and schedules. Tait's frustration is illustrated in a letter he writes to Thomson in June 1864 about halfway through the collaboration where he says:

> I am getting quite sick of the great Book . . . if you send only scraps and these at rare intervals, what can I do? You have not given me even a hint as to what you want done in our present chapter about statics of liquids and gases! I have kinetics of a particle almost ready, nearly the whole of the next chapter, but I don't see the fun of paying 30/- for sending the MSS to you [in Germany] for revision, when in all probability you won't look at it till some indefinite period when you are in Arran, where it would be certain of reaching you—and for 8d. Now all this is very pitiable: I declare you did twice as much during the winter as you are doing now.[3]

The *Treatise* was universally known as 'T & T′' and Thomson and Tait used the abbreviations in their extensive correspondence. One of the formulae in the *Treatise* was $d\rho/dt = jcm$ and as a result their close friend James Clerk Maxwell became known as $d\rho/dt$! Maxwell and Tait had been friends since they first went to school at Edinburgh Academy at the age of 10, and from there to Edinburgh University and then onto Cambridge.

In 1852 Tait had graduated as Senior Wrangler and Maxwell in 1854 as Second Wrangler. After a couple of years in Cambridge, Tait obtained the Chair of Mathematics at the Queen's College, Belfast where he worked closely with Thomas Andrews and William Thomson's brother James. It was during this period that he discovered Hamilton's work on quaternions for which he remained a lifelong enthusiast. Meanwhile, Maxwell had become Professor of Natural Philosophy at Marischal College in Aberdeen, but he had to leave as a result of the 1860 merger of Marischal with King's College because he was junior in the ranking.

Meanwhile J.D. Forbes relinquished the Chair of Natural Philosophy at Edinburgh University in 1859 and Maxwell and Tait both applied. There were five other candidates including E. J. Routh who had beaten Maxwell by being Senior Wrangler in their year. Tait had 19 testimonials including ones from Andrews, Hamilton, Thomson, Boole, Todhunter, and Challis. The deciding factor seems to have been teaching ability, and on this basis Tait was appointed. Indeed the Edinburgh newspaper *The Courant* on reporting his election remarked:

> .... it will be no disrespect to the warmest friends of the successful candidate, and we do not mean to dispute the decision of the curators, by saying, that in Professor Maxwell the curators would have had the opportunity of associating with the University one who is already acknowledged to be one of the remarkable men known to the scientific world. His original investigations ... have well established his name among scientific men; while the almost intuitive accuracy of his ideas would give his connection with a chair of natural philosophy ... a sure and valuable guide to those who came with partial knowledge requiring direction and precision. But there is another power which is desirable in a professor of a University with a system like ours, and that is, the power of oral exposition proceeding upon the supposition of a previous imperfect knowledge, of even total ignorance, of the study on the part of pupils ...

Tait was an enthusiastic, committed, and lucid teacher. According to Professor Andrew Gray, assistant to Thomson:

> Tait's professorial lectures were always models of clear and logical arrangement. Every statement bore on the business in hand; the experimental illustrations, always carefully prepared beforehand, were called for at the proper time, and were invariably successful[4]

And this view was shared by his pupils, including John Flett, a student in the mid-1880s who wrote:

> His lectures were so clear, concise and logical that they made the subject as clear as day, and were illustrated by ... experiments that were exceedingly simple and never failed to

come off with complete success. Yet, strange to say, I found Tait difficult to follow. He made everything so clear that it seemed perfectly easy. Yet he wasted not a word and when I came to repeat his descriptions and arguments I found that it was extremely difficult to make the logical chain of proof as irrefragable as Tait had made it . . . I enjoyed every one of Tait's lectures, though I came to have the idea that he was a sort of juggler who performed many tricks before the audience and left you in a sort of dazed condition afterwards not clearly or fully realising the significance of his words.[5]

J. M. Barrie, more famous as the author of *Peter Pan* but previously an undergraduate at Edinburgh, paints a picture of Tait as a captivating and popular lecturer:

. . . The small twinkling eyes had a fascinating gleam in them; he could concentrate them until they held the object looked at; when they flashed round the room he seemed to have drawn a rapier. I have seen a man fall back in alarm under Tait's eyes, though there were a dozen benches between them. These eyes could be merry as a boy's, though, as when he turned a tube of water on students who would insist on crowding too near an experiment . . .[6]

MOTIVATION FOR THE *TREATISE OF NATURAL PHILOSOPHY*

It was Tait's concern to have adequate textbooks for his students to support his lectures that led to the project that resulted in T&T′. Throughout his academic career Tait produced many textbooks either alone or with a collaborator.

1856	*A Treatise on Dynamics of a Particle* (with Steele)
1857	*Sketch of Elementary Dynamics* (with Thomson)
1867	*A Treatise on Natural Philosophy* (withThomson)
1867	*An Elementary Treatise on Quaternions*
1867	*Elementary Dynamics* (with Thomson)
1873	*Elements of Natural Philosophy* (with Thomson)
1873	*Introduction to Quaternions* (with Kelland)
1875	*The Unseen Universe* (with Balfour Stewart)
1875	*Recent Advances in Physical Science*
1878	*Paradoxical Philosophy* (with Balfour Stewart)
1884	*Heat*
1884	*Light*

1885	*Properties of Matter*
1895	*Dynamics*
1899	*Newton's Laws of Motion*

However it was one of the three books he published in 1867, *A Treatise on Natural Philosophy*, that was to prove the most influential.

Thomson and Tait had three main reasons for embarking on the writing of *A Treatise of Natural Philosophy*. The first was to provide appropriate textbooks to back up their lectures, for Thomson at Glasgow University and Tait at Edinburgh. Indeed Tait would have liked to write a book that he could have used himself when he was learning about natural philosophy and wrote as much in a letter to Thomson at the start of their collaboration in mid December 1861:

> ... I fancy that we might easily give in three moderate volumes a far more complete course of physics, Experimental and Mathematical than exists (to my knowledge) either in French of German. As to English there are NONE ...
>
> I am myself a good example of the want of such a book as we contemplate, having got all my information bit by bit from scattered sources, which often contained more error than truth. The next generation will thank us.[7]

There were also requests by colleagues such as Ludwig Fischer, Professor of Natural Philosophy at St Andrews University for a text to help in understanding the approach to physics based on the concept of energy.

Tait was insistent that the textbooks should be affordable, and in the correspondence later on in December 1861, when Thomson seems to have developed more ambitious plans, he bridled:

> The only objection I see to this, but it is a grave one, is the expense to the students, especially the Scotch ones. We may mulct and bleed Oxford & Cambridge & Rugby &c &c to any extent, but how about our own classes? What we want at once is not the fame of authorship, but the supply of a want in elementary teaching.[8]

Thomson's ambition for the work revealed another motivation for the book which was the balance between experimental demonstration and mathematical deduction. Tait's original conception was to produce a textbook to support his lecture courses which were based on experimental demonstration. The involvement of Thomson quickly led to an increasing emphasis on mathematical deduction. A compromise was reached—they used large print and small print sections with the former being reserved for the experimental part and the latter for the more advanced sections. As they described their aim in the Preface:

> Our object is twofold; to give a tolerably complete account of what is now known as natural philosophy, in language adapted to the non-mathematical reader; and to furnish, to those

who have the privilege which high mathematical acquirements confer, a connected outline of the analytical processes by which the greater part of that knowledge has been extended into regions as yet unexplored by experiment[9]

Their style was to be very different therefore from, on the one hand, the 'feebly descriptive sort that one associates with the name of Lardner,'[10] and on the other, the highly analytical exposition of Lagrange.

Despite this original plan however, over the five years of writing, the 'small print' section expanded at the expense of the 'large print'. The balance and mix nevertheless remained important to them, as they wished to nurture physical intuition and mitigate against a reliance on mathematical manipulation. As they say, again in the preface:

We believe that the mathematical reader will especially profit by a perusal of the large type portion of this volume; as he will thus be forced to think out for himself what he has been

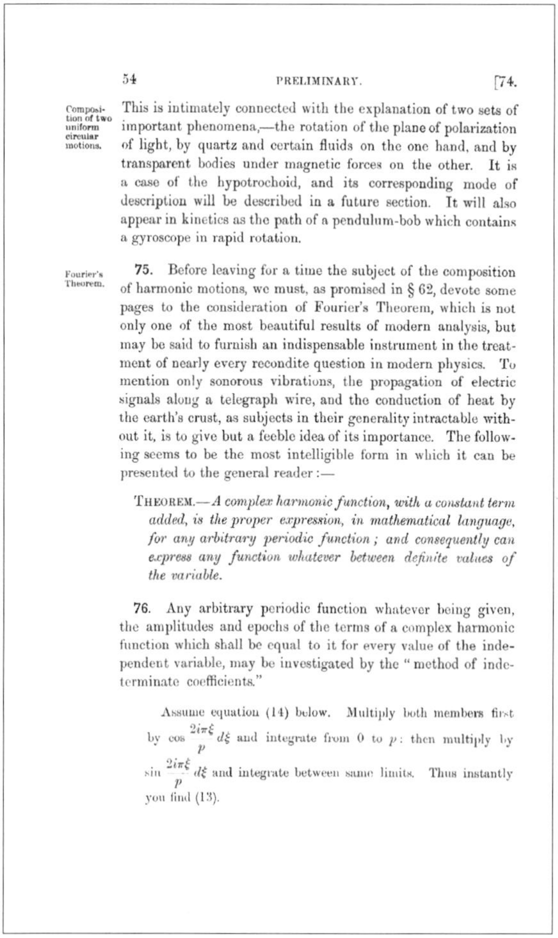

54 PRELIMINARY. [74.

Composition of two uniform circular motions.

This is intimately connected with the explanation of two sets of important phenomena,—the rotation of the plane of polarization of light, by quartz and certain fluids on the one hand, and by transparent bodies under magnetic forces on the other. It is a case of the hypotrochoid, and its corresponding mode of description will be described in a future section. It will also appear in kinetics as the path of a pendulum-bob which contains a gyroscope in rapid rotation.

Fourier's Theorem.

75. Before leaving for a time the subject of the composition of harmonic motions, we must, as promised in § 62, devote some pages to the consideration of Fourier's Theorem, which is not only one of the most beautiful results of modern analysis, but may be said to furnish an indispensable instrument in the treatment of nearly every recondite question in modern physics. To mention only sonorous vibrations, the propagation of electric signals along a telegraph wire, and the conduction of heat by the earth's crust, as subjects in their generality intractable without it, is to give but a feeble idea of its importance. The following seems to be the most intelligible form in which it can be presented to the general reader:—

THEOREM.—*A complex harmonic function, with a constant term added, is the proper expression, in mathematical language, for any arbitrary periodic function; and consequently can express any function whatever between definite values of the variable.*

76. Any arbitrary periodic function whatever being given, the amplitudes and epochs of the terms of a complex harmonic function which shall be equal to it for every value of the independent variable, may be investigated by the "method of indeterminate coefficients."

Assume equation (14) below. Multiply both members first by $\cos \frac{2i\pi\xi}{p} d\xi$ and integrate from 0 to p: then multiply by $\sin \frac{2i\pi\xi}{p} d\xi$ and integrate between same limits. Thus instantly you find (13).

Fig.11.2. A page of the *Treatise on Natural Philosophy* illustrating the use of small and large print.

too often accustomed to reach by a mere mechanical application of analysis. Nothing can be more fatal to progress than a too confident reliance on mathematical symbols; for the student is only too apt to take the easier course, and consider the formula and not the fact as the physical reality.[11]

The third reason motivating their approach was to base their natural philosophy on the principle of conservation of energy and extremum principles, achieving the replacement of 'Newton's Principia of force with a new Principia of energy and extrema'.[12]

And in so doing they wrote a book with a style and emphasis that has influenced physics textbooks to the present day.

COLLABORATION

Tait's original plan was to publish a work in three volumes, and it was for the third in particular that he was delighted to have Thomson's involvement because this would be the one he called the 'unique'[13] one on mathematical physics, and he felt that he could not do it on his own.

Tait was initially very ambitious about the time it would take to produce the volumes, suggesting in December 1861 that the first volume would be ready for the publisher after about six weeks of work if they were able to devote three to four hours a day to the task.[14] This was a very demanding schedule given their teaching commitments. Professors at Scottish universities were the only ones in their subjects who taught. It was not until 1892 that help arrived in the form of lecturers being appointed but even then only in limited numbers. Tait's colleague George Chrystal, who held the chair in mathematics, estimated that Tait had had about 10,000 students during his 40 years at Edinburgh.[15]

But it was not the demands of their university responsibilities that caused delay so much as coming to an agreement on what approach to take, what the books should contain, and in particular the working relationship between the authors.

On the approach and the balance between the experimental content and mathematical content the compromise was, as we have seen, to have two sizes of print, the smaller for the mathematical part, and this was quite quickly arrived at. Thomson would write to Tait about work that he had done in his 'small print hour'[16]

Both the biographies of Smith and Wise and that of S. P. Thompson have analysed the development of the intended content with the former biography, producing an illuminating table that compares the intention with the outcome. They do this by annotating a letter Tait sent, early on in the project, to his former colleague at Belfast, Thomas Andrews. They indicate in square brackets the Chapters that were eventually published.[17]

Section I.
Chap. I. Introductory. [Thomson's introductory lecture—omitted]
II. Matter, Motion, Mass, etc. [II]
III. Measures and Instruments of Precision. [IV]
IV. Energy, Vis viva, Work. [II]
V. Kinematics. [I]
VI. Experience (Experiment and Observation). [III]
Section II. Abstract Mechanics (*Perfect* solids, fluids, etc.).
Chap. I. Introductory (I have written this and will let you see it soon). [V]
II. Statics. [VI and VII]
III. Dynamics (Laws of Motion, Newton. Did you ever read his Latin? Do.). [II]
IV. Hydrostatics [VII] and [HydroDynamics. [Never written]
Section III. Properties of Matter, Elasticity, Capillarity, Cohesion, Gravity, Inertia, etc. etc. (This is to be mine.) [Outlined by Tait; portions expanded by Thomson; never completed]
Section IV. Sound. [Never written]
Section V. Light. [Never written]

Tait continues:

This will give you as good an idea as I yet possess as to the contents of our first volume. All the other physical forces [heat, magnetism, electricity, electrodynamics] will be included in Vol. II, which will finish up with a great section on the *one* law of the Universe, the Conservation of Energy.

As this analysis clearly reveals, only the first section of volume 1 was completed in its entirety and in the second section the emphasis was on statics and hydrostatics. The third section on properties of matter was started but never completed. Section IV on sound and section V on light were never written. When it was eventually published in 1867, *the Treatise on Natural Philosophy*, had two divisions, the first dealing with preliminary notions and covering kinematics and dynamical laws and principles—taking up just under half of the 727-page treatise. The second division dealt with abstract dynamics covering statics of a particle and then statics of solids and fluids, which had 300 pages devoted to it.

The major causes for the six year gap from late 1861, when the collaboration started, to publication in October 1867, were Thomson's delays, diversions, digressions, and inability to keep to deadlines. When Tait wrote just before Thomson's 40th birthday that he was 'getting quite sick of the great Book' he was writing to an internationally renowned scientist with many and varied interests who was driving his science forward on the many fronts described in other chapters as well as being involved with planning for the Atlantic telegraph, designing many instruments, and running his laboratory. He was a tremendously busy person and apart from his academic interests he traveled frequently, adding to the difficulty (and cost) of communicating by post.

Sometimes he did not reply to letters at all; at the other extreme he rewrote entire sections—expanding them considerably, frequently at the proof stage. This was not, for Thomson, a mere consolidation and exposition of existing knowledge, but rather a development of his thoughts on energy, matter, force, and their relation to the practical world. His mind ranged too widely to be constrained by the demands of a textbook. It must have driven Tait mad. In March 1863 Tait writes:

What in the name of goodness are you doing with the proof sheets . . .
SEND THE PROOFS BACK THE SECOND DAY AFTER RECEIPT AT THE LATEST. GOLDEN RULE.[18]

And in May 1864:

Send me my M.SS & and then go and see McMillan and account to him for deficits & extensions &c &c and promise him on *your* part (and then you need have no fears about adding *mine*) that Vol I will be ready in the end of July.[19]

Even so, publication was still three and a half years in the future. Costs were spiralling and it was eventually published by the Clarendon Press at Oxford in conjunction with MacMillan. Tait was midwife to *the Treatise on Natural Philosophy* as Halley was to Newton's *Principia* and it would never have seen publication without him. It sold well, but not to the financial advantage of the authors, who were told by the Clarendon Press in March 1869 that 'the book was still so much in debt to the Press in the actual outlay that there was no balance payable to the authors'.[20]

To support their students during the development of the *Treatise* they had produced various other works. First for the university session 1863–64 a *Sketch of Elementary Dynamics,* a brief discussion of kinematics and dynamical laws and principles which was to be the subject of division I of the *Treatise*. Then the *Sketch* was supplemented by the publication of a portion of Thomson's lecture at Glasgow under the title of *Elements of Dynamics.* However by 1864 they had come to the view that the *Treatise* was not going to be suitable as a textbook for their students and they decided to print some of its large print sections together with parts of the *Elements of Dynamics* separately under the title of *Elements of Natural Philosophy* which appeared in 1873.

A second edition of the *Treatise*, with some additional material mainly in appendices, appeared in two parts, the first in 1879 and the second in 1883, published by Cambridge University Press. Then the authors decided that no further volumes would be written. Sylvanus P. Thompson relates in his biography the

PUBLICATIONS

OF THE

CAMBRIDGE UNIVERSITY PRESS.

A TREATISE ON NATURAL PHILOSOPHY. By Sir W. THOMSON, LL.D., D.C.L., F.R.S., Professor of Natural Philosophy in the University of Glasgow, and P. G. TAIT, M.A., Professor of Natural Philosophy in the University of Edinburgh. **Part I.** Demy 8vo. 16*s*. **Part II.** Demy 8vo. 18*s*.

ELEMENTS OF NATURAL PHILOSOPHY. By Professors SIR W. THOMSON and P. G. TAIT. Demy 8vo. *Second Edition.* 9*s*.

MATHEMATICAL AND PHYSICAL PAPERS. By Sir W. THOMSON, LL.D., D.C.L., F.R.S., Professor of Natural Philosophy in the University of Glasgow. Collected from different Scientific Periodicals from May 1841 to the present time. Vol. I. Demy 8vo. 18*s*. Vol. II. 15*s*. [Volume III. *In the Press.*

MATHEMATICAL AND PHYSICAL PAPERS, by GEORGE GABRIEL STOKES, M.A., D.C.L., LL.D., F.R.S., Fellow of Pembroke College, and Lucasian Professor of Mathematics in the University of Cambridge. Reprinted from the Original Journals and Transactions, with Additional Notes by the Author. Vol. I. Demy 8vo. 15*s*. Vol. II. 15*s*. [Volume III. *In the Press.*

AN ELEMENTARY TREATISE ON QUATERNIONS. By P. G. TAIT, M.A., Professor of Natural Philosophy in the University of Edinburgh. Demy 8vo. 14*s*. [*New Edition. Preparing.*

A HISTORY OF THE THEORY OF ELASTICITY AND OF THE STRENGTH OF MATERIALS, from Galilei to the present time. Vol. I. Galilei to Saint-Venant, 1639—1850. By the late I. TODHUNTER, Sc.D., F.R.S., edited and completed by KARL PEARSON, M.A., Professor of Applied Mathematics at University College, London. Demy 8vo. 25*s*. Vol. II. By the same Editor. [*In the Press.*

THE ANALYTICAL THEORY OF HEAT, by JOSEPH FOURIER. Translated, with Notes, by A. FREEMAN, M.A., Fellow of St John's College, Cambridge. Demy 8vo. 16*s*.

A TREATISE ON THE THEORY OF DETERMINANTS and their applications in Analysis and Geometry, by R. F. SCOTT, M.A., Fellow of St John's College. Demy 8vo. 12*s*.

HYDRODYNAMICS, a Treatise on the Mathematical Theory of the Motion of Fluids, by HORACE LAMB, M.A., formerly Fellow of Trinity College, Cambridge. Demy 8vo. 12*s*.

London: C. J. CLAY AND SONS,
CAMBRIDGE UNIVERSITY PRESS WAREHOUSE,
AVE MARIA LANE.

Fig.11.3. From the second edition of the *Treatise on Natural Philosophy* of 1879 an advertisement for other available texts.

reply he received when he asked Thomson why no more than Volume I was produced:

His reply was that the ground they had proposed to cover had in the years that followed been largely covered by such books as Rayleigh's Sound (1873), Maxwell's Electricity and Magnetism (1873) and Lamb's Hydrodynamics (1879).[21]

The other reason, apart from the existence of other textbooks, was pressure of time, and this was explained in Thomson's own words in a letter he wrote to Professor Simon Newcomb in April 1881 shortly after Newcomb had been elected as an Honorary Fellow of the Royal Society of Edinburgh:

... I am working hard now at reprint of Vol. I., Part II., of Thomson and Tait's *Natural Philosophy* which will contain some considerable additions. Alas, alas! for vols. ii., iii., and iv.; *Ars longa; vita brevis,*

I am afraid neither of us will live to see them. We are both working hard in different branches of our study, and I hope there is some good work in both of us yet before we die. I am bringing out a reprint of all my papers already published. About 70 octavo pages are already in print. It will fill three or four octavo volumes, and will, in occasional different papers, bring out a great deal that I would have written for 'T and T′' volumes ii., iii., and iv.

I look forward also to possibly a separate publication on "Hydrodynamics," and on the 'Equilibrium and Motion of Elastic Solids'. Alas! however; I have been absolutely stopped, for three or four months now, in the work on 'T and T′', Part II., Vol. I., and in the reprint of my own papers, on account of incessant and pressing engagements both here and in London. To-morrow I become freed from my University duties; and the day after I hope to take refuge in the *Lalla Rookh,* where very soon I shall get to work, at least on my reprint—Believe me, with kind regards, yours very truly.[22]

So although the grand project was abandoned in its original format, Thomson felt that much of the material intended for it would appear in other forms and forums.

TREATISE ON NATURAL PHILOSOPHY

The *Treatise* begins by acknowledging the tradition of natural philosophy which Thomson and Tait are continuing and which underlies its structure and its aim. The Preface opens with a quotation from Fourier.

Les causes primordiales ne nous sont point connues; mais elles sont assujetties à des lois simples et constantes, que l'on peut découvrir par l'observation, et dont l'étude est l'objet de la philosophie naturelle.

(Fundamental causes are not known to us; but they are subject to simple and constant laws, which one can discover by observation and whose study is the object of natural philosophy).

This is followed by an appeal to another authority. And what greater authority is there than Newton? He is referred to in the very first sentence of the Preface and it is clear that Thomson and Tait see themselves as following in his footsteps in their investigation of natural philosophy which is 'the investigation of laws in the material world, and the deduction of results not directly observed'.[23] Indeed, as Andrew Whitaker comments in his chapter, this appeal to rediscovering Newton contributed to Kelvin's drift to reputational death. The style they adopt for T and T′ is to introduce material in non-mathematical language and then, when appropriate, to illustrate, develop, and expand using mathematics. However on one occasion even this approach defeats them. When discussing Hamilton's stationary action they state that 'this can scarcely be made intelligible without mathematical language'.[24]

In spite of Tait's enthusiasm for, and interest in, quaternions, they are not used in the *Treatise*. Indeed, Thomson when writing to Chrystal, who was seeking background information for his obituary of Tait said:

We have had a thirty-eight year war over quaternions. He had been captured by the originality and extraordinary beauty of Hamilton's genius in this respect; and had accepted I believe definitely from Hamilton to take charge of quaternions after his death, which he has most loyally executed. Times without number I offered to let quaternions into Thomson and Tait if he could only show that in any case our work would be helped by their use. You will see that from beginning to end they were never introduced.[25]

The underlying concepts of the book have been usefully classified by Smith and Wise into three areas. The first is the concept of energy with other ideas either being derived from or supporting it. They define, for example, the British absolute unit of force, named the poundal by William's brother James, as 'the force which, acting on one pound of matter for one second generates a velocity of one foot per second'[26], thus relating force to the change in velocity which it produces and then showing the relation of this absolute measure to weight which varies with location. They then go on to make explicit the equivalence between the work done by a force and the increase in kinetic energy—in those circumstances when, as they say, 'no other forces act on the body which can do work or have work done against them'.[27] This gives them a kinetic energy measure for work and consequently for force.

The second concept was to understand the nature of matter and 'Thomson's belief in a continuous substance as a substratum for atoms, molecules, and forces, or for matter and energy'.[28] In the *Treatise* there are frequent references to the forthcoming discussion on Properties of Matter—never completed. This concern is one reason why the great book was never finished, because it was also a research program to lay the dynamical foundation of a unified theory of ether and matter and obtain an explanation of the dynamics of continuous media. Without this underlying foundation the *Treatise* had a 'heuristic nature'[29] commented on by Joseph Larmor and attacked by Pierre Duhem. Larmor commented:

It was but rarely that his expositions were calculated to satisfy a reader whose interests were mainly logical...This fluent character, and want of definite focus, has been a great obstacle to the appreciation of 'Thomson and Tait', as it is still to Maxwell's 'Electricity', for such readers as ask for demonstration, but find only suggestion and exploration.[30]

The third feature that is striking on reading the text is their investigation of practical situations, for example the 'motion of governing masses in Watt's centrifugal governor: also of gimballed compass-bowl'[31] treated using Lagrange's generalized equations of motion.

Indeed as a reviewer said in the *Scotsman* of November 6th 1868:

The world of which they give the Natural Philosophy is not the abstract world of Cambridge examination papers—in which matter is perfectly homogeneous, pulleys perfectly smooth, strings perfectly elastic, liquids perfectly incompressible—but it is the concrete world of the senses, which approximates to, but always falls short alike of the ideal of the mathematical as of the poetic imagination. No iron beam is there met with so rigid as not to bend, no sphere of metal equally tense in its parts, no body that does not yield so much as to be incapable of having a fixed centre of gravity. Nowhere is there actual rest; nowhere is there perfect smoothness; nowhere motion without friction.

Also included as an appendix in the second edition is a marvelous discussion on continuous calculating machines collecting together previously published work either by Thomson or his brother James. There are tide predicting machines, a machine for the solution of simultaneous equations, an integrating machine, a machine for calculating the integral of the product of two given functions, and the mechanical integration of linear differential equations of the second order with variable coefficients. The tide predicting machine was to give the depth of water over a period of years for any port for which the 'tidal constituents have been found from harmonic analysis of tide-gauge observations'.[32]

MOTION VERSUS MATTER AND FORCE

A distinguishing feature of the *Treatise* is its separation of the topic of motion from that which causes it. The first chapter is devoted to this kinematics and as Thomson and Tait state in the introduction to the chapter:

There are many properties of motion, displacement, and deformation, which may be considered altogether independently of such physical ideas as force, mass, elasticity, temperature, magnetism, electricity. The preliminary consideration of such properties in the abstract is of very great use for Natural Philosophy, and we devote to it, accordingly, the whole of this our first chapter; which will form, as it were, the Geometry of our subject, embracing what can be observed or concluded with regard to actual motions, as long as the cause is not sought.[33]

They then develop the topic, first considering the motion of a point, then a point attached to an extensible cord leading to the motions and displacements of rigid systems and finally the deformations of solid and fluid bodies.

They give pride of place to Fourier's theorem—the development of a periodic function in terms of simple harmonic terms—not only as a 'beautiful result'[34] but also for its widespread usefulness in 'nearly every recondite question in modern physics'[35] ranging across sound, electricity, and heat conduction and which underpinned Thomson's machine for predicting the tides. There then follows descriptions and analysis of various complicated motions of rigid bodies, of Hooke's joint, of surfaces rolling on surfaces, of screws, and many other combinations, also introducing deformations of the bodies. Many of these situations have practical applications.

Thomson and Tait begin chapter 2 on dynamical laws and principles by reiterating that so far they:

> have considered as a subject of pure geometry the motion of points, lines, surfaces and volumes with or without change of dimensions and form; and the results we there arrived at are of course altogether independent of the idea of matter and of the forces which matter exerts.[36]

Now they are going to consider the action of force: whether it maintains relative rest, or produces acceleration of relative motion.[37] The initial sections parallel the development that Newton followed in his *Principia Mathematica*. In the section on matter they make the unfulfilled promise:

> To our Chapter on Properties of matter we must refer for further discussion of the question. *What is matter?* And we shall then be in a position to discuss the subjectivity of *Force*.[38]

Fundamental to their approach is the concept of energy and the underpinning law of energy which: '... in abstract dynamics, [may] be expressed as follows:

> The whole work done in any time, on any limited material system, by applied forces, is equal to the whole effect in the forms of potential and kinetic energy produced in the system together with the work lost in friction.[39]

They see this approach as already inherent in Newton's third law, and view the first two laws as giving only 'a definition and a measure of force'.[40]

They then claim:

> This principle may be regarded as comprehending the whole of abstract dynamics, because, as we now proceed to show, the conditions of equilibrium and of motion, in every possible case, may be immediately derived from it.[41]

They start off right away to substantiate this claim by first considering the situation where friction is absent and the system is in equilibrium. This is achieved by obtaining a necessary and sufficient condition for equilibrium which is that: '... the work done by the applied forces is equal to the potential energy gained, in any infinitely small displacement from that configuration'.[42]

The development proceeds to obtain 'what posterity regards as the single most important aspect of the treatise'[43] This aspect involved basing their dynamics on a variational principle—the principle of least action. This action they defined as twice the time integral of the kinetic energy, and the principle of least action is:

> Of all the different sets of paths along which a conservative system may be guided to move from one configuration to another, with the sum of its potential and kinetic energies equal to a given constant, that one for which the action is the least is such that the system will require only to be started with the proper General velocities, to move along it unguided.[44]

Smith and Wise argue very convincingly that this approach to dynamics, of considering how a system evolved over time, was informed by, and in sympathy with, the temporal perspective made necessary by the second law of thermodynamics.[45] From the principle of least action and the principle of conservation of energy they could deduce the Langrangian and Hamiltonian equations of motion. Before the *Treatise* this approach was not adopted in textbooks.

RECEPTION

The *Treatise* was well received on publication in 1867 and indeed Sylvanus P. Thompson saw it as a turning point in the production of textbooks on natural philosophy. He wrote in 1909:

> Any one who might doubt the enormous and enduring service rendered to science by this work has only to compare any recent treatise with the best of those in existence before 1867 to be convinced on the point.[46]

That it was widely used seems to follow from his next observation, where he is reluctant to analyse its content because 'the forty-two years that have elapsed since its publication have made the work so familiar to every student of natural philosophy'.[47] However, it could not serve as an introductory textbook, and the syllabus for Tait's 1869 course in Natural Philosophy (which had two divisions) in the *Edinburgh Calendar* had the following reading list: Herschel's *Astronomy* in Lardner's *Cyclopaedia* (for the lower division); Newton's *Three Sections* and Goodwin's *Course of Mathematics* (for the higher division) followed by the comment:

> ... In 1867 there was published the first volume of a 'Treatise on Natural Philosophy' by Sir W Thomson and Prof Tait; and an elementary work on the same subject for less advanced students will soon appear (a portion having been separately printed with the title *Elementary Dynamics*). The latter will of course, so far as it goes, form the textbook for the course.

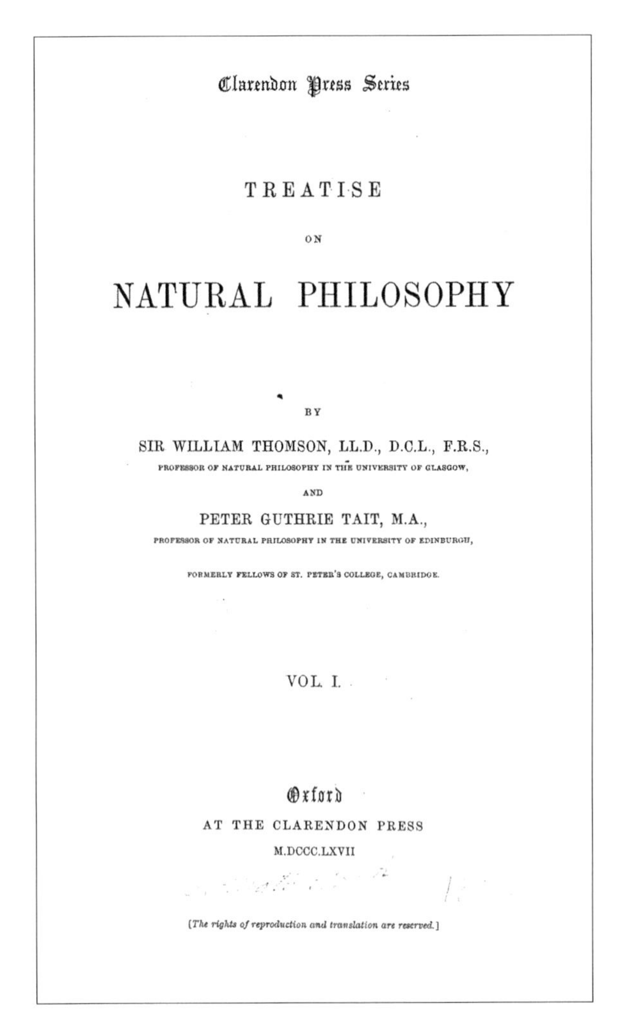
Clarendon Press Series

TREATISE

ON

NATURAL PHILOSOPHY

BY

SIR WILLIAM THOMSON, LL.D., D.C.L., F.R.S.,

PROFESSOR OF NATURAL PHILOSOPHY IN THE UNIVERSITY OF GLASGOW,

AND

PETER GUTHRIE TAIT, M.A.,

PROFESSOR OF NATURAL PHILOSOPHY IN THE UNIVERSITY OF EDINBURGH,

FORMERLY FELLOWS OF ST. PETER'S COLLEGE, CAMBRIDGE.

VOL. I.

Oxford

AT THE CLARENDON PRESS

M.DCCC.LXVII

[The rights of reproduction and translation are reserved.]

Fig.11.4. The work of 'two northern wizards'.

It was the *Elements* and not the *Treatise* that was recommended over the next 30 years for the introductory courses. Indeed, to the undergraduates of the day even the *Elements* was too difficult, with J. M. Barrie in his *An Edinburgh Eleven* saying that it was more familiarly known as the 'Student's First Glimpse of Hades'.[48] This reaction was echoed in a review of the *Treatise* in the *Athenaeum* of 5 Oct 1867 where the reviewer, possibly de Morgan, says:

If anything they have not sufficiently diluted the mathematical part with expanded demonstration. But what of that? The higher class students for whom this work is intended are rats who can gnaw through anything: though even their teeth will be tried here and there, we can tell them.

But, nevertheless, its value was recognized and given the highest praise by the reviewer in the *Medical Times* and *Gazette* of 16 November 1867 who says:

Should the three succeeding volumes at all come up in value to the present one, Thomson and Tait's *Natural Philosophy* will deserve to take place with Newton's *Principia* and Laplace's *Mecanique Celeste*. This is strong language, but not too strong.

A similar view was held by Helmholtz and Wertheim, who in the preface to their German translation of 1871, say:

The present volume will introduce to the physical and mathematical German public the beginning of a work of high scientific significance, which will, in the most excellent fashion, fill in a very perceptible gap in the literature of the subject . . .[49]

Thomson and Tait's *Treatise on Natural Philosophy* was one of the most important scientific works of the nineteenth century, and it is appropriate to conclude with some of Clerk Maxwell's comments in his review of volume one of the second edition in *Nature* Vol xx, 1879, published shortly before his death. It is particularly appropriate to finish with this quote since of the three men—Thompson, Tait, and Maxwell—it is Maxwell whose reputation stands highest today, and the quotation shows Maxwell's view of their achievement:

The credit of breaking up the monopoly of the great masters of the spell, and making all their charms familiar to our ears as household words, belongs in great measure to Thomson and Tait. The two northern wizards were the first who, without compunction or dread, uttered in their mother tongue the true and proper names of those dynamical concepts which the magicians of old were wont to invoke only by the aid of muttered symbols and inarticulate equations. And now the feeblest among us can repeat the words of power and take part in dynamical discussions which but a few years ago we should have left for our betters.

12

Kelvin on Atoms and Molecules

Elizabeth Garber

Lord Kelvin published his first paper on atoms and molecules in 1861 and his last more than 40 years later in 1907. Although he was unsatisfied with the models he developed, such persistence indicates the importance of the search. Most of his work on atoms and molecules developed in a search for a dynamical theory for the interaction of matter and the ether. His ultimate goal was to unite all known phenomena into one theory. However, no one model was ever satisfactory because the molecules' properties depended upon the particular phenomenon under examination. Each phenomenon, or cluster of phenomena, required different molecular properties, some of which contradicted those of other models. This methodology was at odds with his ultimate goal of a unifying theory. Considering specific molecules to examine one, or a range, of similar appearances seems inconsistent with his ultimate objective. Taken in isolation his publications on molecules and atoms do not seem to follow a coherent pattern of argument but rather appear as a series of papers to develop disconnected models of molecules and atoms. It is only in the context of his search for a grand theory connecting these disparate phenomena that his pursuit of these elusive structures begins to make sense. His ultimate goal of uniting all optical, electromagnetic, and even gravitational phenomena in one theoretical net was truly a grand theory of everything physical.[1]

Although a full range of mechanical models, Newtonian centres of force, hard spheres, point masses, systems of springs, and gyrostats eventually appeared in his publications he had, early in his career, rejected the notion that matter consisted of atoms. His commitment to continuum physics became visible in the textbook he co-authored with Peter Guthrie Tait, the *Treatise on Natural Philosophy*. He was supposed to include a section on molecules, but he never completed it.[2] Kelvin hoped to replace the physics of the ultimate parts of matter based on these ill-founded representations by the dynamics of the continuum.

Despite his early successes in the mathematization of electrostatics and electromagnetism, Kelvin adhered firmly to the belief that ultimate explanations in physics were dynamical. And, like his contemporaries, he also believed in the existence and mechanical nature of the ether. His early paper of 1847, that through shared mathematical forms, connected mechanical, electrical, and magnetic phenomena, also spurred his efforts to develop a theory of the ether that would connect all of them in one grand theoretical scheme.[3] Dynamics was the bedrock of his explanations of any and all phenomena. He had an extraordinary ability to visualize phenomena mechanically, but unless he could do so Kelvin could not understand them.[4] Kelvin explained the actions in one physical situation using the mathematics of a theory taken from another domain in physics and built analogies through the replications of mathematical forms. Thus, the rotational motion of a molecule embedded in the ether illustrated directly the mathematics of the rotation of the plane of polarized light in a magnetic field. He went further. The mechanical operations of the molecule in the ether explained the phenomenon of the rotation of polarized light in a magnetic field. From his earliest papers, the electrical and magnetic properties of matter were explained in mechanical language.[5] Atoms and molecules were agents whose complex motions were the link between electricity, magnetism, and the ether, which if successfully tied together would allow him to reach his goal of a general, dynamical theory of everything. Atoms and molecules and their interaction with the ether were also crucial to his confrontations with James Clerk Maxwell's electromagnetic theory of light. Kelvin's challenge was directed to Maxwell's idea of displacement and the displacement current for which he could not visualize any mechanical image.[6]

Kelvin also shared, with other nineteenth-century physicists, the belief that the same physical laws and mathematical forms applied to all scales of phenomena. The behaviour of a model built to human scale, such as a gyrostat, were replicated in the structures and behaviour of molecules on a much smaller scale. The physical structures and laws were the same. Therefore, molecular motions could be described from observations and experiments then applied directly to the invisible, molecular bodies.[7] He also assumed, along with his colleagues, that molecules and atoms were absolutely stable. The evidence lay in their ability to survive chemical change. However, for Kelvin, ultimate authority was vested in experiments, some of which he developed himself, or replicated in his own laboratory at Glasgow University. Complimenting this, Kelvin expended much effort in numerical calculations to secure arithmetically the implications of his theories, as well as calculating the exact fit of theory to experimental results.[8] Because of his commitment to experiment and to understanding exactly how the ether and matter interacted, his modelling had to respond to ever changing experimental results. This occurred in his efforts to replicate spectra with mechanical molecular motions and in his efforts to deny the equipartition theorem because it conflicted with necessary attributes of his models in the face of spectral data.[9] As the focus of experimental research on spectra and molecular structure developed his theoretical efforts were bound to be replete with failed models. His models were challenged, not just by the incorporation of novel numerical results, but by the results from new kinds of experiments. His theoretical models were always temporary. However, with an understanding of what was at stake for Kelvin in his analysis of the behaviour of molecules and atoms over a broad range of phenomena, we can understand his repeated assaults on the same problems and the energy and imagination he focused on the puzzle of the structure of atoms and molecules.

THE SIZE OF ATOMS AND MOLECULES

Kelvin's first foray into atoms and molecules was to estimate their size. Historically we accept John Dalton as establishing the existence of atoms. Yet, in the 1860s chemists in Britain still argued whether their discipline needed such hypothetical entities. In this decade Sir William Brodie avoided them by introducing his 'chemical calculus'.[10] However, this was also the decade in which Maxwell published his first two papers on the kinetic theory of gases and gave molecular meaning to Avogardro's number.[11] Thus, the first, independent estimates of molecular size by Josef Loschmidt in 1865, by G. Johnstone Stoney in 1868, and by Kelvin two years later, were part of an ongoing debate between some chemists and physicists.

In a letter to Joule, reprinted as a paper, Kelvin asked rhetorically why, if atoms were infinitely small, chemical reactions were not infinitely swift. Chemists could not answer this question and other equally important ones because they were 'barred' from seeing atoms as 'real portions of matter occupying finite space, and forming a not immeasurable, small constituent of any palpable body'.[12] He proceeded to use several physical phenomena that led to mutually compatible estimates of the approximate size of an atom. He first used an optical argument, adapted from Augustin Cauchy. The properties of water and glass at distances of small fractions of a wave length of light were quite dissimilar. Kelvin concluded that optics forced us to admit that:

> the distance from the centre of a molecule to the centre of a contiguous molecule in glass, water, or any other of transparent liquids and solids exceeds 1 / 10 000 of the wavelength of light, or, 1 / 200 000 000 cms.

He then deduced a value for atomic size from his own experiments on the contact electricity between zinc and copper, ending with an atomic distance of the same order of magnitude. Using Newton's Rings to estimate the thickness of a soap film he then considered the work done in stretching such a film from 1 mm to

1/20 000 000 mms, in thickness. The work done in stretching the film was the heat equivalent of over a thousand times that needed to raise the same film through one degree Centigrade. Such amounts of heat would vaporize the liquid. The excess must come from the 'contractile force of the film' because the film was several molecules thick. From this Kelvin concluded that several molecules existed in a film 1/20 000 000 mms thick. These experiments put a limit on the 'smallness of the ultimate particles of matter'. Finally, he turned to the kinetic theory of gases and used Maxwell's experiments on the viscosity of gases. He argued that, given the average velocities of molecules of oxygen, nitrogen, or air at normal temperature and pressure were about 50 000 cm/sec, and the average time between collisions, this led to an average path length. In a paper that quickly followed Kelvin used Clausius' work on kinetic theory where he assumed that this average path length was at most 5,000 times the diameter of a molecule. From this result he deduced that, for gases, the diameter of a molecule could not be less than 1/500 000 000 cms. The densities of solids and liquids are 500 to 16 000 times that of air at normal temperature and pressure and hence the number of molecules in a cubic centimeter lay between 3×10^{24} to 10^{26}, hence there was a 'very high probability' that for such bodies the distance between the centres of molecules lay between 1/100 000 000 and 1/200 000 000 cms. He later added further arguments from the dispersion of light and capillary attraction.[13]

VORTEX ATOMS

While Kelvin never completed the section on molecules for his text with Peter Guthrie Tait, he investigated molecular models in a series of papers. The model he developed in most detail was that of the vortex atom. While he had written on vortices earlier, his research into them became serious in the 1860s after he read Hermann von Helmholtz's papers on the same subject.[14] We can sense the importance he attached to this model in his correspondence to Tait, and in his efforts to contact Helmholtz on the subject, while in Germany in the summer of 1868.[15] The vortex fitted his commitment to a continuum model of nature, was stable, and purely dynamical. Initially, he was drawn to Rankine's vortex model for the atom. In the 1840s Rankine used the model to develop ideas on the nature of heat and to deduce his expressions for the laws of thermodynamics.[16] Kelvin's interest was coupled to his analysis of the Faraday effect. His initial conclusions were tentative. Whether matter was continuous, or molecular, or consisted of vortices, 'is impossible to decide, and perhaps in vain to speculate, in the present state of science'. He was only after a 'dynamical illustration'. By 1867 the vortex had become 'the only true atom.'[17] In these early papers his preliminary calculations showed that two linked, vibrating vortices could account for the two D lines in the sodium spectrum.

Helmholtz provided Thomson with a detailed analysis of the stability of this mode of motion.[18] Here was continuous matter in motion that acted as an independent entity; an atom. While Rankine's mathematical exploration of vortices was unsatisfactory, Helmholtz's proof demonstrated that a vortex in a fluid had many of the necessary attributes of atoms and molecules. They were mechanically stable, interacted with one another, yet after such interactions remained as they were before the interaction, as single vortices, or linked together in stable doublets or triplets etc. And there was a simple demonstration of their properties. Tait's smoke box, while not elegant, demonstrated them with simplicity. The vortex generator was a smoke-filled cardboard box, with a circular hole in one side. With a healthy thwack to a wall smoke rings emerged from the hole. They were stable, and interacted with one another causing vibrations in each other as they came close, or, intertwined yet moved apart and then remained single and intact.[19]

Kelvin described the dynamical characteristics of vortices in a series of papers delivered to the Edinburgh Philosophical Society in the same year, yet the formal mathematical properties of such physical entities only appeared in 1869. In these early papers Kelvin went over the same ground as Helmholtz only in more detail, and then went beyond him. Both used the analogy between fluid rotational velocities and magnetic forces, and with the laws of flow (force) around an axis (electric current). Space was a continuum filled with an incompressible, frictionless fluid. Any material phenomena depended solely on the motions within this fluid. Kelvin demonstrated that outside the axis (current) the system of flow was irrotational and involved no 'molecular rotations'. There were no rotational motions within the walls of the vortices that swirled around the central axis. Kelvin's addition to the theory was in his analysis of multiple continuous spaces, that is, knotted vortex tubes, many of which bring to mind knotted Celtic motifs.

He established the uniqueness of any irrotational motion generated by any means after the vortices had been established, and that the same work was done to produce the same fluid motion independent of the pathway to that motion. Kelvin's theory appeared to be supported by experiments when Frederick Guthrie demonstrated that a vibrating tuning fork attracted nearby light objects. Kelvin took these results as an illustration of attraction due to vibration, as expected from his theory of vortices. It was an illustration of the feasibility of a theory of matter without the concept of force.[20]

In the early 1870s it looked as if a grand theory based on vortex atoms might succeed. In statics, forces were banished because 'we know of no case of true statics in which some, if not all of the forces are not due to motion', as well as in liquids, elastic solids, and in the 'hydrostatics' (kinetic theory) of gases. It even seemed to explain gravitation, not by action at a distance, but by the actions 'of intervening matter'. Another attempt in this direction was connected with his resurrection of P. L. LeSage's gravitational theory, developed to avoid the action-at-a-distance theory of Isaac Newton.[21] LeSage's universe consisted of a gravific fluid made up of extremely small particles. Matter consisted of much larger structures, empty except for the bars that defined their shape. Unlike LeSage's system, Kelvin's had to obey the law of conservation of energy. Therefore, when the ultramondane particles collided with ordinary matter some of their translational energy was converted into vibrations and rotations. Only a small minority of such particles would be subject to these conditions even through long periods of time. Essentially, gravity would be unaffected. Kelvin then discussed some of the problems of kinetic theory of gases and claimed that the difficulties of LeSage's theory were no worse than those of the kinetic theory of gases. This excursion into such hypothetical areas demonstrates the lengths to which Kelvin would go, and his boldness of asserting these views publicly, in his attempts to complete his grand theory.

This first foray into a theory that potentially encompassed all phenomena was short lived. LeSage's theory was quietly dropped, although it connected with Kelvin's metaphysical and theological predispositions. The other focus of his attention, the vortex, was also becoming problematic. The analogy between the rotation of a magnetic field about an electric current and the distribution of hydro-kinetic velocity did not work. The magnitude of the mutual action between two magnets and vortex flow was identical in quantity, but the mechanical forces acted in the opposite directions. When magnets attracted the equivalent flow systems repelled each other.[22]

Kelvin was not alone in investigating the mathematics and physics of vortices. They became the subject of the Adams Prize Essay in 1882 in the form of a 'general investigation of the action upon each other of two closed vortices in a perfect incompressible fluid'. J. J. Thomson's winning essay was published in 1883. His investigation was mathematically more complete and more rigorous that Kelvin's[23] He also developed a series of cases to show that Kelvin's analogy between the magnetic and dynamic case was insufficient. Kelvin had taken the experimental demonstration of A. M. Mayer, who had investigated the stability of various configurations of long, thin magnets floating upright on water subject to the attraction of a fixed magnet. Kelvin had developed a mathematical analogy and concluded that if such magnets arranged at the corners of an equilateral triangle, a square, or a regular pentagon were stable, vortex columns arranged in the same ways would also be stable. J. J. Thomson concluded that this was not true for complexes of vortices of seven or more.[24] He also demonstrated that Kelvin's attempts to construct a vortex model of a gas were only valid for monatomic gases, whose energy only increased with an increase in the diameter of its vortex atoms. He then developed a theory of chemical valency, based on such atoms.[25] This gave added impetus to Kelvin in his critique of the kinetic theory of gases.

In the 1870s Maxwell had investigated vortices, along with all other available mechanical models of atoms, finding them all wanting.[26] A gas made up of vortices was as subject to the equipartition theorem as any other molecular model, and to the same problems when faced with the experimental values of the ratio of the specific heats of gases. Such molecules would transform all their energy of agitation into internal energy and their specific heat would become infinite. In addition, a vortex was a mode of motion, inertia a property of matter. However, the vortex model of molecules was too important for Kelvin to relinquish it easily. In a system consisting of an infinite number of such molecules the second law of thermodynamics was never violated. He also connected the stability of his molecules to his cosmogony.[27]

KELVIN CONTRA KINETIC THEORY

Although he used some of its experimental results in his early estimates of the size of atoms, Kelvin now opposed kinetic theory. He had expected that gas theory would be subsumed under his continuum theory. In particular he pointed out that after many collisions between the 'elastic-solid' molecules of Clausius and Maxwell, their translational energy would be transformed into vibrational energy. The kinetic energy and hence the temperature of the gas would plummet. Kelvin's alternative was the continuum vision of vortex atoms. In addition, kinetic theory was based on molecules that were either perfectly hard bodies, or centres of force that behaved as if their collisions were perfectly elastic. Kelvin always built his models of matter from a defensible image of its constituent particles. He could not accept different theories of:

> heat, light, diffusion, electricity and magnetism in gases, liquids, and solids, and the describing precisely the relation of these different states of matter to one another by the statistics of great numbers of atoms, when the properties of the atom itself are simply assumed.[28]

Kelvin began a campaign to undermine the statistical foundations of kinetic theory with a series of 'Test Cases'. These cases appeared through the 1880s, his arguments being countered by other demonstrations showing flaws in his reasoning. He was, for different reasons, as distressed as Maxwell about the equipartition theorem which seemed to doom kinetic theory. Yet, for most physicists in the late nineteenth century, kinetic theory was too useful to discard and the insights that it offered into the molecular structure of gases too important to ignore. Kelvin's reaction to this wanton disregard to the fundamentals of dynamics at the heart of kinetic theory led him, in the 1880s, to propose that the equipartition theorem was wrong. Rather than persuading his colleagues, this simply alarmed the community of theoretical physicists, including Tait. He was met repeatedly with arguments that pinpointed his errors in statistical reasoning. Indeed, in 1891 Lord Rayleigh showed that Maxwell's energy distribution function was correct and in 1900 that the limitations of kinetic theory lay not in its statistical reasoning but in dynamics.[29] The root cause was in the tools of the trade, not the equipartition theorem. Changing the model of the molecules of a gas would not help. Even after he became disenchanted with vortices because they could not explain inertia or gravitation, Kelvin persisted in his campaign against the equipartition theorem and the kinetic theory of gases. He also developed a demonstration that showed that vortices were not stable.[30] However, he continued to try and evade the inevitable results of the equipartition theorem. Finally, he declared that we should simply 'deny the conclusion' and thus lose sight of one of the 'Clouds' that obscured the brilliance of nineteenth-century achievements in the molecular theory of heat and light.[31]

GYROSTATIC MOLECULES

In 1875, at the same time that he was investigating vortex atoms, Kelvin introduced the gyrostatic molecule. He defined it as a 'rapidly rotating flywheel, frictionlessly pivoted on a stiff moveable framework or containing case'.[32] A stretched chain of such gyrostats enabled him to model Faraday's Magneto-Optical results. He used them again to address the same phenomena in the Baltimore Lectures of 1884. However, in the latter case the gyrostats were contained within a spherical cavity in the ether, and the flywheel of the gyrostat rotated with enormous rigidity.[33] With this model he could, again, replicate the Faraday effect. However, the model proved insufficient. The rotation of the plane of polarized light should change as the inverse of the square of the wavelength of the light. It did not, and the gyrostatic model was finally abandoned.

By the time the first edition of the *Baltimore Lectures* went to press it included an 'improved' gyrostatic model. In this later improvement the simple flywheel had become a system of two gyrostats connected together by a ball joint and mounted within the ether cavity. Initially the gyrostats, and the ball joint, were at rest and in a straight line.[34] The new molecule, if embedded 'adhesively' in the ether was more efficient in its coupling with the ether and its representation of the Faraday effect. The gyrostats only experienced translational forces in the line through their centres, and, the whole system would turn as a rigid body. The upshot of the mathematics of the system was that the model could account for the 'anomalous dispersion' in the magneto-optical effect. He used this gyrostat mainly to illustrate the propagation of a circularly polarized wave through the ether, and thus it was useful in thinking about the propagation of a magnetic wave through a medium. In general, gyrostatic molecules interacted with the wave motion passing through the ether which set them in motion. The internal motions of the gyrostats generated vibrations that were of different frequencies from those of the ether, which were then passed back to the medium. Such a system might account for the absorption lines in, say, the spectrum of sodium. In this case the model worked well and Kelvin could report in great detail on the results of such oscillations and compare them to the results of experiments on the sodium D lines.

However, the usefulness of gyrostats did not outlast the 1880s. Indeed, in the 1890s Kelvin's molecular models took a different turn. From this decade onwards there was much more of a threat to Kelvin's dynamical explanation of physical phenomena. Although, introducing Maxwell's work into the Cambridge Tripos had not been easy,[35] his electromagnetic theory had been adopted by a growing number of physicists and electrical engineers in Britain. Electromagnetism began to challenge dynamics as the explanatory language in some domains of physics. While not triumphant in Britain, by 1900 electrodynamics became the explanatory language of choice for some physicists in the German Empire. Not just a particular model, but Kelvin's assumption of an ultimately dynamical explanation for all physical phenomena was being compromised. Even in Britain the reign of the kind of thinking Kelvin's physics represented was coming to a close. However, Kelvin continued with his mechanical vision of nature even as the foundations of theory shifted around him.

THE BALTIMORE LECTURES

In the middle of this decade in which the attention of theorists in Britain shifted towards Maxwell's electromagnetic theory, rather than challenging it, Kelvin was invited to lecture at Johns Hopkins University in Baltimore. The Lectures grew out of a desire to put the fledgling Johns Hopkins University on the academic map, and to give more of a sense of identity to the scattered, yet growing community of physicists in the United States. After some negotiations, the topic of the series of 20 lectures to an invited audience was decided as Molecular Dynamics.[36] Kelvin interpreted this subject matter as the production of all known optical phenomena by the interaction of matter with the ether. Thus he could directly challenge

Maxwell's electromagnetic theory of light. He conducted the proceedings as a research seminar, focusing on the difficulties of the theory, rather than going over known results. The twin foundations of the lectures were the theory of the propagation of waves through elastic solids of various types, and the construction of mechanical molecular systems coupled to the solid ether. The internal motions of the mechanical molecules, passed to the elastic solid, were the source of all the phenomena of light, from simple reflection and refraction to complex spectra. Some lectures were divided into two parts by these topics. In later lectures the discussion of one or other of these topics predominated. He included discussions of the limitations of the theories of elasticity of mathematicians such as Green or Cauchy, and his alternatives to them. The Lectures were a bully pulpit for Kelvin to promote his vision of theoretical physics in general and his ideas on what he considered the most important theoretical issues facing the discipline.[37]

The molecular models he pursued in these lectures were variations on a system of springs within a spherical shell in the ether, gyrostats, and combinations of these two models, depending on the phenomenon under discussion. His discussion of molecular models was less tidy than his development of the theory of the motions of elastic solids of various kinds, because no one molecule could encompass all the known phenomena of light. In the second edition of the *Lectures* the molecular models became more numerous and complex where the electron and radioactivity etc., had to be incorporated into his theoretical net. He assumed that all his molecular or atomic models were far smaller than the wavelength of light.

The molecular model Kelvin introduced in the first lecture, and spent more time discussing than any other, consisted of a spherical cavity within the ether, with either a rigid shell, or coupled elastically to the ether. In the shell were up to three other concentric shells coupled to each other and the lining by zigzag springs. At the centre was a heavy mass, either stationary or movable.

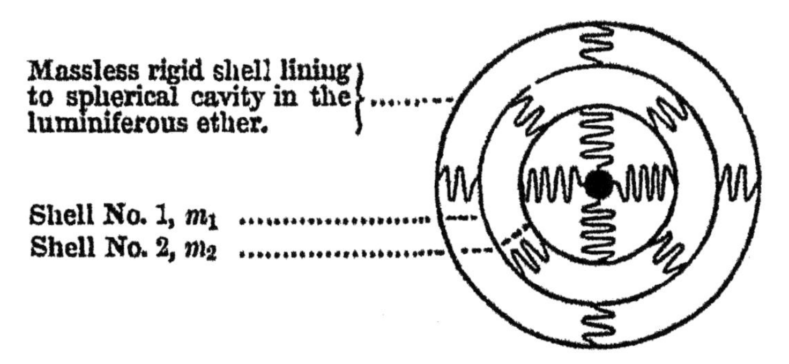

Fig.12.1. Model of a spring-loaded model of a molecule, embedded in the ether. (From Lord Kelvin, *Baltimore Lectures* (1884), Lecture I, p. 13.)

In 1884, he used this model to explain polarization by reflection, double refraction, and anomalous dispersion. The details of the inner structure, and how the model coupled to the ether, depended on the optical phenomenon under discussion. The shells transmitted to the ether the kind of motions that would lead to a particular optical phenomenon. Where available, these motions had to replicate known results. Thus Kelvin used this model to explain refraction. In this case, after he developed a theoretical model of the motions of the ether, Kelvin had a formula that he admitted, in 1904, contained enough disposable constants to replicate the index of refraction 'by an empirical formula, as it were'. His formula was 'ample for representing the refractive index of ordinary, transparent substances'.[38] However, there was a problem. When the vibrations of the ether were close to the fundamental frequency of the molecule, the refractive index of the substance became negative. In addition, the model had to cover the full range of experimental results, from the ultraviolet to the infrared, which did not seem unlikely in 1884. This had changed by the date of publication of the second edition with developments in experimental physics. However, this model could demonstrate the absorption lines in spectra.

In the later edition of the *Lectures*, Kelvin returned to the problem of double refraction and introduced a new molecular model. He considered two overlapping, embedded molecules and a new consideration. The continuum of the ether was traversed by molecules that behaved like those in gases. Spherical shell atoms within the ether overlapped each other and within these overlapping atoms the ether suffered symmetrical condensations and rarefactions. Kelvin argued that in the overlapping volume the force of the ether on atoms brings them together in an apparent attraction, balanced by the repulsion between the atoms themselves. It seemed 'not improbable' that these forces were involved in the equilibrium of two atoms in known diatomic molecules. To justify these assumptions, Kelvin pointed to isotropic crystals with atoms in a cubic array. The forces of the ether pushing the atoms inwards are balanced by the repulsions exerted on the ether by the atoms next to it. If we imagine forces applied outward along the x-direction, and inward forces on the array in the y-direction, atoms would overlap. While this was an interesting mathematical case, physically Kelvin could not explain the forces within the atoms. In addition, any disturbance within the ether, however small, should produce double refraction around the crystal. This should be perceptible, but alas remained undetected.[39] Kelvin did tackle double refraction with a full mathematical investigation of spherical waves originating in the application of force to an elastic solid within a limited space. He traced the physical implications of the application of different kinds of forces to keep the shell moving in simple harmonic motion. From the rigid shell, waves propagated throughout the elastic solid ether. However, for any known optical phenomenon the ether needed to execute many millions of vibrations per second. Yet there must be viscosity in the interactions of the molecules with the ether. If so, molecular vibrations should die too quickly for any optical phenomenon to be visible. The other alternative was 'a theoretic violation of the conservation of energy'. If there was no viscosity, then whatever the energy of vibration of the molecules, they must become vibrations in something else, but what? Somehow there had to be vibrations, with arbitrary periods without viscous consumption of energy. This, in Kelvin's opinion, was possible. Uranium glass fluoresced for several thousandths of a second after the light source was removed.

In the Baltimore Lectures the motions of his theoretical models were themselves modelled by linear vibrations. Initially, a vertically hung series of springs and masses was installed in the lecture hall. Later, this was joined by a 'wave machine'. The linear system of springs and masses yielded the possible vibrations in the system when masses and spring coefficients differed. The 'wave machine' also made visible the motions of the system when certain modes of motion dominated. The operator set the heavy bar, H, in motion and the period of this bar was altered by the positioning of the rings that connected it to the suspension system. The masses would eventually vibrate in their simple harmonic motions. Superimposed upon these was the fundamental motion of the heavy bar, H. The motions of the smaller masses died down more quickly than that of the bar and Kelvin produced the spectacle of the smaller masses vibrating at the natural frequency of the heavier bar. He could thus demonstrate a mechanical system that might set the ether vibrating at particular frequencies

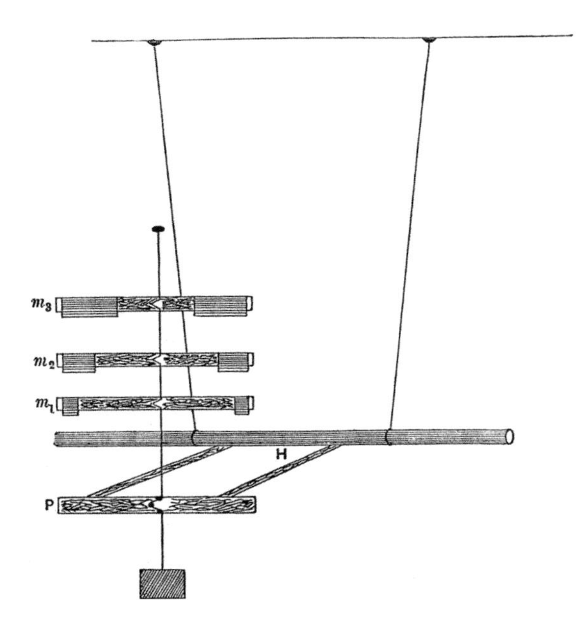

Fig.12.2. Bifilar suspension system, used to illustrate the motions of molecules embedded in the ether. (From Lord Kelvin, *Baltimore Lectures* (1884), Lecture XIV, p. 215.)

to mimic the production of particular spectral lines. Conversely, by suppressing certain frequencies he could demonstrate the production of absorption lines in a spectrum. Kelvin also got the masses in the wave machine oscillating in their natural periods. These motions dying away gradually was an illustration of fluorescence.

While both the vertical system of masses and springs and the 'wave machine' could illustrate molecular motions, the latter system also demonstrated the complex motions of his molecular model that would be of 'enormous difficulty' to obtain algebraically. Visual experience replaced, indeed for Kelvin was more reliable than, theoretical calculations.[40] The linear model was there to help make the mathematics 'plain'.

In the *Lectures* the mathematical modelling of molecules was based on the linear, vertical system of zigzag springs of different coefficients of elasticity separated by bodies of different masses. With this theoretical system various coefficients and conditions were varied to imitate the multiple vibrations of heavy, elastic atoms in, and their couplings to, the ether. Much time was spent on expressing the motions of this linear system algebraically. The motions examined included those that were equivalent to dilations and contractions of the spheres, their oscillations in a straight line, or two spheres oscillating linearly along the line of their centres.

Kelvin moved from a discussion of the wave machine's vibrations, to the possible modes of vibration in the molecular model and back again, then to the vibrations of the ether in the case of the phenomenon under consideration where the modes of vibration of the model and molecule become crucial. With such a back and forth between theory and physically present molecular model Kelvin examined phosphorescence, where the molecules returned vibrations, with changed frequencies, back to the ether. However, energy dissipation remained a problem even after many tens of millions of vibrations. He had to assume that the elasticity of the molecules was perfect, because he could not allow viscous terms. By moving the linear system so that the motions of the largest masses in the system were superimposed on one another, Kelvin could suggest how spectral lines appeared; the D lines of sodium being his particular example. This presented yet another problem. The energy of the sodium molecules must increase enormously as the amplitudes of particular wavelengths overwhelm the others. This kind of motion could also be a perfect illustration of double refraction if only one could visualize how the molecules acted upon the ether. However, since he could not visualize this interaction with his models, he could not explain double refraction. The problem here lay, not in the vibrations of the molecules, but in those of the ether. The illustrations of the molecular motions were only partial, and seemed to highlight the difficulties of theory, not its strengths.

Another problem discussed at length was that of dispersion. In the second edition of the *Lectures* (1904) Kelvin introduced a simplified spherical shell model, a central mass connected to the shell by springs. In this case the mass at the centre had to be 'enormous'. Molecules collide and vibrate, and the central mass moves in one direction, the shell in another. The motion of the shell is passed on to the ether, and Kelvin treated this as a spherical wave travelling outwards from a source. However, there was another problem, the model failed in just the visible section of the spectrum.[41]

All the models shared a fatal flaw. There should be a, never detected, condensational wave as well as transverse waves radiating out from the molecules. Eventually Kelvin abandoned the general model of a sphere embedded in the ether. The frequencies necessary to produce the phenomena required a ratio of the mass of molecule to that of the ether which seemed highly unlikely, and the range of frequencies which had to be covered by the model required improbable molecular spring structures. Even so, while the difficulties of a model might be discussed as insurmountable in one lecture, they would reappear in the next.

BOSCOVICHEAN ATOMS

None of the molecular models Kelvin proposed during the Baltimore Lectures led to any long term results. In the following decades he did continue to develop other molecular models that were influenced by the experimental confirmation of the existence of the electron, and the growing importance of Maxwell's electromagnetic theory. His final forays into molecular structures included action-at-a-distance models. He was also looking for a mechanism to explain radioactivity. In the 1890s he focused on molecular properties as well as their interactions with the ether. In this search he returned to a long neglected centre-of-force model of matter developed by Father Roger Boscovich in the mid-eighteenth century. At the centre of the atom the force was repulsive. At a certain distance from the centre the repulsive force changed to attraction. Thus the atoms acted as solid particles. At an even greater distance from the centre the action returns to repulsion and so on.[42] Kelvin had condemned Newton's centre of force atom early in his career. Boscovish's model was more defensible, and he used it to visualize the equilibrium of atoms in crystals. Matter was made up of Boscovichean atoms arranged geometrically. His discussion of crystal types was based on principles laid down by Bravais. Any homogeneous solid was seen as a multiple of crystals, that is, had lines of symmetry distributed randomly in space, equally in all directions. Kelvin distributed points homogeneously throughout space, surrounded by cells, each cell being a hollow polyhedron. Bravais' protocol for such arrangements regarded the plane of the symmetrically distributed points as networks. Slippage along such planes could account for the twining of Iceland Spar and changes in crystalline structure.[43] Promising a full mathematical treatment later, Kelvin demonstrated with models and mechanical reasoning how physical crystals could be accounted for by Boscovichean atoms in various periodic structures. He went on to build crystals of different structures that reflected light selectively, and explained how they might sheer along certain planes etc. Three years later, and with more geometric details, he elaborated on these ideas in his Boyle Lecture at Oxford. He included a discussion of the closest packing of cells, a polygon of fourteen faces (a tetrakaidekahedron in Kelvin's nomenclature). Kelvin explored Boscovichean centres of forces further to tease from the geometrical structure the pyro- and piezo-electric properties of quartz. These atoms gave Kelvin a new foundation for a theory of matter in all its forms, including gases. Using them as free particles he once again challenged the Maxwell-Boltzmann distribution of energy in gases. The challenge did not stand.[44]

Kelvin extended this work on the centre of force atoms into their possible interactions with the elastic solid ether where the atoms and the ether occupy the same space. Atomic forces only act immediately on the ether within the atom where the density of the ether increases and decreases radially, but on average the density of the ether within and outside of the atom is the same. To explain the refractive indices of isotropic bodies such as glass or water he assumed a particular distribution of ether within the atoms when at rest. Kelvin traced the patterns of density changes and the possible density patterns in the ether through and around such atoms. He was able to deduce refractive indices, although not quite on target, for oxygen, nitrogen, and argon. However this could all be adjusted by changing the constants in his formulae. He also recognized that the results of the Michelson-Morley experiment were fatal to his theory. But even here he could argue that their results indicated that the ether, at least close to the Earth, was at rest with respect to it. Anyway, the Lorentz contraction made these results moot.

ELECTRIONS

Kelvin resurrected the Boscovichean atom for the last time in 1902. Again the ether was both within and around his atoms, although the forces of repulsion and attraction came from 'neutralizing electrions'. The 'electrion' was a charged particle that at times had the properties of the electron, sometimes it had a positive charge. The ether was compressible, but the inertia of crystals was no longer isotropic, a condition that was formally rejected as inconsistent with the conservation of energy.[45]

Other physicists in the 1890s also incorporated the electron into a grand theory of the ether and matter, yet still maintained an ultimate, mechanical explanation of such phenomena. Between 1893 and 1897 Joseph Larmor developed a new theory of the ether, whose ultimate structure was a gyrostatic, adynamic, solid with rigidity and with the free motion of matter through it. His atoms of matter contained electrons and he explained magnetism as the rotations of such electrons in the atoms, not in the ether. In his mechanical explanation Larmor considered the electron as a centre of strain within the homogeneous ether.[46]

Kelvin also returned to the motion of a body acting on the ether by attraction and repulsion.[47] The density distribution of the ether within the atom was replaced by 'a special class of atom'. Unlike electrons Kelvin's 'electrions' could be positively as well as negatively charged. Clearly they are not part of the atoms but somehow reside within them. He considered a positive electrion as a unit that attracts the ether 'into the space occupied by its volume'. The negative electrion rarifies the ether within itself through repulsion. However, the final explanation of these repulsions and attractions were stresses in the ether. If different portions of the ether of different densities repel each other, the law of repulsion depends on the difference in the density of the ether within the atom from that outside the atom. The ether consists of electrions moving freely through atoms embedded in it. The atoms of the ether are much smaller than those of matter. Ordinary matter consisted of spherical molecules with positive electricity uniformly distributed throughout its volume as well as negative electrions. He then considered the case of two overlapping atoms, containing one electrion each. The electrions are displaced from their previous, central positions because of their mutual repulsions and the attractions from neighbouring atoms. The displacement depends on the relative radii of the atoms. Without mathematical details, Kelvin argued that this atomic model and the displacement of electrions, would suffice to describe oxygen, nitrogen, hydrogen, carbon monoxide, and sodium chloride, if each single atom had one electrion. Kelvin then described the behaviour of atoms containing two or more electrions, and finally atoms full of them. Within the atoms the electrions are distributed uniformly, unless in the assemblage of molecules, such as an equilateral triangle, the electrions at the surface would rearrange themselves to reestablish equilibrium. He then connected his models to the electrical properties of finite bodies. The external electrostatic force depended upon the number of electrions and the radius of the atom. This led to an expression for the electric moment per unit volume in terms of the number of electrions and the radius of the atom. Kelvin then applied this to crystals. Most of his argument is verbal, or in simple algebraic terms, not detailed examinations of each case. He deals with what seems physically, and on straightforward electrostatic arguments, likely.[48]

He also used electrions in atoms to describe the propagation of waves through the ether. When electrions, one at the centre of each atom, moved, equilibrium was reestablished by the motions of electrions and the ether. The displacements of the ether and electrions replaced the motions of the mechanical models discussed in the Baltimore Lectures of 1884. Rather than the motions of complex spring systems and masses, an electrion acted directly on the ether through its electric charge and thus, while the motion of the ether was still mechanical, its underlying cause was now electrical.[49]

EXPLAINING RADIOACTIVITY

In some of the final papers in his long career Kelvin used electrions to account for radioactivity. This connected with a longstanding objective of his grand theory, to understand the interaction of matter and the motion of the ether. The vibrations in atoms, whether they contain electrions or not, produce a continuous spectrum, without the bright lines seen in the spectra of ordinary matter. The collisions of atoms of a 'glowing' gas must give rise to the latter through the vibrations of the molecules of the gas which are not available from its atoms. These spectral vibrations occur via collisions between the atoms of the gas and the motions of the electrions within those atoms. Radium atoms are special because they store up far more energy than other atoms from the electrions within. He reused an earlier atomic model where the electrion is attracted to the atomic centre. The atom is stable when the electrion with small potential energy is close to the boundary of the atom, or at its centre, where its potential energy is greatest. When 'loaded' in this way the electrion is within a region of stability for the atom. He assumed that in crystals of some compounds of radium, atoms are initially 'unloaded' then quickly become 'loaded'. This requires work, taken as heat from the crystal. When the atoms become unloaded electrions are projected out of the atoms with enormous velocities, so high in fact that they leave the crystal and enter the space around it where they are detected. This process continues to a maximum, then decreases to zero. The discharge of electrions forces unloaded atoms out of the crystals where they appear as 'alpha radiation'. This will not produce much heat compared to the production of beta particles. The latter produce the heat that fuels the continuation of the process. This could continue indefinitely without violating the laws of conservation of energy or storing potential energy in the radium atoms.[50]

In a further attempt to understand the radioactivity of radium, Kelvin abandoned the uniform distribution of positive electricity throughout spherical atoms.

The distribution was now a concentric strata of positive and negative electricity of equal electrical density. The whole had a positive charge. The process of loading, unloading, and the energies involved were the same. The results of experiments with polonium required slightly different explanations, but the electrions remained as bodies within, but not of, the atom. The molecules of polonium were represented by two overlapping spheres that contained four electrions each, with an additional one encapsulated within the overlap. This equilibrium was unstable and if one atom moved away from the other even slightly the encapsulated electrion would vibrate, sending waves out of the atom into the ether. The two atoms, now under the influence of a net repulsion, would separate with an extra electrion in one of them. The atoms' zone of stability was so narrow that only a very small disturbance could set atoms flying apart, one with an extra electrion. He also postulated mechanisms by which the atoms might once again become electrically neutral. However, this model could explain only the known results from polonium; the fate of radium atoms under the same theory was somewhat different. These were spheres with only ¼ electrion's quantity of positive electricity at their centres. At opposite sides of the atom on the same radius are one electrion's worth of negative charge. To neutralize this atom simply take a positively charged atoms of very large volume, 7/4 electrion's worth, and place it concentrically with the initial atom. Out of this we get two atoms and two electrions in a stable position with electrions within the boundary of the smaller atom. Here are two neutral atoms which 'if put together in any substance would give it the alpha and beta properties of Radium'.[51] Kelvin's engagement with radioactivity and electrions suggest that he felt the need to encompass new phenomena within his explanatory net. Thus his theory was never a finished product but always in process. Models long discarded could be resurrected and developed in other directions. This did not mean that he discarded his ultimately dynamical explanation of such phenomena.

CONCLUSIONS

At the Jubilee celebration of his professorship at Glasgow University Kelvin described these efforts at a grand theory as a 'failure'. In 1896 he claimed that he knew no more about electricity, magnetism, or of the relationship between ether and matter, or of chemistry, than when he had begun to teach, and this despite attempts to develop models of molecules and atoms for over 40 years. For all the energy and intellectual passion that he channelled into solving the problem of the structure of matter and its interactions with the ether, his ultimate explanation of their properties remained in the same set of images and explanations that he began with—the language and imagery of dynamics. The electron remained a foreign entity, both inside and outside the atom, even after the confirmation of electrically charged particles within atoms.

This mindset was fixed early in his career. We repeatedly find the same phenomena are crucial in the development of his ideas on atoms and molecules. He did not stray far from the research program he sketched out in his notebook of 1845. Faraday's experiments on magnet fields and the rotation of polarized light appeared most often as a crucial phenomenon that required explanation. All his molecular and atomic models had to model this experimental result. This was coupled to his early conviction that dynamics was literally the only conceivable theoretical language and imagery for the ultimate explanation of all phenomena. His intellectual agility lay in his dynamical understanding of mechanical systems and the analogies he drew between them and electricity, magnetism, and the flow of heat. He, together with Peter Guthrie Tait, had shown their mastery of these mechanical concepts and the development of this theoretical structure with their text on the subject.

Within these limitations, however, Kelvin displayed an enormous range of intellectual agility, and an even deeper understanding and ability to visualize how mechanical systems work. Thus, he saw through his models, and as they failed he conjured up others with which he pushed explanations that bit further, but none to his satisfaction. He would treat one subject within a short period of time using more than one model. While not alone in his quest, his was the most persistent and longest of any nineteenth century physicists engaged in such endeavours.

While brilliant, he had his blind spots, the most durable being kinetic theory and the electromagnetic theory of light. He could not think statistically, nor could he accept the idea of electric displacement. The individual molecule or atom was the carrier of properties that became visible through experiment. The kinetic theory ignored the attributes of individual, ultimate parts of matter. In general, he voiced admiration for the electromagnetic theory of light, but could not accept it because he could not visualize displacement mechanically. Neither of these theoretical developments, which his colleagues and historians see as turning physics in fundamentally new directions, changed the nature of his quest. They well might present him with new phenomena to discuss and new theoretical directions to explore, but only within the language of matter in motion. In the last decades of his life one has to wonder just how many of his colleagues agreed with his assessment of the most urgent research problems of physics. Probably they were closer to Henry Rowland who, as Professor of Physics at Johns Hopkins, hosted Kelvin's Baltimore Lectures. He had already decided that the electromagnetic theory of light was the only one of use in explaining just the phenomena Kelvin was discussing. Whatever physicists thought of his solutions, with his lectures at Baltimore Kelvin left physics and physicists in the United States in very different condition. American physicists at the Lectures created a sense of collegiality and of belonging to a profession, even though they returned to Colleges scattered across thousands of miles of territory. Kelvin had confirmed their sense of professional identity by the very nature of his lectures, discussing problems to be solved, not the results of research done elsewhere. In these lectures and throughout his professional life Kelvin could always be counted on to confront difficulties within his chosen field, and not get diverted into secondary problems.

13

Kelvin and the Development of Science in Meiji Japan

Colin Latimer

INTRODUCTION

The modern era in Japan is normally considered as beginning in 1868 when the feudal age, or Edo era, finally ended. Under the rule of the Edo Shogunate headed by the Tokugawa family, the whole country had existed in a state of peace and virtual isolation, essentially sealed off from the rest of the world, for over 250 years. The country was ruled by the *bakfu* government in Edo (Tokyo) while the Emperor continued to reside in Kyoto. Japanese who left and returned were executed to prevent the introduction of outside ideas. Society was highly organized with clearly defined classes and a similar fate awaited those who failed to carry out orders.

Under the isolationist policy, foreign trade and literature were strictly prohibited. Japan was an agrarian non-industrialized nation and the construction of ocean-going ships was banned until 1853. The traditional shipbuilding technology was suitable only for the construction of small wooden boats required for coastal navigation[1]. Financial difficulties exacerbated by natural disasters, including large scale famine, followed by foreign pressure for Japan to open its doors to trade, eventually resulted in the Shogunate stepping down. The Emperor Meiji declared the restoration of Imperial rule in January 1868 (the Meiji restoration) and he and his entourage transferred from Kyoto to Tokyo (Edo was renamed Tokyo) in September 1868. The following year the Diet was established and a constitutional monarchy headed by the Emperor was formed.

The new and globally ambitious Meiji government quickly realized the importance of science and technology. They introduced a programme of importing expertise on short term contracts from the west, called *oyatoi gaikokujin,* or 'honourable foreign employees', generally referred to simply as *oyatoi*. This programme was especially important in shipbuilding which was urgently required for globalization, and became the leading sector industry through which the Japanese gained access to the advanced technology of the day[2]. They also recognized from the beginning that applied science and engineering were essential prerequisites for success, and here again the *oyatoi* programme played a vital role. One of the first schools set up in 1873 was *Kobu Diagakko* (The Imperial College of Engineering), run by the Ministry of Public Works to promote industrial education, which 14 years later was amalgamated with the Science School of the University of Tokyo to form the Imperial University of Tokyo.

A major problem was the shortage of teachers capable of teaching advanced courses. Scottish merchants, businessmen, and bankers had, however, fostered links with Scottish universities which had become highly regarded by the Japanese[2]. Furthermore in the realm of pure and applied science, Sir William Thomson (later Lord Kelvin) was clearly considered an intellectual giant of international renown and:

> around Thomson clustered a galaxy of professors, as pupils, teachers and business partners—Lewis Gordon, Fleeming Jenkin, P G Tait, Alfred Ewing—active in university science in Scotland and entrepreneurship in the electrical industry. Products of the Scottish universities, and usually of the Cambridge Mathematical Tripos also, were amongst the finest flower of Scottish Victorian culture, and in their presence the Scottish universities were confident enough to hold to the traditions[3].

Kelvin's applied research interests included those related to shipbuilding and navigation. He had 11 patents including his mariner's compass, depth sounding device, and his tidal machines. His 45 others included topics such as telegraphy and electrical instrumentation. He had formed companies to exploit his inventions. All this was exactly what the Japanese wanted to foster and it was only natural that they should seek Kelvin's involvement in the appointment of teaching staff. As we shall see he was happy to oblige and essentially all the *oyatoi* appointments in natural philosophy and engineering were made on Kelvin's recommendation. In addition to their teaching duties, these *oyatoi* founded a number of research programmes. After 1879 when the first Imperial College of Engineering students graduated, some were selected to go abroad for further study. Not surprisingly a significant number ended up in Glasgow to work with Lord Kelvin, and on their return home many were to take up important positions in the new Japan.

KELVIN'S PROTÉGÉS IN TOKYO

The two days of celebrations convened by the Royal Society in London in 1924 to celebrate the centenary of Kelvin's birth 'in a manner befitting the memory of one to whose achievements mankind are so greatly indebted', included addresses commemorative from all over the world[4]. The presentation that came from Japan stated:

The National Research Council of Japan specially desires to recall with gratitude the interest he took in developing physical science and in encouraging research in this part of the world. At the time when Japan was remodelling her education on modern lines, she was fortunate to have his eminent disciples as organisers. Dyer, Gray, Ayrton, Perry, Ewing and Knott, personally recommended by Lord Kelvin himself, came to the Far East, and by their personal examples, inspired the young students with the spirit of research and love for the pursuit of truth, a spirit which sprang from the soul of the Great Master.

The first named in this role of honour, Henry Dyer (1848–1918), had been appointed founding Principal and Professor of Engineering of The Imperial College of Engineering at the remarkably young age of 24 with the brief from Prince Ito, the Japanese Prime Minister, 'to train men who would be able to design and superintend the works which were necessary for Japan to carry on if she adopted Western methods'[5]. Dyer was born in Bothwell near Glasgow and was an engineering apprentice from 1863 for five years at James Aitken and Co., Foundrymen. During this time he attended evening classes at Anderson's College in Glasgow which ran scientific and technical classes for artisans to augment their practical skills with theoretical knowledge. There he met Yozo Yamao, who was gaining work experience at Napier's shipyard before returning to Japan to join the Ministry of Public Works where he helped draw up plans for what was later to become The Imperial College of Engineering.

In 1868, Dyer entered the University of Glasgow and obtained the Certificate of Proficiency in Engineering Science which enabled him to proceed on to the new BSc degree which Kelvin and John McQuorn Rankine, the Professor of Civil Engineering and Mechanics, introduced in 1872. About this time Ito was in the UK with the Iwakura Mission to recruit staff for the new Imperial College of Engineering in Tokyo. Rankine, just before he died suddenly—probably of diabetes—at the end of 1872, recommended Dyer, his most brilliant recent engineering graduate, for the position of Principal and Professor of Engineering. Testimonials in support of Dyer's application were prepared early the next year and came from Kelvin and several other colleagues at the University of Glasgow. It has been claimed that Kelvin had some reservations about Dyer's suitability for the position[6]. Nevertheless the 24 year old Dyer was appointed and special dispensation given for him to graduate early in 1873, having passed exam papers specially set for him.

Dyer spent the voyage out to Japan planning the curriculum which 'combined the best of the British and continental systems for training engineers'[7]. This was immediately agreed to by his former fellow student, Yamao, and teaching began in August 1873 with 56 students embarking on a six-year course. This included, in the later years, practical classes along the lines pioneered in Glasgow by Kelvin, who had introduced the first experimental facilities for undergraduate students in Britain almost 20 years earlier in 'a deserted wine cellar of an old professorial house'[8]. This highly successful educational scheme introduced by Dyer and later colleagues was eventually to return with them to Britain where it was replicated at, amongst other places, Finsbury Technical College, later to become Imperial College London. As Brock has noted,[9] how far Dyer chose his fellow professors is unclear'. However, they were all remarkably young men in their 20s and early 30s with Dyer the youngest, and the majority of them had clear links with Kelvin.

The first to arrive, in 1873, was William Edward Ayrton who had read mathematics at University College London, followed by a short period of service in the Indian Government before spending a year working on electrical research with Kelvin in Glasgow. He later described his year with Kelvin as 'the inspiration of his life'[10]. He arrived in Japan, having been recommended by Kelvin, accompanied by his wife Matilda. 'She, in contrast, was a refugee from Edinburgh University's expulsion of women medical students forced to qualify at the Sorbonne and retrain as a midwife in London'[11]. Thus they each brought new forms of expertise to Japan, although sadly Matilda was dogged by ill health and returned to Britain a year before William in 1877. In addition to his other teaching duties, Ayrton was given the task of supervizing the design and construction of the new laboratory facilities.

Fig. 13.1. John Perry (1850–1920) studied under James Thomson in Belfast, and acted as William Thomson's assistant in Glasgow before arriving in Tokyo in 1875. (From Proceedings of the Royal Society A, volume 111 (1926), opposite page i)

The next to arrive, in 1875, was John Perry who was an Irishman from Garvagh in Ulster. In 1868–70 he had attended the engineering classes of Professor James Thomson (brother of Kelvin) at the Queen's College, Belfast. During the summer months he worked at the Lagan Foundry to supplement his theoretical instruction with practical work—an educational experience which in Tokyo, and later life, he was to promote with great enthusiasm. He then became a lecturer in physics at Clifton College, Bristol where he established a physical laboratory and workshop in 1871. He left there in 1874 to become the Honorary Assistant to Kelvin in Glasgow for a year, before Kelvin recommended him for the Professorship of Civil Engineering at the ICE in Tokyo. Upon arrival it was the teaching laboratories of Ayrton that impressed him most. In later life, 35 years on, he wrote:[12]

> When I arrived in Japan in 1875, I found a marvellous laboratory, such as the world had not seen elsewhere. At Glasgow, at Cambridge and at Berlin, there were three great personalities; the laboratories of Kelvin, and of Maxwell, and of Helmholtz, however, were not to be mentioned in comparison with that of Ayrton. Fine buildings, splendid apparatus, well chosen, a never-resting keen-eyed chief of great originality and individuality: these are what I found in Japan.

Today, however, general opinion is of the view that he was not being strictly accurate and that he was eliding the early laboratory of 1873 with that opened in 1877[11]. Nevertheless, Ayrton and his laboratories clearly impressed the new arrival. The two Kelvin protégés obviously gelled and formed a working partnership to the benefit of both, which was to continue even after their return to England. The partnership was so congenial that for a time they shared equally all their receipts, even the lecture fees that either of them earned. They became known as 'the Japanese twins'. Ayrton 'was the worldly practical member of the firm, Perry the dreamer'[13]. Together they worked tirelessly to improve teaching methods in Japan and Britain. This included stressing the importance of practical work and spending time working in industry. However, as Hirayama has recorded, 'special mention should be made of the fact that the professors occupied themselves ardently in researches besides their teaching duties'. Together they researched a wide range of topics from the electric and magnetic properties of common materials such as beeswax, stone, and lead chloride to measurements of gravity. They also investigated the Japanese 'magic mirrors' which were a great fascination to Europeans. These mirrors were thin pieces of bronze coated with mercury with figures invisibly etched on the back. When the mirror was illuminated with intense light, the figures could mysteriously be projected on to a wall[14]. Following the Kelvin approach they involved their students, here mainly former *samurai* unused to practical work, in these research activities.

Ayrton and Perry wrote approximately 26 papers during the 4 years they were together in Japan, and their first, on 'Studies on ice as an electrolyte' (*Philosophical*

Magazine Series v4 (1877)114), was the first piece of physics ever reported from Japan[14]. The volume of their work was so great as to draw from Maxwell the jest that 'the centre of electrical gravity seemed to have shifted to Japan'. However, perhaps Ayrton and Perry's most noteworthy work was their scientific investigation of earthquakes in 1877 using a seismometer (a device that indicates the occurrence of an earthquake but does not produce a record) that they had developed themselves. In the history of seismology, Hudson has singled out Ayrton and Perry as, 'so far ahead of their time that they scarcely influenced their contemporaries'[15]. It is worth noting that at that time seismology was also receiving considerable attention in Britain. There had been a substantial earthquake in Comrie, near Perth, in Scotland in 1839, which had caused a large dam to burst, and as a result the British Association for the Advancement of Science had formed a Committee for Seismological Investigations which included Lord Kelvin. Ayrton and Perry left Japan in 1878 and 1879 respectively, both returning to teach at Finsbury, later the City and Guilds' College, and eventually at Imperial College London. In the 1890s, Perry would become one of Kelvin's leading opponents in his arguments over the age of the Earth, as described by Patrick Wyse Jackson in this book.

James Alfred Ewing was appointed Professor of Mechanical Engineering and Physics at the Imperial University of Tokyo in 1878—a position he held until 1883. He had been a student of P. G. Tait and Fleeming Jenkin, Professor of Engineering at Edinburgh University. Both these men were close collaborators of Kelvin. Tait and Kelvin were writing their *Treatise on Natural Philosophy* (see Chapter 10 in this volume) while Jenkin collaborated on Kelvin's inventions in connection with submarine telegraphy. The two collaborators competed for Kelvin's time and attention when he came to stay with one or other of them when in Edinburgh. Tait took the view that it was a prostitution of Kelvin's great abilities to devote himself to engineering and invention[16]. Jenkin arranged for Ewing to assist in three of Kelvin's submarine cable laying expeditions to Brazil and the River Plate during the summer vacations. In Japan, Ewing, with a group of his best pupils, took up those researches which were soon to make his name famous—namely magnetism and seismology. Ewing had a noteworthy group of students which included Ryintaro Nomaro who later became the director of the South Manchurian Railway, Rynsaki who became a distinguished mining engineer, Tetso Tsuchida, Shohei Tanaka, and Aikitsu Tanakadate. In 1881 Ewing, aided, as he says in one of his papers, by his band of students, discovered magnetic hysteresis which he named, for posterity, from the Greek verb 'to be behind'. Alas, he had later to cede precedence for discovering the phenomenon to the German physicist, Warburg, who had published his work the previous year, calling it 'Elastische Nachwirkung'[17].

Ewing also devoted much time to the construction and use of seismographs, and developed a new seismograph (an instrument which produces a permanent continuous record of earth motion—a seismogram) based on a long horizontal pendulum with which he succeeded in recording the magnitude of an earthquake for the first time. Small earthquakes in Japan are of course quite common—Ewing claims to have experienced over 300 in 5 years. However a severe earthquake struck Yokohama on 22 February 1880 and this was to provide a great stimulus to the study of seismology in the country. Ewing teamed up with two colleagues at the Imperial College of Engineering, Thomas Gray and John Milne, to develop his seismograph further.

Thomas Gray had been appointed to the position of Demonstrator of Physics and Instructor of Telegraphy at the Imperial College of Engineering in 1878 for a period of three years. He was a graduate of the University of Glasgow where he had studied under Kelvin who had recommended him for the position. John Milne, a Liverpudlian who was a graduate of King's College and the Royal School of Mines in London, had been appointed Professor of Geology and Mining in 1876—the same year Perry arrived. However, it was not until after the 1880 earthquake that he found his true scientific niche in seismology. First Gray, and then Milne, made considerable improvements to the Ewing seismograph, which became known as the Ewing-Gray-Milne model, and eventually became the standard instrument used to observe earthquakes. This instrument had some similarities with Kelvin's siphon recorder for receiving telegraph messages, and Milne arranged for it to be manufactured for global use by Kelvin's instrument maker James White, the firm which eventually became Kelvin and White in 1884. This is perhaps the earliest example of technology transfer from Japan to the West[11].

In 1880, Milne was one of the main proponents of the Seismological Society of Japan, the first such society in the world. Over the next 12 years he published 20 volumes of transactions and other journals in which he, himself, contributed two thirds of the articles. The Japanese scientists, Seiki Sekiya and Fusichi Omori, also helped develop the new quantitative approach to seismology. Sekiya was able to produce a 'wire diagram'—a four dimensional (time and space) picture—of the motion of a particle on the ground moving during an earthquake[18]. Upon his return home with his Japanese wife to the Isle of Wight in 1895, he established his own private earthquake observatory and continued to collect worldwide earthquake data. In 1901 he established the International Committee of Seismology.

In January 1883, the Rector of Tokyo University (*Tokio Diagakko*) wrote to Kelvin seeking a replacement for Ewing, asking him to nominate 'one of as high scientific talent and standing as possible'[19]. Kelvin recommended Cargill Gilston Knott, a research assistant of P. G. Tait in Edinburgh, for the post of Professor of Physics. It is perhaps worth noting that Kelvin would not himself have used the term 'physics'—a neologism that he (and Faraday) objected to strenuously[20]. Ewing, who had worked alongside Knott in Edinburgh, also supported this appointment. In Tokyo, from 1883 to 1891, Knott's main research interests involved magnetism,

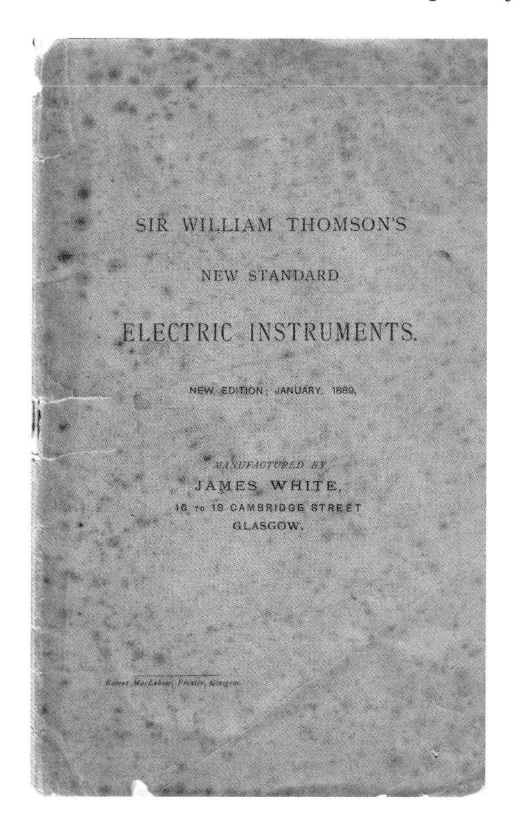

Fig. 13.2. James White's January 1889 catalogue of Sir William Thomson's Electrical Instruments. White also produced the Ewing-Gray-Milne seismograph which became the standard instrument used to observe earthquakes.

where he benefited from Ewing's legacy, and seismology where Milne was still the driving force. His published work concerned magnetostriction—the study of how magnetism varies under material stress and strain. In these studies he was assisted by a Japanese student, Hantaro Nagaoka, who was to become one of Japan's first physicists of repute—he was the first to suggest a planetary model for the atom, based on the rings of Saturn. In geomagnetism, Knott and Aikitsu Tanakadake (a former student of Ewing) organized a major magnetic survey of Japan. In seismology, Knott concentrated on the analysis of earthquake records and was able to give these a sounder physical and theoretical basis based on Fourier analysis—a favourite approach of Kelvin's. J. J. Thomson[4] records that:

> I have heard him say, though I think few will agree with the first statement, that his [Kelvin's] work on problems connected with the Earth was his most important contribution to physics, and that whenever he had done anything with which he was particularly pleased, Fourier's Theorem was always at the bottom of it.

It is interesting, therefore, to note that Kelvin, although he was clearly aware of the problem, did not contribute to the topic of the propagation of earthquake waves, leaving the problem to Knott.

With the departure of Knott in 1891 and Milne in 1895, the *oyatoi* programme drew to a close. It had served its purpose and scientific and technological self-reliance could now be achieved without further assistance from the West. Japan now had a generation of able scientists which included Tanakadate, Nagaoka, Sekiya, Omori, and Shida who were able to pick up the baton. For Kelvin's protégés in Tokyo—Dyer, Ayrton, Perry, Gray, and Ewing—and also for Milne, the Japanese experience was rewarding and pleasurable. They all displayed a great capacity for developing friendships with their students, many of whom were not much younger than themselves. In later life they all kept in contact with many they had taught, warmly welcoming those who visited them in Britain. On leaving Japan they received numerous expressions of thanks. On Perry's departure for example, 56 students presented him with a photograph of themselves, with him in their midst, along with an affectionate farewell address in which they express regret that so many of his pupils are 'scattered about in different parts of the country' and were therefore 'exceedingly sorry for their absence'[21]. When leaving Tokyo, Ewing was presented with personal gifts by the Mikado and later received the Japanese Order of the Precious Treasure. Knott and Milne were both awarded the Order of the Rising Sun by the Emperor. The future achievements of this group have been summarized by Craik[22] and Brock[9].

JAPANESE SCHOLARS IN GLASGOW

The early Meiji Government, as part of their national strategy to import Western applied science and technology, launched a programme to sponsor academic studies abroad. Between 1885 and 1912 nearly 1000 students were sent abroad, of whom two-thirds studied science, engineering, or medicine[1]. On their return to Japan they were required to work for the Government and most of them took up university teaching or research positions. Scottish universities, especially Glasgow, proved a major attraction. As Checkland has noted:

> The academic advantage which made Glasgow a desirable centre for the Japanese was the presence of teachers, including Kelvin, whose research interests lay in the problems of the Clyde industries especially shipbuilding—applied science and technology"[2].

The students enrolled in the university established a base (an Honorary Japanese Consul was created in 1890) which encouraged others to come and work as apprentices in the local industries such as Lobnitz and Co., G. & J. Weir, and later Barr and Stroud. It is said that men from Beardmore's taught Japanese to operate a forge hammer and 'for long the cries of "Awa Parkheid" and "Awa Camlachie" were used in Japan when a workpiece was being moved'[23]!

Between the years 1878 and 1898, approximately 20 students studied with Kelvin (exact numbers are difficult to ascertain as not all of Kelvin's visitors matriculated at the university), of whom 8 studied natural philosophy, thus 'Kelvin's Japanese connection was apparently directed as much or more towards engineering than physics'[24]. In the academic year 1882–3 members of the natural philosophy class voted 'for general eminence' three Japanese—in first, second, and seventh places. They were Rinzaburo Shida, who was awarded the Cleveland Gold Medal for the 'best experimental investigation of magnetic susceptibility', Naomoto Takayama, and Kiyoshi Minami. The care and concern with which Kelvin looked after these young Japanese, not only in the laboratory but also by entertaining them in his home, were greatly appreciated. Lord and Lady Kelvin were very hospitable to all their foreign visitors, although their entertaining was very correct and formal—something, however, which the Japanese could appreciate. As Agnes King, Kelvin's niece, has written, 'Numerous distinguished guests enjoyed their hospitality, and many of all nationalities who would otherwise have been very lonely, were ... invited to the house'. A Japanese scientist had unfortunately to refuse an invitation to dinner because, as he said, he had 'no night dress' with him and 'in this country I understand men must always dine in night dress'. Another was also prevented from accepting an invitation through being 'concealed in bed by a bad cold'[25].

During the celebrations for his 80th birthday in 1904, Kelvin received a telegram expressing 'hearty birthday greetings' signed by six Japanese who had all worked with him: Masuda, Taniguchi, Watanabe, Mano, Goto, and Tanadake. Kelvin carefully preserved this telegram, keeping it in his famous green notebook that went everywhere with him. A similar telegram was also sent for the *Celebrations on the Occasion of Lord Kelvin's Jubilee as a Professor* in 1899. In his reply to the numerous addresses on this occasion, Kelvin[26] highlighted the fact that:

> I have had interesting and kindly addresses from my old Japanese students of Glasgow University, now professors in the University of Tokyo, or occupying posts in the Civil Service and Engineering Service of Japan.

Kelvin's work on behalf of his Japanese students was also officially acknowledged.

The commemorative address at the Kelvin Centenary celebrations in 1924 from the National Research Council of Japan presented by 'Professor Tanakadake [who] has come all the way from Japan to do honour to his master Lord Kelvin'[27], already quoted above[4], continued:

> And when these students came to Glasgow to receive direct instruction from Sir William, he was their father and friend guiding them by hand through the untrodden realms of physical research.

Aikitsu Tanakadate became one of Japan's most renowned physicists and the foremost of his generation[28]. He had originally intended to study and train for the traditional *samurai* duties of governing the country. At the University of Tokyo, he was a student of Ewing and the American Professor of Physics, Thomas Medenhall, and after much heart searching as to whether science was an honourable calling he decided 'to study physics, which is the basis of all sciences so as to make up in full measure for our country's deficiencies'[29]. After he was appointed to a lectureship, he collaborated with Knott on the magnetic survey before going to work with Kelvin in Glasgow in 1888 for two years. There he developed an affinity with Kelvin. As Koizumi records, 'to the end of his life Tanakadate spoke of the deep impression [Kelvin] had made on him'[29]. Indeed so deep was the impression that his own students in later life nicknamed him 'Lord Kelvin'. Kelvin's own respect for Tanakadate ensured that these students always received a warm welcome in Glasgow. Before returning to Japan, where he became Professor of Physics at the Imperial University of Tokyo and Director of the Physical Institute, Kelvin gave Tanakadate a collection of his personal cards to use as introductions to other scientists in Europe. His research interests were naturally related to those of Kelvin and consisted of developing electrical instrumentation for geomagnetic research. Japan issued a series of commemorative stamps featuring Tanakadate in 2002, 50 years after his death.

Rinzaburo Shida was a former student of Ayrton and Dyer and was the first graduate of the Imperial College of Engineering to be sent to work with Kelvin—on electromagnetism and telegraphy. He too, like Tanakadate, was able to travel widely in Europe carrying introductions from Kelvin. On his return to Japan in 1883 he was appointed to Ayrton's old position as Professor of Natural Philosophy at the Imperial College of Engineering. In contrast to wired telecommunications which were imported, radio technology was developed autonomously in Japan. The first transmission experiments, conducted in 1886, were due to Shida and used the conduction method across the River Sumida in Tokyo by immersing electrodes in the water[30]. Shida founded the Institute of Electrical Engineers of Japan in 1888, and was given the title of Doctor of Engineering. Sadly, he died of tuberculosis in1892 at the relatively young age of 37—a not uncommon occurrence in Japan at that time.

CONCLUSION

During the Meiji era, Japan, with the aid of the *oyatioi* and other programmes, made astonishingly rapid progress in modernizing from a feudal to a modern industrialized society. And, as the *Japan Weekly Mail* stated in an editorial of 1878, 'in no direction has Japan symbolised her advance toward the assimilation of the Western world more emphatically than in that of applied science'[31]. The aim, to emulate late Victorian Britain and indeed to become 'the Britain of the East', was substantially achieved over the period of a few decades. Lord Kelvin, who never visited Japan, played an important indeed unique role in the transfer of educational ideas, science, and technology. A letter from the President of 'The Lord Kelvin Association' in Utsunomiya in 1906, which describes a meeting at which 'Your Lordship's portrait was hung over the platform and some speeches were made in praise of your scientific attainment'[32], provides clear evidence of the reverence with which Kelvin became regarded in Japan.

ACKNOWLEDGEMENTS

It is a pleasure to be able to thank all my Japanese friends and collaborators, especially Professors Nobuo Koboyashi, Hiro Tawara, and Nori Kouchi, who have over the years made my visits to Japan so enjoyable and patiently educated me in the history and culture of their country. I am also grateful to Professor Alex Craik for his helpful advice and comments on the manuscript.

With the exception of Kelvin's class lists at the University of Glasgow, kindly provided some years ago by their Archives and Business Record Centre, this chapter has been compiled from secondary sources. Its aim has been to provide an easily readable rather than academic account of this under-appreciated aspect of Kelvin's life. Much of the information it contains, and further details, can be found in the scholarly articles dealing with related topics by Brock[9], Craik[22], Gooday and Low[11], Koizumi[29], and especially the book by the late Olive Checkland[2] which is a *tour de force.*

14

Kelvin, Maxwell, Einstein and the Ether: Who was Right about What?

C. W. Francis Everitt

ETHER MYTHS, HUMAN REALITIES

The history of physics—by which I mean the version of history physicists purvey and historians too often swallow—is full of myths, many of nationalistic origin. Among such is the myth of the ether, which runs something like this. Our Victorian forebears, dry souls empty of metaphysical reflection, held that all space is filled with an invisible medium, the ether. This ether came into being in the 1800s with the wave theory of light; it worked, but as time went on some people, of whom Lord Kelvin was the most extreme, went too far. They became obsessed with trying to invent mechanical models of it. Then, in 1887, came disaster. Ether theory (according to the myth) requires that as the Earth moves through space, the measured velocity of light parallel and perpendicular to its motion will vary. Two American physicists, A. A. Michelson and E. W. Morley (Morley, in fact, was a chemist), devised a test of 'ether drift'. To their and everyone's surprise, it didn't exist. Crisis. In desperation, two new explorers, one Irish, G. F. FitzGerald, the other Dutch, H. A. Lorentz, separately advanced the weird ad hoc hypothesis that as Michelson's apparatus traversed the ether it somehow shrank lengthwise by just the right amount to hide the effect. Various physicists devised further experiments to see the shrinkage. All failed. Then, from that rare source of high romance, the Swiss Patent Office, emerged in 1905, the young prince-magician Albert Einstein, a new King Arthur. From the stone Physics, with a mighty heave, he drew the mystic sword Relativity, stronger than Arthur's Excalibur. It flashed and whirred; there was smoke and flame; when the smoke had cleared, there vanquished on the ground lay the dark dragon Ether, slain never to return and trouble scientific Camelot.

Every one of the foregoing claims—that the ether worked, that Kelvin was naïve, that the Michelson-Morley result was a shock, that the FitzGerald contraction was an ad hoc guess, and even that Einstein abolished the ether—is a myth, and not alas mythic in any of the ways Carl Jung and Joseph Campbell have pointed to as myths of enlightenment.

In treating Kelvin—or, as he was known to Maxwell and others in the pertinent years, William, then Sir William, Thomson—I shall argue that his insight into ether theory was outstanding, far superior to that of any of his famed predecessors, Poisson, Cauchy, MacCullagh, etc. (with one exception) and that most of the criticisms of him by twentieth century writers were merely silly. The exception is Stokes. Stokes's influence, personal and intellectual, is fascinating. Five years older, vastly knowledgeable, he comes across for Thomson as the wise elder brother and the voice of calm reflective reason to Thomson's impulsiveness. The point sharpens when one learns that the formidable Lord Kelvin was a second son whose older brother, the engineer James Thomson, doggedly overbore him in argument well on into mid-life. One part of this discussion will be to see where Kelvin bent to Stokes and where, sometimes cavalierly, he went beyond him.

To praise Kelvin is not to say that he got it right. The man who more nearly did that was the second in our title, James Clerk Maxwell, with his electromagnetic theory of light. What kind of theory was this, looking to electromagnetism not mechanism for explanation? The name implies something radically new but not whether it was or wasn't an ether theory. We are led to an odd finding. The issue seems strictly scientific but works in a cultural frame. Maxwell saw layers of explanation.

Fig.14.1. James Clerk Maxwell. (From R.T. Glazebrook, James Clerk Maxwell and Modern Physics, Cassell & Co., London, 1896, frontispiece.)

Fig.14.2. Albert Einstein. (Courtesy AIP Emilio Segre Visual Archives.)

Kelvin judged that the theory breached ether principles and, worse, didn't explain anything. Einstein thought that it and special relativity made the ether unnecessary. Lorentz held to ether tenaciously. We reach three broader questions: what is an 'ether theory', what is 'scientific explanation', and, what is meant in science by 'getting it right'.

Mythic history tends to nationalism. Einstein once said, 'If I am right, the Germans will say I am German and the French that I am a Jew; if I am wrong, the French will say I am German, and the Germans that I am a Jew'. To which pre-1914 remark one must add with despair that in the set of papers, *Das Relativitätsprinzip*, assembled with his blessing in 1922 there was, save for a footnote by the ever-careful Lorentz, not one word honouring the great Frenchman who began it, Henri Poincaré. (As Poincaré, in 1913, had a version with seven notices of Lorentz and none of Einstein.) Nor is it just Germany vs. France. The venturer from England who moves to the US finds there, with surprise, a different history of physics and going deeper discovers how much the American one was rewritten after 1940 by those who, while fleeing *German science*, brought with them an inherited German outlook. How, for Kelvin and ether, are we to

Fig.14.3. Lord Kelvin (From Silvanus P. Thompson, The Life of William Thomson Baron Kelvin of Largs, 2 volumes, Macmillan & Co., London, 1910, opposite p. 880.)

transcend myth? National bias must go, but for him especially one must ask which nation and whose bias?

In one thing the Germans and French could agree, the English were an unsophisticated lot. As Wilhelm Wien, the East Prussian farmer's son, put it in 1911 to the New Zealand flax-miller's son, Rutherford, 'No Anglo-Saxon can understand relativity'. Or hear Pierre Duhem on Kelvin and his 'English school' treating atoms and ether as machinery: 'nous pensions entrer dans la demeure paisible et soigneusement ordonnée de la raison déductive; nous nous trouvons dans une usine' [we thought to enter the peaceful and neatly ordered abode of deductive reasoning; we find ourselves in a factory]; or, more sympathetic but still baffled, Poincaré on Maxwell:

All our [French] masters, from Laplace to Cauchy, proceeded along similar lines. Starting with clearly enunciated hypotheses, they deduced from them all their consequences with mathematical rigour, and then compared them with experiment. It seemed to be their aim to give to each of the branches of physics the same precision as celestial mechanics.

What does a Frenchman expect?—'a theoretical whole, as logical and as precise as the physical optics founded on the hypothesis of the ether'. What does 'this English scientist' offer in his Treatise on Electricity and Magnetism?—'a large number of provisional and independent constructions, between which communication is difficult and sometimes impossible'.

The cliché—French logic, English muddle—embodies three large mistakes. First, to make Kelvin a simpleton, ignorant of French mathematics, is preposterous. This was the youth who, enchanted at 15 by that pinnacle of French elegance, Fourier's Théorie Analytique de la Chaleur, had in a fortnight 'mastered it—read it right through', and who then at 20, during a five-month stay in Paris met and impressed all the great French mathematicians, not least Cauchy. Second, neither Kelvin nor Maxwell was English. Third, Poincaré's image of the physical optics founded on the hypothesis of the ether as a logical and precise theoretical whole is grotesquely wrong.

OF IRISH ANGLES AND A SCOTTISH ACCENT

Kelvin was not English. Nor, despite living and working in Glasgow for 70 years, was he a Scot. He was Irish, born in Belfast from a County Down family, and all his life felt the call of the land he had left when he was eight, which leads us to that other Irishman from Skerry, County Sligo: Stokes. The sad breach in modern Ireland makes it hard now to catch the bond between the two men, both Protestant, one North, one South (but it is no Irishism to observe that North is east and South is west and the northern parts of the southern county lie north of the southern parts of the northern one), meeting expatriate at Cambridge in 1841 with Stokes at 20 the most brilliant man of his year and Thomson a 17-year-old prodigy. Friendship was immediate, mathematical and Irish. Each saw Cambridge admiringly through non-English eyes; each knew that Ireland, in Trinity College, Dublin, had its own mathematical line with Hamilton, the first mathematician since Newton to be knighted, then at the height of his powers. They had duality of vision. When the 21-year-old Thomson took over from R. L. Ellis as editor of the Cambridge Mathematical Journal in 1845, his first step (preplanned with Hamilton and Ellis) was to widen its scope and rename it the Cambridge and Dublin Mathematical Journal.

Those wry words of Einstein on Frenchmen and Germans say something about pre-1914 Europe and more about himself. He was Jewish and honoured it, and we now tend to make this his total non-scientific identity. It wasn't. He was, however he wished to transcend it, also and always a German. Maxwell too knew duality. Far from the 'English scientist' of Poincaré's dreaming, he was a Scot of ancient family, educated and intellectually formed in Edinburgh before entering England and Cambridge at 19 in 1850. Two clues, as different in kind as were the two men's gifts to physics, illuminate him and Kelvin: in Maxwell accent, that this man of classically graceful written English spoke with a Scottish accent, in Kelvin angle, his Irish knack of coming at every subject from an unexpected direction. Culture, science, and the personal are entwined; Maxwell's Edinburgh heritage, Kelvin's Irishness, Einstein's training in the Swiss Patent Office, each influenced their science but to follow how we must link the cultural inquiry to the scientific one of who was right about what.

Alongside the ether, there was for Kelvin and Maxwell another unfolding story: the story of electromagnetism. The hinge-point in each was the electromagnetic theory of light, which, while it neither made nor unmade the ether, did, as Maxwell saw and Kelvin didn't, alter the frame of the debate. Was that divergence personal or cultural? I claim both; meanwhile it is fitting, given his influence, that the most acute appraisal of Kelvin's genius was by Maxwell in a review from 1872 of 'Sir William Thomson's Papers on Electrostatics and Magnetism'.

The publication, a year before Maxwell's Treatise, of that assemblage of Thomson's writings in the two fields since 1843, is a curious event. Was he, one wonders, dismayed at being outrun? Be that as it may Maxwell, generous with the generosity of truth, was full of praise, extolling Thomson's mastery of 'truly scientific or science-forming ideas'. The line of conceptual ascent in electromagnetism is usually drawn as a single one, Faraday-Maxwell. A glance at the index of Maxwell's Treatise—36 citations of Faraday, 47 of Thomson—and his thanks in the preface 'to Sir William Thomson, to whose advice and assistance, as well as to his published papers, I owe most of what I have learned on the subject' opens a wider view. Reminiscing in the 1890s to his biographer Silvanus Thompson, Lord Kelvin recalled the time when he, at 19, through another Thomson, David, Faraday's nephew, was 'inoculated with Faraday fire'. That was one half, with Faraday starting it and Thomson by insights into Faraday and new concepts of his own, laying the trail for Maxwell. The other, more baffling, was the connection—if any—to ether. We may allow Kelvin in a letter of 1896 to FitzGerald, his own words of self-irony:

> I have not had a moment's peace or happiness with respect to electromagnetic theory since November 28, 1846 (see vol. i , p. 80 of M[athematical and] P[hysical] P[apers]). All this time, I have been liable to fits of ether dipsomania, kept away at intervals only by rigorous abstention from thought on the subject.

The mighty Lord Kelvin felt need of EA (Etherholics Anonymous). Ether dipsomania, Faraday fire, what wild words are these? Well, Kelvin was Irish with the Irishman's urge to rhetoric and self-drama. Not blarney, which those of us who have kissed the Blarney stone may admit, but something better, a vivid unexpectedness with words to match that unexpectedness of idea I call angle. We reach a world too little known, the intellectual world of Irish Protestantism, and with it young Thomson's intense bond with his very Irish father, James Thomson, who too had the gift of creative surprise—as when in 1815 on his way from farm-hand in Ballynahinch, County Down to Presbyterian minister he transmuted to mathematician.

To enter Glasgow University at 24 in 1810, graduate in two years not four, then stay two more, all the while working each summer in Ireland and facing a twice-yearly crossing by sailboat of a notoriously wild sea, takes a fierce inner drive. Then, back in Belfast, James Thomson becomes not a minister but, at 29, a professor of mathematics. He marries; he works; he writes books. In 1831, his wife dies. Left with seven young children, with iron Calvinist will he goes on, and by a supreme leap gains the Chair of Mathematics at Glasgow. It is the time of his sons. William, his favorite, is 8, James 10; within two years they enter the university, graduating in 1839 with William at 15, Gold Medalist of the year, and James, second. He takes the family to Paris to learn French, then to Frankfurt to learn German; William will face Cambridge and its traditions with a European outlook. A brilliant undergraduate career, five more months in Paris meeting the great and mastering experiment in Victor Regnault's laboratory, and then the climax: at 22, justifiably but not without heavy parental manoeuvering, William is made Professor of Natural Philosophy at Glasgow, two years younger than James had been as a beginning student there 36 years earlier.

Such is the story of an Ulster farm-boy and his son's 'Faraday fire'—a fire Maxwell also knew. Visitors to his austere office in the Cavendish Laboratory, carpetless and with bare walls, saw just one adornment, a portrait of Faraday, but the tone of his appraisal was different, about Faraday not himself, cooler than Kelvin's yet ultimately even more self-revealing:

> Faraday shews us his unsuccessful as well as his successful experiments and his crude ideas as well as his developed ones, and the reader, however inferior to him in inductive power, feels sympathy even more than admiration, and is tempted to believe that, if he had the opportunity, he too would be a discoverer.

Kelvin and Maxwell, two men with much in common: mathematics, admiration for Faraday, a dual education, Scottish then English, but how different in temperamental colouring and in vision of how science is to advance. For Kelvin, it was by engagement, close furious involvement in some previously unaddressed subject. For Maxwell, it was by distancing: thoughts, facts, and ideas in steady accumulation and then a moment of dawning when it fell into a new pattern—which brings us to another matter, his accent.

How often when men or women enter new cultures, they gain new accents. Maxwell did not. A comparison with his friend, Lewis Campbell, who went from

Edinburgh via Glasgow to Oxford, and the young Benjamin Jowett, is telling. An acute observer (Oscar Browning) saw him years later as 'Oxford within and Oxford without', an exemplar of all that was best from that refined culture. With Maxwell, heir to two cultures, an Edinburgh that was still the 'Athens of the North' and Cambridge at a high period, it was not so; he spoke the strong Gallovidian dialect, weird to his Edinburgh classmates, gained democratically from the servants and farmhands and the boys he played with on his father's estate, Glenlair, in Kirkcudbrightshire, southwest Scotland. His nine years' schooling in Edinburgh, six at Cambridge, his parents' Edinburgh tones, had no effect. Picture, if you are American, a student from the Tennessee back-country admitted with acclaim to Harvard, holding always his Tennessee accent. One recalls the world of diglossia studied by the late Charles Ferguson in visits to Switzerland in the 1960s. Articles in German-language newspapers were in high German but the jokes and cartoons, aimed to feeling rather than intellect, were invariably in one or other of the Schweitzerdeutsch dialects. Not by chance did Maxwell do his main work at Glenlair; emotion was at rest and intellect could be brought to bear.

What of Duhem's 'English school' obsessed with reducing physics to machinery? At stake are two issues, whether physics altered in crossing the Channel, and whether two non-Englishmen, Maxwell and Kelvin, fit Duhem's pattern. Maxwell, with his philosophical mind did not, but neither, in any easy way, did Kelvin. Consider his Papers on Electrostatics and Magnetism, 30 years' work with a running commentary from the viewpoint of 1872. The mythmakers would have us see ether and machinery. Neither is visible. The one mention of ether (not by name) is in a letter to John Tyndall about magnetism. As for machinery, much confusion exists that is unfair to Kelvin, between analogy and explanation. He was a master of analogy as in his breathtaking one of 1842 between the flow of heat and the laws of electrostatics, but crystal clear from these early papers is that fact that for him, then, analogy was a device not of explanation but of discovery. Later it seems different, witness in his Baltimore Lectures in 1884 the ringing cry, ... the test of "Do we or do we not understand a particular subject in physics?" is "Can we make a mechanical model of it?" , but that remark, delivered with Irish rhetoric, needs context. Part of its aim was to expose the ruinous state ether theory was in.

We must still ask whether Duhem and Wien and Poincaré were on to something. I believe they were, and that in ether theory the intellectual divergence was there in its two founders, Thomas Young and Augustin Fresnel. Rather than 'machinery', let us use the constructively vague term 'thinking physically' and hear not European philosophizing but words from another Cambridge non-Englishman, Rutherford, in a letter to W. H. Bragg, following that exchange in Brussels with Wien four years after Kelvin's death. The subject—to Wien no doubt even more beyond reach of the Anglo-Saxon mind—was not ether or relativity but Planck's quantum.

> I was rather struck... by the fact that the continental people do not seem to be in the least interested to form a physical idea of the basis of Planck's theory. They are quite content to explain everything on a certain assumption, and do not worry their heads about the real cause of the thing. I must, I think, say that the English point of view is much more physical and much to be preferred.

Some will take Rutherford as naïve. A note to Niels Bohr, in 1913, on the Bohr atom dispels that illusion. After due praise, he said:

> ... the mixture of Planck's ideas with the old mechanics makes it very difficult to form a physical idea of what is the basis of it all. There appears to be one grave difficulty in your hypothesis... namely, how does an electron decide what frequency it is going to vibrate at when it passes from one stationary state to the other? It seems to me that you would have to assume that the electron knows beforehand where it is going to stop.

By thinking physically, Rutherford had seized the key point missed by Bohr, not met until the 'old mechanics' had been replaced at the atomic level by a suitable new one.

Minds of the power of Kelvin, Maxwell, and Einstein transcend cultures. Einstein—the young Einstein—had a gift for 'physical' thinking not always given to his followers. With Kelvin and Maxwell, the Scottish-English bridge entered their beings in very different—creatively different—ways. Maxwell's theory involved a profound change in the nature of scientific explanation. Maxwell could make that change; Kelvin saw it as evasion (which it was, but there can be good evasions), and that is why he spent so much effort trying to rescue the ether. Views here may diverge. Some have thought Kelvin misguided; others, that by so remorselessly exploring the case he helped lay ground for Einstein. I prefer the latter, but with the caveat that we should not assume, as so many seem eager to assume, that Einstein was in all respects right.

YOUNG, FRESNEL AND A METAPHYSICAL QUESTION

The ether—the optical medium invented in the early 1800s by Young and Fresnel—began so well that one's first wonder, before recalling the intimidating shade of Newton, is why it's invention took so long. By making light a compressive wave like sound in an air-like medium, Young from 1800, and Fresnel with greater mathematical sophistication from 1814 on, were able to account for diffraction, interference and other previously obscure phenomena, and predict many new ones. Young's 'physical' view of the aberration of starlight was charming. In a telescope moving sideways across the line of sight, the apparent positions of stars are displaced away from the motion. Young pictured the ether, far rarer than air, passing between the molecules of the telescope wall 'as freely perhaps as the wind through a grove of trees'.

In 1808 (after Young, before Fresnel) a French engineer, Etienne-Louis Malus, made a discovery that brought triumph and catastrophe. Raising a crystal of Iceland spar to light reflected from the windows of the Luxembourg Palace in Paris, he noticed the intensity varying with orientation; he had discovered polarization by reflection. The strange double refraction in such crystals, two emergent rays from a single entering ray, had long been known; Malus made it more compelling. A raft of discoveries followed but the theory was wrecked; nothing like this happens with sound. The answer from Young in 1817, and then Fresnel, is that light waves are not longitudinal but transverse as in a vibrating rope. To Fresnel, it brought triumph but to Young, with his acute intuition, it was, however true, 'horrible'. It was a case of geometry vs. physics. Experiments by Brewster in Scotland, Fraunhofer in Germany, and himself and Arago in France, gave Fresnel laws of diffraction and formulae for the reflection of polarized light and its propagation through crystals. It was an intellectual achievement of the first rank, but to devise an ether that would account for it was quite another matter. To have Young's gas-like medium permeating all space seems reasonable; one can picture it and believe in it. For transverse waves, the medium cannot be a gas or a liquid, it must be in some sense solid. If, like Young, we are physically minded, we ask questions: obvious ones such as how bodies like the Sun and Earth glide freely through this universal solid, but soon more awkward inquiries.

One, answered neatly by Fresnel, was this. A solid will convey transverse waves but also, with a different velocity, longitudinal ones making, with an ether, strange things happen that don't happen. At the surface of a sheet of glass, for example, the extra non-light waves generate light and vice versa. Fresnel made the ether incompressible. The longitudinal velocity became infinite; the problem vanished. Or did it? Hardly, for an incompressible ether had, as time went on, other increasingly evident troubles. It is not my intent to give all that occurred in the 40-odd years from Fresnel to the day in 1862 when Stokes produced for the British Association a lethal and fundamentally correct 'Report on Double Refraction' which demolished every one of the 11 competing ether theories advanced to that time. The issues were of three kinds:

- *Dynamical explanation 1: general.* No one knew at first whether Fresnel's two sets of laws were correct. As people began to devise ethers, they found a curious thing. Over and over again, theories of ether and its interaction with matter would emerge, fitting the laws of reflection but not double refraction or conversely. Experimentalists found this delightful, a lead to work of ever increasing accuracy culminating in R. T. Glazebrook's beautiful investigation in 1879 which confirmed the Fresnel wave surface in doubly refracting crystals to parts in 10^5. Ether builders were less happy.

- *Dynamical explanation 2: plane of polarization*. A more specific question was this: the two reflection formulae found by Brewster and Fresnel gave intensities of rays polarized in and at right angles to the plane of reflection. Geometrically, there was no ambiguity but in relating them to ether vibrations, one had a choice; the vibration could be parallel or perpendicular to the plane of polarization. Correct results could be obtained either way through proper supplementary conditions but the choice affected other matters, and in particular the treatment of double refraction.
- *Ad hoc vs physically based models*. Two routes to explaining Fresnel's laws by an ether seem possible. One is to start with a model—say, an elastic solid ether with properties more or less similar to the solids we know—and deduce the laws. The other is to start with the laws and deduce an ether. Both failed. The theory produced in 1839 by George Green belongs to the first: elegant, logical, physically-based, but with one small problem, it was refuted by experiment. Among the second were two 'theories' from that preeminent analyst Cauchy, who to get Fresnel's laws made a succession of moves that in a less esteemed mathematician would be called fudges.

The story to 1860, six long decades, is an absorbing one, vastly more so than the myth: at one level, wave theory, total explanatory and experimental success; at the deeper conceptual level, failure, no answer in sight. A break came around 1840 when Green, Cauchy, Franz Neumann, and others were done; then with Stokes in the 1840s and 1850s, hope and a run of discoveries, turning slowly, inexorably, to despair. Where to go must await Maxwell and the heroic efforts by which Kelvin, in the 1880s and 1890s, very nearly rescued the ether; meanwhile, in Cauchy and Green, the two most powerful intellects before 1840, one notes the same unbridgeable gulf between 'continental' and 'Anglo-Saxon' mindsets captured by Rutherford and Wien.

We are into a strange region with a wrong excess of choice. There is the longitudinal wave and two ways to get rid of it: Fresnel's, with the wave velocity infinite; another, noted by Green and explored by Cauchy and Kelvin, to give the ether a contractile property making it zero. There are two choices for plane of vibration: parallel or perpendicular to the polarization. There is an awkwardness at the boundaries of transparent media even with the longitudinal wave suppressed. Above all, is the almost ultimate metaphysical question of what is to explain which. Take the velocity of light, c. In elastic bodies, waves are transmitted with a velocity proportional to $\sqrt{(E/\rho)}$, where ρ is density and E the relevant elastic constant. With ordinary solids or air, E and ρ are measurable; with the ether, they aren't. The velocity of sound can be calculated, or if you will, 'explained', not the velocity of light. For it, one seems in the odd position of explaining one unknown, c, in terms of two even-less-knowns, the E and ρ of the ether. Maxwell and Einstein took one line, making c a fundamental; Kelvin, in an argument to which we come later, neatly turned the case on its head to obtain a lower bound on the ether's density.

STOKES TO THE RESCUE?

Stokes came to the ether in 1845 after having written two masterworks on hydrodynamics, and—this was characteristic of him—one in pure mathematics drawn from them. His leap to optics was in an unexpected place, aberration, for which he offered an explanation very different from Young's. While his was not a theory we now have any belief in, it was beautiful and interesting and historically important in the story of the Michelson-Morley experiment. The mythbuilders make the null result of that great measurement a shock. It was not; Michelson's first thought, with good reason, was that he had proved Stokes's theory.

To anyone who thinks physically, as Stokes did, Young's vision of ether flowing through matter like the wind through a grove of trees loses charm when the wind is a mass of glass. Fresnel had no qualms but the more Stokes thought about it the less he liked it. His new explanation had two interlocking principles: (1) having the ether not rigid but viscoelastic, i.e. solid for the high frequency vibrations of light, yielding like pitch to the slow motions of the planets; (2) the neat Stokesian observation that if the motion of the convected ether is 'irrotational' (as one would expect mathematically), light waves approaching the Earth will follow a curved path. The angle of deflection proved to be v/c, the ratio of the Earth's velocity to the velocity of light—the old correct result from a new premise. It was a brilliant start, and more came with it. In the 40 years after Fresnel, one man only approached him in combined mastery of theory and experiment (and far surpassed him as a teacher), Stokes. Fluorescence was his greatest discovery—we still speak of Stokes and anti-Stokes lines—but greater in conceptual power was his 'Dynamical Theory of Diffraction' (1850) where one of those unsettleable choices, whether the plane of vibration was parallel or perpendicular to the polarization, was settled. With consummate ingenuity Stokes saw that diffraction, unlike reflection and double refraction, yields a test, a different variation of intensity with polarization angle in the two cases. His experiments set the planes at right angles. The result carries over to Maxwell's theory, making the electric rather than the magnetic vector in it perpendicular to the plane of polarization, as is indeed the case.

A man far ahead—to many, possibly even himself, the one who might master the ether, such in the 1850s was Stokes. Then, in 1862, came that 'Report on Double Refraction' where, one by one, every ether yet invented was weighed in the balances and found wanting. Larmor and Rayleigh, and more recently the historian Jed Buchwald, have held Stokes's evaluation of one theory, MacCullagh's second, as unjust, with attempts—misguided in my view—to assimilate his work to Maxwell's. In the context, Stokes was correct. Reading his Report with its austere farewell, 'the true dynamical theory of double refraction has yet to be found', is a powerful and moving experience, that rarest of pleasures, meeting someone who knows what he is about.

And knew he was stuck. That, rather than pressure of other work, is why Stokes never wrote the commanding treatise on light expected of him. It is hard to be commanding in an impasse. Yet, if his Report was an ending, it was also in its impact on Maxwell a new start, coming as it did midway between the first hint of his new theory in 1861 and the paper of 1865 where it took definitive form. How Stokes helped will appear in the section after next.

KELVIN, ENERGY DENSITY, AND ETHER DENSITY.

In William Thomson at 36 in 1850, we meet a figure unlike any other. Where 1849 had papers on hydrodynamics, electricity, the heating and cooling of buildings by circulating air, and above all, the second law of thermodynamics, this year would bring ones on steam engines, geometry, electrolysis, magnetic properties of crystals, regelation, and (one of his few truly comprehensive investigations), his 'Mathematical Theory of Magnetism'. Here, he developed the whole field from two equal standpoints, magnetic poles and Ampèrean currents, introduced the two vectors H and B that accompany them, and in what seems at first no more than an elegant mathematical transformation gave the integral $\int H^2 dV$ through the space surrounding a magnet, from which came the concept central to any true field theory, energy-density.

The idea that forces, like armies, have fields of action, was Faraday's, but Faraday had no clue about energy. Thomson, who, following Thomas Young, brought the word energy into physics, replacing the awkward use of force in English and Kraft in German for this fundamentally new concept, had every clue. Its relevance may be put in terms either of energy or superposition. A field theory locates energy not at source bodies but in the surrounding space, and it is there that superposition occurs. Each separately modifies the state of the space, and the combined action acts back to create the force. This dual vision, so casually present in Thomson's $\int H^2 dV$ was to be the heart of his and Maxwell's theories. Where energy resides is, of course, distinct from (though easy to confuse with) whether energy-density has to do with an ether.

In 1854, Thomson published, simultaneously in English and French, a paper 'On the possible density of the luminiferous medium and on the mechanical value of a cubic mile of sunlight' [Note sur la densité possible du milieu lumineux et sur la puissance mécanique d'un mille cube de lumière solaire]. It exemplifies his divine boldness with numbers, crude estimates better than no estimate at all, or, as he put it later:

> When you can measure what you are speaking about and express it in numbers, you know something about it; but when you cannot measure it, when you cannot express it in numbers, your knowledge is of a meagre and unsatisfactory kind; it may be the beginnings of knowledge, but you have scarcely in your thoughts advanced to the stage of science.

His three for ether theory were (1) the energy-density of sunlight, (2) ether density, (3) the mass of ether in a sphere the radius of the Earth's orbit. The starting point was the heat output from the Sun for which there existed by 1854 a quite good estimate.

In our terms, which were not then his (though it was he in 1866, following a suggestion from the engineers Charles Bright and Latimer Clark, who led the naming and defining of the watt), the power from the Sun is 1400 watts/m^2. Multiplying by the time light takes to go a metre, one has for the energy density of sunlight at the Earth, 4.7×10^{-5} joules/m^3; Thomson's figure transposed is close, 3.8×10^{-5} joules/m^3. With superb confidence he went for a lower limit ρ_{min} on the ether density by the following chain of reasoning. The energy in the wave will be half potential and half kinetic, with kinetic energy-density $\frac{1}{2}\rho v^2$, v being the vibrational velocity. There must be some limit on wave amplitude; Thomson guessed 2% of the wavelength. Relating this to the velocity of light and vibration frequency, he found for v an upper limit $v_{max} < 2\pi c/50$ giving, by $E = \frac{1}{2}\rho v^2$, a density limit ρ_{min} of $253.2\, E/c^2$ or 4×10^{-23} gm/cc. A sphere of ether the radius of the Earth's orbit weighs 10^{12} tons. A first thought may be to dismiss such numerology as futile. That would be immodest but also, more importantly, wrong, for sunlight has mass. Put W for power, and τ for the time light takes to travel from the Sun to the Earth; the solar energy E in a sphere the radius of the Earth's orbit is $4\pi r^2 W\tau$; through $E = mc^2$ the mass in it is $M_\phi = 4\pi W\tau^3$ or 2×10^9 tons, 0.2% of Thomson's limit. More, his ether mass M_k ties to our sunlight mass M_ϕ; their ratio M_ϕ / M_k is $3/W\rho c^3$, that is, $3v^2/2c^2$, a natural ratio for Newtonian to relativistic terms giving just that 0.2%. One might quibble about Thomson's $\lambda/50$ but some lower limit on ρ was as inevitable for a material ether as $E = mc^2$ is for us.

To have caught and worked this point, undreamt of in 50 years of ether theorizing, was genius. Thomson, without planning it, has brought us to where it would become evident that mass and mass-density are not solely mechanical. The story is a complex one, still unfinished. It began in 1882 with another Thomson, J. J., who proved Maxwell's theory that a moving charge has an electromagnetic mass and later daringly conjectured that the entire mass of the electron might have this form. Einstein, in his first discussion of $E = mc^2$, inverted the point by surmising that the huge heat output from radioactive substances came through annihilation of mass. The mysteries have, if anything, deepened, at one end no exact theory of the masses of fundamental particles, at the other dark energy and dark mass. Against Kelvin's limit 4×10^{-23} gm / cc may be set a current estimate for dark energy, seldom called ether, of 10^{-29} gm / cc.

A separate revolutionizing set of events came in the 1880s and 1890s, the progressive discovery through Maxwell and experimental invention of the full electromagnetic spectrum from radio waves to γ-rays. Kelvin found this new world fascinating. We, looking back, may ask whether the existence of diverse radiations over 15 orders in frequency, all having the same velocity, should, like the new understanding of mass, be unsettling. But that brings us to Maxwell and his findings and evasions, which included evading some Kelvinesque questions.

MAXWELL AND SCIENTIFIC EXPLANATION

In pondering Maxwell, one could begin with Faraday or Kelvin or William Hopkins, his—and Kelvin's—superbly organized Cambridge tutor; one could stress that among mathematicians, he was with those who are by nature geometers rather than algebraists; but best is a word borne to him in his student days which he, 30 years later, introduced into physics: the now magical term relativity. For this, we look to the man who, though he did not quite invent it, first gave the word weight, the Professor of Logic and Metaphysics at Edinburgh, Sir William Hamilton (not to be confused with his Irish namesake William Rowan Hamilton). Philosophers are of diverse kinds. Hamilton, no system-builder, had three large gifts: erudition, the ability to inspire, and an eye for ratiocinative detail. Maxwell at 16, eager in class and welcomed with others to evenings of discussion in the master's house with its library of 10,000 books, was enchanted.

Among this immensely knowledgeable man's grand proclamations was 'the relativity of human knowledge'. To follow, forget Einstein (and the dream that everything is relative): Hamilton's focus was epistemology. Drawing on Kant, he held that we know nothing of the thing in itself (Ding-an-sich), our knowledge is of relations, of ourselves to the object or objects to each other. Coupled in metaphysical release, was a distinction between knowledge and belief. There were philosophers who, assuming perhaps too easily that they knew what knowledge is, insisted on putting it before belief; belief without facts (their facts) was irrational. Opposite was the post-Kantian idealism of Fichte who sought in ego direct knowledge of the absolute. Both were adrift. To have beliefs, and take them as in some sense absolute, is rational, but knowledge in them is, and has to be, relative.

From Kelvin on numbers and Faraday winding coils in his laboratory to Fichte on the absolute seems a far journey, yet to Maxwell their double heir in 1863, relativity of knowledge—scientific knowledge—was liberation. More than a new theory, his was a new kind of theory, new in what it did, new in how it was done: a union of three realms, electricity, magnetism, and light, based not on axiom and deductive chain, but a system of relations. So we speak of Newton's laws but Maxwell's equations. Of all who touched it, nearest was Poincaré with his dawning realization of 'something rather artificial' in the French deductive style. But if Poincaré caught the what, he hardly reached to the why, which lay not in oddities of the English, or Scottish, mind but in the facts of the case—the conceptual situation Maxwell found himself in.

I proceed now first to the growth of Maxwell's thought in five papers, then to the discomforts on the one side of Kelvin—the man who started it—and on the other of Heinrich Hertz—the man who, by discovering radio waves, did more than any to establish the theory. The papers are:

- 1855–1856 'On Faraday's Lines of Force'
- 1861–1862 ' On Physical Lines of Force'
- 1863 'On the Elementary Relations of Electrical Quantities'
- 1865 'A Dynamical Theory of the Electromagnetic Field'
- 1868 'A Note on the Electromagnetic Theory of Light'

Pivotal was the fourth, of which he wrote on 5 January 1865 to his cousin Charles Cay, 'I have also a paper afloat containing an electromagnetic theory of light, which, till I am convinced to the contrary, I hold to be great guns'. While not the first gleam—that came in the remarkable third section of 'On Physical Lines of Force'—it is where Maxwell recognized and resolved the explanatory crisis. The five together reveal one uniquely Maxwellian quality: each is a total view, but the view changes markedly from one to the next. The Treatise hewed the same course. To Robert Andrews Millikan looking back in 1950, it ranked with Newton's Principia, 'the one creating our modern mechanical world and the other our modern electrical world'. To Maxwell, in a conversation recorded by Larmor, it was simply a view of the stage he had reached.

Three further remarks first: that while the titles of the two earliest papers point to Faraday, in each Maxwell was almost equally indebted to Thomson; second, that he came to ether, in 1861, by way of Faraday and magnetism rather than optics; third, that his new understanding of 1865 hinged on a Hamilton-like distinction between knowledge (about electromagnetism) and belief (in an underlying undescribed ether). The 'Elementary Relations of Physical Quantities' in the third title is another Hamilton touch. As for the first paper, having nothing about ether it may seem irrelevant; in fact, beginning as it did Maxwell's lifelong regard for Faraday and lifelong journey into the nature of scientific explanation, it gave direction to the rest.

Curved lines of force, revealed by sprinkling iron filings on a sheet of paper over a magnet, long preceded Faraday. He saw them anew in three ways, geometric, electromagnetic, and physical. Geometrically, they set direction and strength (the closer the lines, the stronger the force). With electromagnetism, there was his finding that in it electric currents, magnetism, and motion are always mutually perpendicular—a fact to him completely at odds with Newtonian push-pull forces. Finally, most daringly, he assigned physical properties to the lines, like stretched elastic bands with a sideways repulsion, acting in space as the actual cause of the force. Maxwell's first paper explored the first two; his second, began as an attempt to model, and therefore explain, the supposed stresses. It was a journey in both. In both, he learned much from Kelvin—or as we must revert to calling him, Thomson—and first that one can do excellent science without looking to explanation.

In a short note written when he was 17, Thomson made an astounding discovery. Two utterly disjoint physical problems, distribution of charge and the flow of heat are in exact mathematical analogy. Learning next that Faraday doubted Coulomb's law of electric force, his first reaction was the contempt of young theory towards slow-witted experimenters. Then the light of his own analogy dawned. For static electric or magnetic forces, the irreconcilable views of Faraday and the all-conquering French mathematicians were complementary, each aiding the other. This was the message 'young Clerk Maxwell' heard from him with due Faraday fire almost the day they met, and returned to, in 1854, to produce 'On Faraday's Lines of Force'.

The curiously different titles of its two parts, I 'theory of the motion of an incompressible fluid', II 'On Faraday's Electro-tonic State', reflect that in them Maxwell used, in different ways, two powerful analogies drawn from Thomson. Part I developed geometric properties of lines of force by analogy with fluid flow in a porous medium, a variant on Thomson's heat flow analogy, much of it already given by Thomson. Part II centred on a quantity to which Maxwell gave successively three names—electro-tonic function here, reduced momentum in 'Physical Lines', vector potential in the Treatise—set in a frame of six mathematical relations, the start of 'Maxwell's equations'. The title catches one of Faraday's heavier passages, from which it took a Maxwell to win gold—the clue being yet another even more surprising analogy of Thomson's, between lines of magnetic force and strains in an elastic solid. Comparing it with the ingenious but problematical action-of-a-distance theory of electromagnetism advanced in 1845 by Wilhelm Weber, Maxwell's closing words on his own structure were:

I do not think that it contains even the shadow of a true physical theory; in fact, its chief merit as a temporary instrument of research is that it does not, even in appearance, account for anything.

Analogy in physics has two aims. One, Thomson's in 1841, is gained technique, transporting the mathematics of heat theory to electricity. Another, Maxwell's here, is gained understanding. With it, met by him in an introductory essay as well as one read to a small circle at Cambridge, is an inquiry, critical to a reader of Kant, from analogy to explanation. In Kantian terms, knowledge is a complex of elements, some within the human mind, others out there in Nature. Analogies like Thomson's bring a sudden feeling of illumination, relevant to mind but not so obviously about Nature. Some distinction must lie between understanding and explanation. Addressing it in 'Faraday's Lines', Maxwell prefigured, without the nationalism, that later clash of arms, Anglo-Saxon vs continental between Rutherford and Wien. Like them, he distinguished two ways in physics, physical and abstract, but unlike them saw both as falling short. Too abstract, we 'lose sight of the phenomena' and miss 'more extended connexions of the subject'; too physical, we 'see the phenomena only through a medium' and risk 'that blindness to facts and rashness of assumption which a partial explanation encourages'. Analogy, by allowing us 'at every step to lay hold of a clear physical conception, without being committed to any theory founded on [it]', is a creative middle way. The passage has received wide attention but cannot, I think, be a resting place; it was for Maxwell, a holding action, his hope being to move beyond analogy to 'a mature theory in which physical facts will be physically explained'. That was the task he set himself in 'On Physical Lines of Force'.

Faraday's idea that magnetic action originates in stressed lines of force is a physical explanation rich in the charm of the unexpected; only later does one say, like Byron of Coleridge explaining metaphysics to the nation, 'I wish he would explain his explanation'. Maxwell's story has romance, influences from the engineer Macquorn Rankine as well as Thomson, four parts, and a surprise ending. The first two, treating magnets and currents, I 'On the Theory of Molecular Vortices applied to Magnetic Phenomena' and II 'The Theory of Molecular Vortices applied to Electric Currents', appeared in March and May 1861. Then to his own complete surprise he went on during the long summer vacation at Glenlair to electric forces and the amazing discovery that the vortex machinery he had set up to explain Faraday's stresses would transmit waves which just happened to have the velocity of light.

What of 'molecular vortices'? In studying heat-engines and gases in the 1840s, Rankine had built a theory of matter with tiny molecules spinning fixed in space in an ethereal medium. It gave many results but was, through the work of Clausius and

Maxwell, soon superseded by kinetic theory. It is one of the high ironies of physics that Maxwell, having given it the death blow for gases in 1860, in 1861 adapted it to magnetism. Vortices shrink axially and expand sideways like tiny spinning Earths, just Faraday's pattern. Working mechanical and magnetic concepts together, he derived, in Part I, results for magnets, steady currents, and diamagnetic bodies. As always, Thomson is there. Having refereed one of Rankine's papers in 1851, he suddenly saw in 1856, that magneto-optical rotation—Faraday's discovery to which Maxwell would come in Part IV—necessarily implies local rotation around the lines of force.

How curious, till one knows him, that Maxwell the philosopher-physicist could meet and master an engineering problem that had escaped Rankine. Lines of force around a current-carrying wire form vortex rings, all turning the same way. Surfaces of adjacent rings move in opposite directions: awkward unless, like Maxwell in the high age of Victorian machinery, one pictures large gearwheels with idle wheels between them and makes the mental leap to layers of tiny idle particles between the vortices. In space, the particles stay fixed, in conductors they are free; the vision comes of a stream of particles in a wire engaging the inmost vortex layer and it, through the idle wheels, engaging the next layer until the whole intricate chain of machinery is at work. Maxwell obtained all the correct equations. He had explained Faraday's explanation.

Pausing with him between Parts II and III, let us weigh what he has done in physics and in analogical reasoning. Physically, he has combined Faraday's stresses with Thomson's concept of energy density, identified the magnetic term with kinetic energy of rotation, and provided connecting equations. For analogy, I examine three uses, here and in Part III: (a) parallelism, (b) extension, (c) counteranalogy, noting that the vortex model has two functions, explanation and discovery, and further that it is not to be taken in all respects literally. Of the idle wheel image Maxwell says, 'I do not bring it forward as a mode of connexion existing in nature'. So, how does it work, or, more exactly, how does he work it?

(a) *Parallelism*: His method throughout is to work two lines in parallel, the vortex machinery to define relations, magnetic or electric results to give them relevant form. It is a powerful mode of discovery already in some degree used in 'Faraday's Lines', made more powerful by the greater level of explanation.

(b) *Extension*: How to extend the theory? By analogy. Magnetic energy is disseminated in space; so is electrical energy. Also, energy has two forms, kinetic and potential, and the magnetic energy is kinetic. Maxwell's guess by extension is that electrical energy will be potential, an elastic distortion of the vortex medium.

(c) *Counteranalogy*: Human thought is full of unthought-out analogies. Electric current and with it electric fluid, a term universal in Maxwell's day, is one: an

> image of water in a pipe (or in the rival two fluid theory, two opposite streams in one pipe). Counteranalogy, a second image to upset the first, is one way of fighting the 'mental inertia', as Maxwell following Faraday called it, that takes image for truth.

Extension made the medium elastic; parallelism translated mechanical to electrical equations; counteranalogy, by making electricity not a fluid but idle particles, led Maxwell to his amazing concept of a displacement current. In the model, a steady current in a wire again resembles water in a pipe. With an oscillating current, however, elastic yielding of the vortices near the wire allows the particles there to oscillate. The current penetrates into the space around it by a distance depending on frequency; an extra term enters. Also the medium can transmit waves. Having built the electromagnetic equations, including the new term, Maxwell calculated their velocity from electric and magnetic quantities. This, in turn, took him back to Weber, in whose theory also appeared a velocity v though with no thought of waves. In 1858, he and Kohlrausch had determined v experimentally. The result in Weber's special units (his being a two fluid theory), was ~ 70% of the velocity c of light.

To Duhem the English mind had exchanged the calm abode of deductive reasoning for the pounding din of a factory. Maxwell, in the quiet of his study at Glenlair—morning, larks rising, cattle lowing in the fields beyond—pictured the factory, reread Faraday, and brought to bear the full power of a mind trained in Cambridge mathematics and Hamiltonian philosophy. By October 1861 he had a wave equation but no numbers. Back in London, he put in Weber's ratio and out came the velocity c: 'We can scarcely avoid the inference that *light consists in the transverse undulations of the same medium which is the cause of electric and magnetic phenomena*'. Letters to Faraday and Thomson, dated 19 October and 10 December 1861 respectively, inform Faraday that the theory is based on his ideas, and Thomson that it is based on his. Part IV returns to Thomson's entry-point, the magneto-optical effect.

The discovery made, Maxwell drops his machinery, setting his next paper 'On the Elementary Relations of Electrical Quantities', on the strictest of empirical grounds, dimensional analysis. Behind, as always, is Thomson, he among physicists—unlike Macavity among cats—inevitably, visibly, there. In 1861, the British Association set up under his chairmanship, a Committee on Electrical Standards. The admirable pioneer had been Weber, but Thomson, by invoking energy principles, raised the subject to a new level. Physicists using the expressions $\frac{1}{2}$ Li^2 for the energy in an inductance and $\frac{1}{2}$ CV^2 for the energy in a capacitance may be curious to learn that they originated in his 1861 Report. Maxwell joined in 1862. The first task, under his leadership in a laboratory at King's College, London, with an apparatus designed by Thomson, was to determine the ohm in absolute units. Meanwhile, another of Thomson's profound clarifications had been a definition of electric current free from Weber's arbitrary two-fluid assumption. The stage was set. In 'Elementary Relations', written with the telegraph engineer Fleeming Jenkin, Maxwell created our nowstandard notation with dimensional relations as products of powers of M, L, T, enclosed in square brackets with separate dimensionless multipliers. For every quantity, two definitions were possible, from forces between charges or forces between magnetic poles. Their ratios proved to be some power of a constant with dimensions $[LT^{-1}]$ and magnitude $\sqrt{2}$ times Weber's constant—in other words, the velocity of light.

Before Maxwell, the velocity of light was one velocity among many. Now it was privileged, central to the scheme of things, granted a certain absolute character. Optical experiments measured it at high frequency. Electrical ones, such as Kohlrausch and Weber's experiment, and another by Maxwell and Charles Hockin in 1868, were at zero frequency. That they should agree to within the limits of a quite small experimental error was extraordinary. In 1863, a track had been laid out leading to Einstein and 1905. It is hard to imagine an empirical result more important to physics than this.

Looking again at the vortex ether, one cannot but feel its peculiar explanatory status. Forces require Faraday stresses, stresses require vortices, vortices require gearwheels, gearwheels require ... what? We are into a regress of ever-more-remote explanation. Yet known relations exist. Already Maxwell had assembled a system of equations between electric and magnetic quantities and deduced wave propagation. Whether or not he had explained electromagnetism or light, he had manifestly connected them, and the dimensional argument assured that in no way was the connection speculative. There remained how best to express it and how to relate it to the rest of physics.

Stokes with his deadly Report in 1862 had exposed the ether crisis. In proposing an alternative theory Maxwell might escape that, but still had two arguable paths. He took both. One, delayed to 1868, was to make the theory purely electromagnetic. The other was to make it dynamical, connected to principles expressed, for example, in Lagrange's equations of dynamics. This he did in 1865. In 'Physical Lines' he had used mechanical analogy to guide the electromagnetic equations; here he used electric and magnetic knowledge to shape the dynamics. The differences between the dynamical and electromagnetic approaches are somewhat technical. The dynamical form included eight groups of equations; the electromagnetic form had only the four now identified as 'Maxwell's equations'. The eight included the incorrectly named 'Lorentz force law', limited at this stage to currents (Maxwell treated moving charges in later writings), and also the vector potential A, now often labelled an unphysical quantity but from Maxwell's standpoint incorrectly so because, as he rightly demonstrated,it is related to the distinctly physical concept of momentum.

From an explanatory standpoint, what Maxwell had now achieved was a hold on regress of explanation. The theory reached just the level needed and no more. To quote Poincaré, 'Maxwell does not give a mechanical explanation of electricity and magnetism; he confines himself to showing that such an explanation is possible'. It was a hybrid theory. The equations were dynamical but the quantities entering them were electric and magnetic ones; it worked because they could be written as energy terms. One of the sweet ironies of Thomson's later doubts, is that it was in considerable degree through his work, including the expressions $\frac{1}{2}Li^2$ and $\frac{1}{2}CV^2$, that the incorporation became possible. One further point is vital. Maxwell had set up a theory for which, unlike the mechanical ethers, no special assumption is needed to get rid of the longitudinal wave. In doing so, however, he had not followed a dynamical argument; the evidence is electrical: 'The equations of the electromagnetic field deduced from purely experimental evidence, shew that the transversal vibrations only can be propagated'. The assumption, a gigantic one, is that experiments at very low frequency can be applied to signals of high frequency like that of light (or x-rays or γ-rays). It is an assumption that works, and one perhaps justifiable—by analogy—with the equally astonishing fact about the common velocity c.

Millikan's words quoted earlier are no exaggeration. Nothing comparable to Maxwell's theory in range of explanation and discovery had appeared since Newton's. Yet people had difficulty with it. In examining the opposite discomforts of Kelvin and Hertz, it helps to recall that questions about scientific explanation had also affected Newton. So used are we to his theory that we forget how controversial it once was. Criticism came from two directions. Older philosophers saw direct action at a distance as delusive, a reversion to medieval 'occult forces'; his disciple Roger Cotes, more Newtonian than Newton, made it an ultimate, a truth beyond need or reach of explanation. The opposed comments of Kelvin and Hertz on Maxwell afford a parallel. To Kelvin, an electromagnetic as against a mechanical theory of light was 'rather a backward step', a betrayal of the path physics was on. Hertz was equally trenchant. To him, all efforts—including Maxwell's own—to 'explain' Maxwell's equations were wasted: as he put it with Cotesian impatience:

> To the question, What is Maxwell is theory? I know of no shorter or more definite answer than the following:– Maxwell's theory is Maxwell's system of equations. Every theory which leads to the same system of equations, and therefore comprises the same possible phenomena, I would consider as being a form or special case of Maxwell's theory; every theory which leads to different equations, and therefore to different possible phenomena, is a different theory.'

That men of such powers could reach opposite opinions is a measure of the rethinking required.

The hybridism of Maxwell's dynamical theory transports us from relativity, in its Hamiltonian sense, to that other principle of Hamilton's: the distinction between knowledge and belief. Very simply Maxwell formulated his theory in such a way that he could believe in an ether without having to know anything about it. Over and over again in his writings one finds evidence of the belief, and some physicists who read him fail, in their post-Einstein euphoria, to observe what he actually did. The striking fact about his theory of light is that while it assumes an ether, it is not about it. It deals with knowable, measurable electrical quantities and the dynamical relations between them.

Hertz, among the greatest of experimenters, was also a theorist, and one held by many—but not I confess me—as a great philosopher of science. Reading him on Maxwell, as well as in that work Principles of Mechanics, which so influenced Wittgenstein, one finds everywhere an extreme wish to remove unnecessary assumptions. The exercise can be a useful one but is itself an assumption. Devotees of Ockham's—sometimes less than sharp—razor should notice a remark of Maxwell's in the Treatise that 'to eliminate a quantity which expresses a useful idea would be a loss rather than a gain in this stage of our inquiry'. Parsimony is not always creative.

But with that we rightly return to that volcano of scientific ideas, nearly all useful, Kelvin.

THE BALTIMORE LECTURES AND AFTER

From Maxwell in remote Galloway, to the all-powerful Sir William Thomson lecturing to an enthralled audience at the new research university, Johns Hopkins, in Baltimore, Maryland, is a distance greater than the 3,400 miles between them. The request had been transmitted two years earlier and engaged an exercise in which characteristically Thomson managed to up the fee from $1,000 to $ 2,000, but then no less characteristically gave more than was asked by extending the number of lectures from 12 to 20. There were two lectures a day; the group, 21, calling themselves by a recondite joke, 'coefficients', included Rayleigh, Rowland, Michelson, Morley, and S. P. Langley. Rayleigh's account to his son is the best: 'What an extraordinary performance that was! I often recognized that the morning's lecture was founded on the questions which had cropped up when we were talking at breakfast'. His son supposed the hosts would have been dismayed by this but Rayleigh's reply was: 'On the contrary, they were very much impressed and he got some of them to do grinding long sums for him in the intervals'. With extreme modernity the lectures were taken down shorthand and reproduced by the 'papyrograph' method, and it is from these pages more than anywhere that one hears the man's authentic voice. Right from the beginning—on page 2—he makes it clear that this would be about 'the outstanding difficulties' in the theory, of which he saw many. It was a phase of ether dipsomania he would be on for another 12 years until April 29, 1896, when he wrote to FitzGerald 'I now abandon everything I have ever thought of or written with respect to the constitution of the ether.'

In the first lecture, Thomson expresses the opinion that it is 'rather a backward step from an absolutely definite mechanical notion that is put before us by Fresnel and his followers to take up the so-called Electro-magnetic theory of light'; reveals a suspicion amounting to rejection of Maxwell's displacement current; and continues by 'insisting upon the plain matter of fact dynamics and the true elastic solid as giving what seems to me, the only tenable foundation for the wave theory of light in the present state of our knowledge.' It sounds entirely negative and we may come to think of him as having passed beyond the realm of critical thought, but nothing could be further from the truth. This was a first shot across the bows. Thomson knew better than anyone, except Stokes and Rayleigh, that the ether was in a mess. Coming to Lecture XVII (delivered after Rayleigh had drifted off to Philadelphia), we find the following:

> Mathematical literature has been loaded with a great deal of bad writing on this subject. A great number of investigations and statements called theories have been made, in which a piece of dynamical work is gone through, and then a condition is arbitrarily introduced, and that is called Cauchy's theory, and something else is called Neumann's theory, and something else is called MacCullagh's theory.

Then, after a not very strong admission of possible unfairness, he goes on to quote Rayleigh's dissection of the principles in two theories and adds:

> One of these principles is just an arbitrary assumption, absolutely inconsistent with the dynamical conditions of the problem. If you want to put not too fine a point on it, you may call it MacCullagh's mistake or Neumann's mistake.

It is physics as knock-about comedy mingled with Irish provocation, which is at the same time totally serious.

Toughest among questions was the interaction between matter and ether. Among the assaults Thomson made on it was a model of molecules as gyrostats interacting with an ether jelly. After a graceful farewell to the coefficients, he returned to Glasgow despairing of the elastic solid ether. A solution he then worked on with considerable success transferred the gyrostatic structure to the ether. Over the next ten years, he tried a succession of ether models, one of the most ingenious being his 'labile' ether of 1887 where instead of giving the longitudinal wave infinite velocity, he made it zero, stabilizing the collapsible medium by attaching it to rigid boundaries of the universe. The comments of Gibbs on the 'happy inspiration of genius' that could look for the solution to a problem in a place where no other physicist would look, balanced in Gibbs characteristic understated style the fact that the point of the paper in which they appear was the complete successfulness of Maxwell's theory.

Not least in Baltimore was the sound of Thomson's voice urging Morley that the Michelson experiment must be repeated to high accuracy as probably the most important investigation we can undertake at this time.

MYTH, TRUTH AND THE MICHELSON–MORLEY EXPERIMENT

Among myths in the alleged history of the ether, the most egregious may be that the null result of the Michelson–Morley experiment came as a surprise. Michelson did not find it so. Read his first account in 1881, before Morley joined him,and you will find that what he believed himself to have done was to establish Stokes's, rather than Fresnel's, theory of aberration. That, in 1881, was an eminently reasonable conclusion.

The myth hides a more interesting truth, not to be put absolutely for it was as much atmospheric as factual, that physics during the 1870s was changing. The elegant 'physical' picture Stokes had set against Fresnel's decidedly unphysical one was fading in elegance. Such transitions in science have been schematized in different ways by Thomas Kuhn and Imre Lakatos and their followers and critics. Rather than speak of paradigm shifts and research strategies, I prefer to think atmospherically with some, but not all, of the atmospheric change being the rise of a new theory of light. To Michelson, all this was irrelevant. As an experimentalist he was instantly taken in 1880 by a letter of Maxwell's, published posthumously in Nature, stating that the difference in measured velocity of light between mirrors parallel and perpendicular to the Earth's motion through the ether, being of order v^2/c^2 in Fresnel's theory, would be beyond the reach of experiment. The challenge was enough; he invented the interferometer.

Three figures, Stokes, Maxwell, and Thomson stood as inspiration. In 1858, the French physicist, Hippolyte Fizeau, had performed an experiment confirming one aspect of Fresnel's theory, by proving that the velocity of light in a moving column of water was greater downstream than upstream. Some time in 1862, Maxwell read his paper and was inspired to devise an ether drift experiment. Since refraction is caused by differences in velocity of light in different media, the Fresnel drag measured by Fizeau could modify the refraction of a glass prism moving through the ether. Maxwell calculated the additional deflection and arranged a chain of three prisms with a return mirror behind them, setting up what would now be called an autocollimator to measure deflection. The predicted effect was 2½ arc-minutes, easily measurable: he found nothing. He submitted a paper concluding that 'the result of the experiment is decidedly negative to the hypothesis about the motion of the ether in the form stated here'. The referee was Stokes, and the subject one of many that Stokes knew better than anyone. Maxwell had made an error in boundary conditions; the effect was not first order in v/c but second order v^2/c^2, completely immeasurable. Maxwell withdrew, though a short account appeared in a note to the astronomer William Huggins, but he did not forget.

Michelson's first apparatus at Helmholtz's laboratory in Berlin was subject to limitations from vibration, also Maxwell's calculated effect was a factor of two too high. We come to the strong voice of Thomson at Baltimore when Michelson and Morley were both 'coefficients'. Morley was the drive behind the enormously improved apparatus floating on a pool of mercury with which the definitive experiment was performed in Chicago.

Steadily the theoretical situation evolved. In 1885, Lorentz raised questions about Stokes's circuital conditions. Meanwhile FitzGerald and the two Olivers, Heaviside and Lodge, progressively increased their grasp of Maxwell's theory to show—in a theme fascinatingly near the one Einstein would start from—that the force between two electric charges depends on their motion with respect to a common frame. A letter from FitzGerald in 1889 to the American Journal of Science made the point. The negative result of the Michelson–Morley experiment could be explained by assuming that intermolecular forces obey the same laws as electromagnetic forces. Michelson's apparatus would shrink by exactly the right amount to compensate for the expected velocity difference. Lorentz independently reached the same result in 1893.

RELATIVITY, MAXWELL, POINCARÉ, AND EINSTEIN

Relativity, supposedly Einstein's word, may be traced backwards or forwards, with an explosion of use best caught, and dated, in the wonderful title of a rather boring book from 1921, The Reign of Relativity, by the widely read Germanophile philosophic statesman, J. S. Haldane. Haldane had been influenced by exchanges with the philosopher, Ernst Cassirer; that and Einstein might suggest German origin, but that is not the case. Certainty is difficult; its first use appears to have been by

Coleridge, in 1833, but only once, and in a characteristically obscure way that was mainly important as a stimulus to the young Hamilton in his reading of Kant.

Behind the many uses of 'relativity' in nineteenth century philosophizing is a trail leading to Hamilton and the stimulus he had on young minds, not least Maxwell's. So we come, in 1877, to Maxwell's short book Matter and Motion, the outcome of a plan he had laid down for his friend C. J. Monro 20 years earlier for 'a general summary of mechanical principles ... into the smallest number of independent truths'. In section 18 'Absolute Space' comes the Hamiltonian remark 'all our knowledge, both of time and place, is essentially relative'. Then at a culminating point in the book, appear sections 102 'Relativity of Dynamical Knowledge' and 103 'Relativity of Force'. The first sentence of 102 reads: 'Our whole progress up to this point may be described as a gradual development of the doctrine of relativity of all physical phenomena', leading, after suitable exposition, to this poetic paragraph, where science flows over to a feeling about life:

> There are no landmarks in space; one portion of space is exactly like every other portion, so that we cannot tell where we are. We are, as it were, on an unruffled sea, without stars, compass, soundings, wind, or tide, and we cannot tell in what direction we are going. We have no log which we can cast out to take a dead reckoning by; we may compute our rate of motion with respect to the neighbouring bodies, but we do not know how these bodies may be moving in space.

Section 103 addresses the equivalence of gravitation and inertia in a way never before done, virtually identical with the one Einstein would come to; sections 104 to 106, decorated with a beautifully apt quotation from Milton's Paradise Lost, cover, without being able to solve any more than Einstein could, the bafflements of absolute rotation.

Poincaré was the bridge. His reading of Matter and Motion well is evident, not only in relativity but in his famous work on dynamical instabilities (chaos theory) with its connections to Maxwell's section 19 'The general maxim of physical science'. As for 'relativity', word and concept, the huge public success of Poincaré's Science and Hypothesis, appearing in 1900, was definitive. From there, directly or indirectly, it flowed to Einstein, who was, however, coy in making acknowledgements, especially to Frenchmen. His first paper, 'On the electrodynamics of moving bodies', embodies it with admirable lucidity in the appropriately Germanic mouthful Relativitätsprinzip. Turning to two works of Cassirer's, Substance and Function (1910) and Einstein's Theory of Relativity Considered from the Epistemological Standpoint (1921), we find in both, relevant, well-discussed quotations from Matter and Motion, as well as a passage, in 1921, on 'the relativity of knowledge' that might have come from Hamilton or from the essay that Maxwell, at 24, had read 65 years earlier to the Apostles Club at Cambridge.

Except, of course, that while Maxwell and Einstein were right in their knowledge of the centrality of the velocity of light and the relativity of motion, there was also a point on which, as we shall see, both of them were wrong.

A ROSE BY ANY OTHER NAME?. . .

We are universally told that Einstein, in 1905, made the ether unnecessary. In what sense is this correct? The question requires reflection in light of: i) the almost totally bogus accounts of the history of the ether; ii) the fact that few people who toss the word ether around bother to say, or possibly think, what they mean by it. Einstein's own words are:

> The introduction of a 'luminiferous ether' will prove to be superfluous inasmuch as the view here to be developed will not require an 'absolutely stationary space' provided with special properties, nor assign a velocity-vector to a point in the empty space in which the electromagnetic processes take place.

The passage has unintended irony. For Einstein—and for Maxwell before him—there was no need to 'assign a velocity-vector to a point in the empty space in which the electromagnetic processes take place', but the 3K cosmic blackbody radiation, discovered in 1965 by A. Penzias and R. Wilson, does assign such a velocity. I now discuss six possible senses in which Einstein may or may not have been correct.

- **Sense 1 Yes**—in the framework of the kind of measurement represented by the Michelson-Morley experiment there are, as Maxwell and Einstein both recognized, 'no landmarks in space'.
- **Sense 2 No**—in field theories, energy is disseminated in space, an 'etherish' thing to have. Credit here rests squarely with Kelvin with his and Maxwell's expressions for energy density ½ B.H, ½ D.E. The anti-ether side of the matter is that these do not correspond to Kelvin's initial expectation that they must be kinetic and potential energies $\frac{1}{2}\rho v^2$, $\frac{1}{2}kx^2$ within a mechanical framework.
- **Sense 3 Yes**—our difference from Kelvin in Sense 2 is traceable to the masslessness of the photon, balancing which, however, is the observation that sunlight has mass (2×10^9 tons in a sphere the radius of the Earth's orbit as compared with Kelvin's 2.4×10^{12} tons).
- **Sense 4 No**—the richness of the vacuum: the existence of virtual particles in quantum field theory conveys a very different sense of the vacuum from Einstein's in 1905.

- **Sense 5 No**—whether one calls them ether or not, dark matter and dark energy have mass. The density of dark energy is $\sim 10^{-29}$ gm/cc, 10 orders of magnitude below Kelvin's ether density but still there as surely as he believed the ether to be.
- **Sense 6 No**—the supreme irony, and one of the most surprising of all discoveries of twentieth century physics, mentioned already, the 3K cosmic background radiation means that there are landmarks in space.

It is appropriate to end on a personal note. Kelvin died in 1907. In the following year, Silvanus Thompson, the Principal of the City and Guilds Technical College in Finsbury, began in earnest the biography he had already discussed with Kelvin. My father, Robert Everitt, was a student at Finsbury in 1908 to 1909 and heard many anecdotes of Kelvin from Thompson, some of which reached me as a young boy. Having from my own direction long been interested in Maxwell, it is good finally to be able through this study to come to an appreciation of the man of whom I heard so much during discussions about life and its meaning at the family dinner table. As Thompson observed, we shall not see Kelvin's like again. Nor, it might be added, Maxwell's or Einstein's like either.

ACKNOWLEDGEMENT

This article began as a talk to the Kelvin Club of Peterhouse, Cambridge. I am grateful to Professor Christopher Calladine for arranging the invitation and Lord Wilson of Tillyorn, the master, for the gracious hospitality and the intellectual exchange.

15
Kelvin and Statistical Mechanics

Oliver Penrose

INTRODUCTION

Besides being one of the founders of equilibrium thermodynamics, Kelvin played a big part in the development of statistical mechanics, both for equilibrium and non-equilibrium. This chapter will review these developments, taking a particular interest in Kelvin's own contributions.

KELVIN AND THERMOELECTRICITY

In the year 1821, three years before Kelvin's birth, Thomas Johann Seebeck from the city that is now called Tallinn (Estonia), discovered that if an electric circuit is made by joining both ends of two wires made from different metals, and the junctions are at different temperatures, then an electric current will flow. Seebeck's effect is the basis of the thermocouple, a method of measuring temperatures which is useful in places such as the furnace of your heating system, where the usual mercury-in-glass thermometers would melt. It would be nice also to be able to use the Seebeck effect to generate electricity, perhaps putting one junction in the sunshine and the other in a cool river; but unfortunately the voltages the effect produces are very small and there are much better methods of turning sunlight into electricity.

In the Seebeck effect temperature differences produce electricity. One might ask whether the causation can also go the other way: can electricity produce temperature differences? In 1834 the French scientist Jean Peltier showed that the answer is 'yes'; he found that when a current flows through a junction where two different metals meet, heat is either emitted or absorbed (depending on the direction of the current).

Thomson (as Kelvin was still called at that time) looked for a quantitative connection between the Seebeck and Peltier effects. As one of the founders of the science of thermodynamics, the 27-year-old Thomson was well placed to apply thermodynamic arguments to the problem of finding such a connection, and he did this in his theory of thermoelectricity (Thomson 1851, 1854).

Let us denote by Π the Peltier heat of a particular metal, that is, the amount of heat carried by unit electric charge moving from one place to another in that metal, and by σ the metal's Seebeck or thermoelectric coefficient, that is, the electric voltage between two places in that metal whose temperatures differ by one degree, when no current flows. Thomson's argument led to the conclusion that Π and σ are related by the formula

$$\Pi = T\sigma$$

where T is the temperature, measured on the Kelvin temperature scale, of course.

To arrive at this formula, Thomson went beyond the boundaries of thermodynamics as it was then being formulated by him and others; that is to say, of equilibrium thermodynamics. Thermoelectricity involves two processes which do not happen at equilibrium: electric conduction and heat conduction. In a given experiment one can make either one of these negligibly small by making either the electric current or the temperature differences small enough, but one cannot get rid of both without also reducing the effect one is looking for to zero. Thomson himself (1851) puts it thus:

> As it cannot be shown that the thermal effect[1] of [heat conduction] is infinitely small compared with that of the electric current, unless [the current] be so large that the term [expressing the effect of electrical heating] cannot be neglected, the conditions for the application of [the second law of thermodynamics] . . . are not completely fulfilled.

Thermoelectricity is therefore an essentially non-equilibrium phenomenon: the thermoelectric effects are inevitably accompanied by at least one of the irreversible processes of electric conduction and heat conduction. Equilibrium thermodynamics did not give a clear recipe for dealing with such situations.

To get around the difficulty Thomson (1854) formulated the following hypothesis, which he adjoined to the usual principles of thermodynamics:

> The electromotive forces produced by inequalities of temperature in a circuit of different metals, and the thermal effects of electric currents circulating in it, are subject to the laws which would follow from the general principles of the dynamical theory of heat if there were no conduction of heat from one part of the circuit to another.

Of this he said 'it must be distinctly understood that it is only a hypothesis, and that, however probable it may appear, experimental evidence in the special phaenomena of thermoelectricity is quite necessary to prove it'.

Thomson's own application of his hypothesis worked brilliantly, and later on the same type of argument was successfully applied to other phenomena including the production of electricity by differences of chemical concentrations, such as occurs in electric batteries (Helmholtz 1878); but the question remained whether the hypothesis is true in general, and if not how to identify the conditions where it is true and why it works. Thermodynamics as it was developed by Kelvin and others in the nineteenth century could not answer these questions. They could not be answered until thermodynamics had evolved into the twentieth century subject of statistical mechanics.

The main new ideas that contributed to this evolution were (i) the use of microscopic models, in which matter is represented as a system of interacting particles rather than as a continuous medium, and (ii) the use of mathematical probability theory. The rest of this chapter will show how these ideas were put together to make a theory that was ultimately capable of handling the tricky questions raised by Thomson's work on thermoelectricity.

A GAS MODELLED AS A COLLECTION OF MOLECULES

The idea that matter might be composed of tiny particles is more than two thousand years old, certainly much older than thermodynamics; but it was only after Newton's formulation of the laws of mechanics in 1687 that the idea could be turned into a quantitative theory explaining things like pressure in terms of the motion of atoms or molecules. This was done by Daniel Bernoulli (1738). In Bernoulli's model, the gas is represented as a system of point particles which interact with the walls of the container but not with each other, so that they travel with constant velocity in straight lines except when they bounce off the walls like billiard balls off the cushion.

This model was developed further by various other scientists and leads to the following simple relation between the pressure p of a gas, its volume V, and its temperature T, measured on the Kelvin temperature scale:

$$p = (N/V)\,kT$$

where N/V is the number of particles (molecules) per unit volume and k is a constant known as Boltzmann's constant, which is necessary to allow for the fact that the temperature and the mechanical quantity pV are measured in different unit systems. The numerical value of Boltzmann's constant is extremely small, 1.4×10^{-23}

in the standard (metre-kilogram-second) system of units, and correspondingly the value of N/V is normally very large, around 3×10^{25} molecules per cubic metre in air at sea level[2].

Bernoulli's model can be improved by including interactions between the molecules. At low densities, such as that of our atmosphere, these interactions have only a small effect on equilibrium properties, but they are essential for any theory of non-equilibrium properties such as heat conduction and viscosity. To deal with the interactions quantitatively, people needed to know how big the molecules were (and, indeed, some people even needed to be persuaded that molecules existed.) Thomson (1870) made a major contribution in a paper estimating the diameter of a molecule by four independent methods. His estimate that 'the diameter of the gaseous molecule cannot be less than 1/500,000,000 of a centimetre' (i.e., in modern language, 0.02 nanometres[3]), is consistent with today's more accurate information, according to which the diameter of a simple molecule such as oxygen, nitrogen, or argon is around 0.3 nanometres.

The quantitative theory of the effect of these interactions in gases that were not in equilibrium was worked out mainly by James Clerk Maxwell and Ludwig Boltzmann. It is now called the kinetic theory of gases. In their model it is assumed that the molecules are point particles which move most of the time in straight lines, but from time to time a pair of particles bounce off one another; the collision changes the direction of motion and the speed for both particles, while conserving their total energy and their total momentum. The effect of these collisions, in general, is to change the distribution of velocities in the gas. This work culminated in the formulation of Boltzmann's kinetic equation (1872), which tells us how the distribution of velocities in a gas changes with time as a result of collisions.

By the distribution of velocities we mean a function f, depending on a vector $\mathbf{v}$ whose three components we write v_x, v_y, v_z, such that (for a suitably chosen small number h) the number of particles whose x, y, and z components of velocity lie simultaneously in the ranges $v_x - 0.5h$ to $v_x + 0.5h$, $v_y - 0.5h$ to $v_y + 0.5\,h$, and $v_z - 0.5h$ to $v_z + 0.5h$ respectively is approximately $h^3 f(\mathbf{v})$. As time progresses the distribution function f changes, and Boltzmann's kinetic equation tells us how it changes.

Boltzmann was able to show that, according to his kinetic equation, the distribution of velocities would approach a particular distribution called the Maxwellian distribution, which characterizes thermal equilibrium. The Maxwellian distribution function is

$$f(\mathbf{v}) = \text{const.}\ \exp\left(-\text{const.}\ \mathbf{v}^2\right).$$

where $\mathbf{v}^2$ means $v_x^2 + v_y^2 + v_z^2$, and the constant multiplying $\mathbf{v}^2$ depends on the temperature (which is related to average energy per molecule). Boltzmann's method of

Fig. 15.1. Ludwig Boltzmann. (Courtesy AIP Emilio Segre Visual Archives.)

proof depended on a quantity he called H, which depends on the velocity distribution function f, and which (according to his kinetic equation) has to decrease with time so long as the distribution function f is non-Maxwellian, that is, so long as the gas is out of equilibrium. This celebrated result came to be known as the H-theorem.

THE REVERSIBILITY PARADOX AND THE NEED FOR PROBABILITY

A characteristic feature of all models describing matter microscopically in terms of molecules is that the assumed laws of motion do not explicitly include any frictional forces. Energy is conserved, which it would not be if there were frictional forces in the microscopic model. If there were microscopic friction, the molecules would soon come to rest, but one of the fundamental ideas of thermodynamics is that heat is regarded as a form of molecular motion and this motion therefore continues for ever. But how is the microscopic lack of friction to be reconciled with the frictional effects we observe in everyday life at the macroscopic level? Or, as Thomson might have put it, how is the lack of energy dissipation in microscopic mechanics to be reconciled with the dissipation which is embodied in the second law of thermodynamics—what Thomson (1852) had called in his paper formulating that law 'the universal tendency in Nature to the dissipation of mechanical energy'?

The problem can be expressed more precisely if we notice that the mathematical laws (equations) used to describe frictionless motion are 'reversible', that is to say invariant under time reversal, whereas those used to describe motion with friction are not. Thus if a film of some frictionless motion, for example the motion of the planets, were to be shown in reverse, the reversed motion would also be possible; but a film showing a motion involving friction, such as a rolling ball coming to rest, would not have this symmetry, since the reversed film would show something impossible, namely the ball starting to move spontaneously.

Thomson (1874) was one of the first to discuss this problem, which came to be known as the reversibility paradox. His paper arose out of discussions between Thomson, Maxwell, and Peter Guthrie Tait with whom Thomson had collaborated on the extremely original *Treatise on Natural Philosophy* (Thomson and Tait 1867). In the 1874 paper Thomson contrasts 'abstract' (frictionless) dynamics with 'physical' dynamics (as actually observed). He says:

> In abstract dynamics the instantaneous reversal of the motion of every moving particle causes the system to move backwards, and at the same speed as before, when again at the same position. That is to say, in mathematical language, any solution [of the mathematical equations describing the motion] remains a solution when t is changed to $-t$. In physical dynamics this simple and perfect reversibility fails, on account of forces depending on friction of solids; imperfect fluidity of fluids; imperfect elasticity of solids; inequalities of temperature . . .

. . . and so on. In a poetic passage, Thomson illustrates the reversibility of the microscopic motions by imagining that:

> If, then, the motion of every particle of matter in the universe were precisely reversed at any instant the course of nature would be simply reversed for ever after. The bursting bubble of foam at the foot of a waterfall would reunite and descend into the water; the thermal motions would reconcentrate their energy and throw the mass up the fall in drops re-forming into a close column of ascending water. Heat which had been generated by the friction of solids and dissipated by conduction, and radiation with absorption, would come again to the place of contact, and throw the moving body back against the force to which it had previously yielded. Boulders would recover from the mud the materials required to rebuild them into their previous jagged forms, and would become reunited to the mountain peak from which they had formerly broken away . . . .

But of course such things do not happen in reality; the reversed motions imagined by Thomson, although consistent with the equations of microscopic mechanics, are totally implausible. How to reconcile the possibility of these extraordinary motions with the fact that they do not happen in practice?

Thomson goes on to consider the example of 'a bar of metal ... varnished with a substance impermeable to heat'. The bar is imagined 'to be first given with one-half of it at one uniform temperature, and the other half of it at another uniform temperature'. Thomson argues that:

> If no selective influence ... guides individual molecules, the average result of their free motions and collisions must be to equalise the distribution of energy among them in the gross ... Suppose now the temperature to have become thus very approximately equalized at a certain time from the beginning, and let the motion of every particle become instantaneously reversed. Each molecule will retrace its former path, and at the end of a second interval of time, equal to the former, every molecule will be in the same position, and moving with the same velocity as at the beginning [but in the opposite direction]; so that the given initial unequal distribution of temperature will again be found.

But, he argues:

> It is a strange but nevertheless a true conception of the old well-known law of the conduction of heat, to say that it is very improbable that in the course of 1000 years one-half of the bar of iron shall of itself become warmer by a degree than the other half ... This one instance suffices to explain the philosophy of the foundation on which the theory of the dissipation of energy rests.

So, in effect, Thomson is saying that although the reversed motion is theoretically possible, it is highly unlikely to happen spontaneously; it will only happen if the system is started off in some very peculiar way. To obtain a complete theory of irreversible processes we need some rule starting the system off in a sensible way. The important point at this stage is that microscopic mechanics alone cannot provide such a rule. It must be supplemented by including some additional information besides the laws of mechanics into the mathematical model. The next question is what form this additional information should take. The information should be of such a nature that it will tell us that certain types of mechanically possible motion, including those envisaged by Thomson in his reversed waterfall, are 'highly unlikely'. That is to say, it will be information about probabilities.

MATHEMATICAL PROBABILITY MODELS

In mathematical models involving probability, the probability p of an event is a number (either 0 or 1 or something in between) such that if the circumstances where the event might occur are repeated a large number of times the fraction of occasions on which the event actually does occur is approximately p; thus, the statement that the probability that a spun coin will come up 'heads' is $\frac{1}{2}$ implies that if the coin is spun many times then about half the spins will give the result 'heads'. The mathematical theory is based on rules for calculating the probabilities of different events which are related in some way. The link between the mathematical theory and real life is provided by a *rule of interpretation* along these lines: if the mathematical theory tells you that the probability of some future event is very very small (less than 10^{-12}, or whatever other small number you think suitable), then it makes sense (if you trust the theory) to live your life on the assumption that the event will not happen. Similarly, if the theory says that the probability is very close to 1 (greater than $1 - 10^{-12}$, say), then it makes sense to live your life as though the event will happen. Insurance companies and the owners of casinos run their businesses very successfully by following such rules of interpretation.

To formulate a theory that distinguishes the microscopic motion describing a commonplace macroscopic event, such as water falling down a waterfall, from the reversed motion describing a preposterous macroscopic event, such as water going up a waterfall, we want a way of attaching probabilities to different motions in such a way that a much higher probability is given to the commonplace than to the preposterous. Such theories are formulated by combining microscopic mechanical models with plausible assumptions about probabilities. The branch of science that deals with such theories has come to be known as statistical mechanics.

Thomson's 1874 paper contains what is perhaps the first statistical mechanics calculation:

> Let a hermetically sealed glass jar contain 2,000,000,000,000 molecules of oxygen and 8,000,000,000,000 molecules of nitrogen. If examined at any time in the infinitely distant future, what is the number of chances against one that all the molecules of oxygen and none of the nitrogen shall be found in one stated part of the vessel equal in volume to 1/5 th of the whole? The number expressing the answer in the Arabic notation has about 2,173,220,000,000 of places of whole numbers.

This result comes from a mathematical model in which specific assumptions are made about the numerical values of certain probabilities: Thomson's statement of this assumption is 'Let A,B denote any particular portions of the whole containing vessel and let *a,b,* be the volumes of those portions. The chance that at any instant one individual molecule of whichever gas shall be in A is $\frac{a}{a+b}$'.

Thomson's calculation illustrates what we might call the quasi-determinism of large systems. The theory he is using is not deterministic, it does not say that the event he is considering (finding all the oxygen and none of the nitrogen in the stated part of the vessel) is impossible, yet in any practical situation we can safely assume that the event will not happen, just as safely as if the theory were deterministic. To pursue this example a little further, it is obvious that on average about 20% of the oxygen atoms will be in the stated part of the vessel, but what is the probability that at the time when the observation is made the actual percentage of atoms in that fifth of the vessel lies between 19 and 21? For the model used by Thomson it can be shown that this probability differs from 1 by less than 10^{-9}, so close to 1 that it would normally be quite safe to assume that the actual percentage will lie between 19 and 21, even though there is no deterministic theory to tell us what the actual percentage will be. Every time you take a breath, you rely on the fraction of oxygen atoms in the breath you have taken being close to 20%, and the (non-deterministic) mathematical model used by Thomson predicts that you will not be disappointed.

BOLTZMANN'S RESPONSE TO THE REVERSIBILITY PARADOX

When the reversibility paradox first attracted attention, it seemed to pose a grave difficulty for kinetic theory. This difficulty was drawn to Boltzmann's attention (see Brush (1976)) by his friend Josef Loschmidt, who was aware of Thomson's 1874 paper. The problem was to reconcile the irreversibility of Boltzmann's kinetic equation, which is manifest in the aforementioned *H*-theorem derived from that kinetic equation, with the reversibility of the microscopic mechanics from which the kinetic equation is supposed to be derived.

Boltzmann (1877a, b, c) responded to Loschmidt's criticism in one of his most original and important papers (1877c). He dealt with the reversibility paradox by putting aside the deterministic thinking that appeared to lie behind his kinetic equation and replacing it by a more detailed model in which probability was central. Not only did changing to this new model make the new theory compatible with the reversibility of the equations of motion, but it also enabled him to explain the physical significance of the previously mysterious quantity *H*. In fact he was able to provide two physical interpretations of *H*: one mechanical and one thermodynamic, and the two interpretations together provided a quantitative link between mechanics and thermodynamics.

To understand Boltzmann's new conception, consider a simplified model in which the velocity of each particle must be chosen from one of a discrete set of possible values; let us call these possible velocities **a, b, c, d,** etc. As a further simplification, we ignore the positions of the particles. Doing this is an over-simplification because it does not take into account the motion of molecules between collisions, but it does highlight the role of probability in describing the effect of collisions.

The velocity distribution at any moment can be specified by giving the number of particles having each of the possible velocity values; we denote the number of particles having velocity **a** by n_a, the number having velocity **b** by n_b, and so on.

These numbers can be called the *occupation numbers* for the different possible velocity values. Their sum is equal to the total number of particles in the gas. Each time there is a collision the occupation numbers change, for example if two particles collide, their pre-collision velocities being **a, b** and their post-collision velocities **c, d,** then the collision decreases the values of n_a and n_b by 1 while increasing those of n_c and n_d by 1.

In this model, each collision is treated as a separate random event: two particles are selected at random, and once they have been selected their post-collision velocities are selected by a further random mechanism. Once the pre-collision velocities are given, there are well-defined probabilities for the various possible post-collision velocities; the details of the formula for these probabilities are unimportant at this stage, except that all collisions violating energy conservation must have zero probability.

At each collision, the occupation numbers change in a random way, and so their time dependence cannot be predicted in advance. What can be predicted, however, is the time dependence of the *expectations* of the occupation numbers. In probability theory, the expectation of a random quantity means a probabilistic average; thus, to say that the expected length of life of a particular newborn baby is so many years is the same thing as saying that this number of years is the average of the lengths of life of some large collection of newborn babies whose start in life is essentially the same as that particular baby, but whose fortunes will vary from one baby to another as decreed by Fate. In the same way, the expectation of n_a means the average of n_a over an imaginary collection of copies of the gas, each started off in the same way, but suffering independently of one another the random decisions of whatever chance mechanism determines the collisions. Such an imaginary collection of copies of a physical system is known in statistical mechanics as an *ensemble*, and the expectation of a dynamical variable is often called an ensemble average. The symbol commonly used in statistical mechanics for an expectation or ensemble average is $\langle\ \rangle$ so that, for example, $\langle n_a \rangle$ denotes the ensemble average of the occupation number n_a.

The distribution function f appearing in Boltzmann's kinetic equation can be defined as the function such that

$$f(\mathbf{a}) = \langle n_a \rangle,\ f(\mathbf{b}) = \langle n_b \rangle,\ f(\mathbf{c}) = \langle n_c \rangle, \text{ etc.}$$

Boltzmann's kinetic equation is a formula giving the time evolution of the function f, based on plausible assumptions about the average number of collisions of each type at each moment in time. It does not tell us how the actual occupation numbers for any particular sample of gas will vary with time, but it does tell us how their expectations will vary with time. The kinetic equation is not symmetric under time reversal, but there is no conflict with the time-reversal symmetry of the laws of mechanics because the random collision mechanism involves probabilities, and probabilities are not, in general, symmetric under time reversal.

For the same reason, there is no conflict between the monotonic decrease of H predicted by Boltzmann's H theorem and the reversibility of the laws of mechanics. The H appearing in Boltzmann's theorem is defined by

$$H(f) := f(\mathbf{a}) \log f(\mathbf{a}) + f(\mathbf{b}) \log f(\mathbf{b}) + \ldots$$

and so it is hardly surprising that the time variation of H, which is determined by the irreversible time variation of f, is also irreversible.

Boltzmann was able to provide a physical interpretation of H which makes it much easier to understand why H should decrease with time. To describe this interpretation, we need the idea of a *complexion;* this means a way of allocating the particles to velocity states. For example, suppose there are just four particles, numbered 1,2,3,4, and four possible velocity states, say **a,b,c,d.** One of the possible complexions is to allocate particle number 1 to velocity state **a,** particle 2 to state **b,** and particles 3 and 4 to states **a** and **c** respectively; this complexion could be described by the 'word' **abac.**

Imagine the gas started out in any complexion whatever, and to evolve by collisions according to the random mechanism described above. It will pass through many complexions, and the probabilities for the various complexions will change with time. Eventually the change will slow down: the probabilities for the various complexions will approach equilibrium values. It turns out that these equilibrium probabilities are given by a very simple rule; the rule is that (assuming a natural time-reversal property of the collisions) all the accessible complexions—that is, all the complexions that can be reached from the given initial complexion by some sequence of possible collisions—have the same probability. Thus, in the equilibrium ensemble, the collision taking the system from a given complexion C_1 to another given complexion C_2 takes place just as often as the one taking it from C_2 to C_1, a state of affairs known as detailed balancing.

Boltzmann's equation is directly concerned not with the complexion of the gas but with its velocity distribution, that is with its occupation numbers. These occupation numbers can be written as a list; thus (with just four velocity states) the list (2,1,1,0) means $n_a = 2$, $n_b = 1$, $n_c = 1$, $n_d = 0$. Every list of occupation numbers corresponds to a definite collection of complexions; for example, the list (2,1,1,0) (meaning $n_a = 2$, $n_b = 1$, $n_c = 1$, $n_d = 0$) corresponds to 12 different complexions: in alphabetical order they are: **aabc, aacb, abac, abca, acab, acba, baac, baca, bcaa, caab, caba, cbaa.** For each occupation number list, we may define its *multiplicity* to be the number of complexions that are compatible with it; so the multiplicity of the occupation number list (2,1,1,0) is 12. It is customary to denote the multiplicity by W, so we can write $W(2,1,1,0) = 12$.

At equilibrium each accessible complexion has the same probability, and therefore the probability of each accessible distribution (occupation number list) is proportional to the number of ways that distribution can arise—that is, to its multiplicity. For this reason the multiplicity is sometimes called the thermodynamic probability, but this name is confusing because the multiplicity is usually greater than 1 whereas probabilities are never greater than 1.

There is a general formula giving the multiplicity of any occupation number list, which will not be given here. When the occupation numbers are large, the formula can be approximated as

$$\log W(n_a, n_b, \dots) = N \log N - H(n_a, n_b, \dots)$$

where $N = n_a + n_b \dots$ is the total number of particles and $H(n_a, n_b, \dots)$ means $n_a \log n_a + n_b \log n_b \dots$. Now, Boltzmann's H theorem tells us that $H(f(\mathbf{a}), f(\mathbf{b}), \dots)$ decreases with time, that is to say that

$$H(\langle n_a \rangle, \langle n_b \rangle, \dots)$$

decreases with time; so on average it is to be expected that $H(n_a, n_b, \dots)$ will also decrease with time and hence that $\log W(n_a, n_b, \dots)$ will increase with time. Thus Boltzmann's H theorem is a statement that, on average, the velocity distribution (occupation number list) will change in such a way that its multiplicity increases. We can interpret this by saying that, starting from any given velocity distribution, the number of possible collisions that increase the multiplicity is greater than the number of collisions that decrease it; correspondingly the probability of a collision that increases the multiplicity is greater than the probability of a collision that decreases it, and so it is to be expected that the multiplicity will increase with time. The velocity distribution having the highest multiplicity of all is the Maxwellian distribution; that is why the equilibrium distribution is Maxwellian.

To this insight, Boltzmann added an even more valuable one. He calculated that for a gas in equilibrium the entropy S is related to the multiplicity W of the equilibrium velocity distribution by a relation of the form

$$S = k \log W$$

This equation, sometimes known as Boltzmann's principle, is engraved on Boltzmann's tomb in Vienna. Boltzmann's principle is at the heart of both equilibrium and non-equilibrium statistical mechanics. It provides a direct link between the thermodynamic quantity entropy and the mechanical quantity *W*, it indicates how entropy may be defined for systems that are not in equilibrium, and it gives a statistical mechanics interpretation of the thermodynamic principle of increasing entropy. Thus what Thomson (1852) called 'A universal tendency in nature to the dissipation of mechanical energy', and which later on would have been called a universal tendency for entropy to increase, could now be re-interpreted as a universal tendency to move from a state of lower multiplicity to a state of higher multiplicity.

SOME FURTHER DIFFICULTIES WITH BOLTZMANN'S EQUATION

Although Boltzmann's 1877 paper was a big advance, it did not entirely satisfy his critics. In 1890 Edward Parnall Culverwell of Trinity College, Dublin questioned Boltzmann's derivation of the H theorem; his memoir (Culverwell 1890) is entitled 'Note on Boltzmann's kinetic theory of gases and on Sir W. Thomson's address to Section A, British Association 1884'. The widespread interest in such problems, which included three papers by Thomson doubting certain aspects of the Maxwell equilibrium distribution, led the BA to set up a committee of two, Joseph Larmor and George Hartley Bryan, to report on the subject. It is a little surprising that Thomson himself was not on the committee; possibly the reason was that he had been one of the protagonists in the preceding discussion. The problem identified by Culverwell was essentially the reversibility paradox again but, as Brush (1976, p. 620) puts it, attention was now fixed not on the general question of compatibility between the reversibility of mechanics and the irreversible behaviour of macroscopic objects, but to the particular one of identifying the stage in Boltzmann's original (1872) apparently deterministic derivation of the *H* theorem where irreversibility had sneaked in.

The culprit, apparently first identified in 1895 by Samuel Hawksley Burbury, a barrister who had turned to mathematics after becoming deaf, is an assumption which he called 'molecular disorder' and which Boltzmann (1877) had called the *Stosszahlansatz* (hypothesis about the number of collisions). At each collision the molecules approach one another with certain velocities, say $\mathbf{v}_1$ and $\mathbf{v}_2$, and recede from one another with different velocities, say $\mathbf{v}_1'$ and $\mathbf{v}_2'$. The post-collision velocities depend on the initial velocities and also on the way the molecules approach one another. Boltzmann had calculated the number of collisions of each type (i.e. with each pair of approach velocities) by assuming that the pairs of approaching molecules are independent of one another[4], that is, that if you could sit on any molecule and await the next collision the collection of potential future collision partners would look exactly the same no matter what the velocity of the molecule you were sitting on. This would not necessarily be the case with respect to past collision partners: as an example, suppose the gas had been started in a special way that gave all the molecules exactly the same energy, then if after a collision your own molecule is going faster than the average, then the chances are that the molecule it just collided with is going slower than the average (since your molecule took some energy from it) whereas if your molecule is going slower than average the chances are that its previous collision partner is going faster—a correlation between the energy of your particle and that of its previous collision partner. This example shows that there can easily be a correlation between the energies of molecules that have just collided, whereas one never expects to find any correlation between molecules that are about to collide.

In fact, since the molecular disorder assumption is made afresh at each collision, Boltzmann's original derivation made not one probability assumption but a large number of separate probability assumptions, one for each pair of approaching molecules. However, these particles are part of a larger system whose parts are interacting with each other by collisions. Could it be that, over a period of time, these interactions might cooperate in some way so as to produce a correlation between the approaching particles which Boltzmann's derivation assumes to be uncorrelated? In other words, is Boltzmann's equation really consistent with the mechanics of large systems of interacting particles?

It was not until nearly a century later that Oscar Lanford (1975) was able to produce a mathematical derivation of Boltzmann's equation (in a certain limiting case) which is fully consistent with the laws of mechanics. Lanford's theorem still uses a probability assumption—some such assumption is inevitable, as we have seen—but his probability assumption refers only on the way the overall motion is started, not separately to the individual collisions. Nevertheless, even Lanford's theorem is not a complete solution to the problem, since it only proves that Boltzmann's equation holds for a rather short time (about 1/5 of the average time between the collisions of a given molecule) after the initial time when the system is started off; it says nothing about what happens after that time.

Another objection to Boltzmann's kinetic equation was raised by Ernst Zermelo (1896), a mathematician who later became well-known as a set theorist. He cited a mathematical theorem which says that a conservative mechanical system left to itself for long enough will eventually return close to its initial state. This contradicts the prediction of the original (1872) version of Boltzmann's *H* theorem that the velocity distribution of the system will go to equilibrium and stay there. But, like the reversibility objection of Loschmidt, this objection is also answered by the probabilistic (1877) model: in this model the *H* theorem refers to the expectation of the velocity distribution, an ensemble average. It does not prevent the velocity distribution of any individual system in the ensemble from deviating occasionally from the Maxwellian average, nor from returning, very very occasionally, back to what it was right at the beginning.

FROM KINETIC THEORY TO STATISTICAL MECHANICS

By the end of the nineteenth century the limitations of classical mechanics were beginning to show. Kelvin's turn-of-the-century lecture to the Royal Institution (Kelvin 1900) was entitled 'Nineteenth century clouds over the dynamical theory of heat and light'. One of the clouds was the prediction of kinetic theory that the energy of a gas whose molecule consisted of two atoms, such as oxygen (O_2) or nitrogen (N_2), would be (essentially) $3kT$ per molecule; experiment gave a result much closer to $2.5\,kT$ per molecule. The other cloud was the (obviously false) prediction of classical theory that electromagnetic radiation in thermal equilibrium with a body at any temperature, no matter how cold, would have an infinite energy. In both cases, classical (Newtonian) mechanics was over-estimating the energy. When he gave his lecture, Kelvin perhaps did not anticipate that these clouds would only be dispelled by a complete revolution in physics thinking, which began with Max Planck's radiation law (1901) and culminated a quarter of a century later in the mathematical formulation of quantum mechanics. But despite this revolution, classical mechanics remains a good approximation for many purposes, and in the present article, for simplicity, I shall concentrate on the part of the theory where the classical approximation is useful.

One of the limitations of the nineteenth-century dynamical theory of heat, even within the classical approximation, was that it was restricted to low densities and therefore to a particular state of matter, namely gases. What can be said about liquids and solids? The assumptions made in the kinetic theory of gases no longer apply: we cannot treat the interactions between the molecules as occasional 'collisions' interrupting their otherwise free motions, since each molecule is interacting with its neighbours the whole time. A big advance, making it possible to incorporate such systems into the theory, was made by Josiah Willard Gibbs of Yale University, USA. His achievement was to put the theory into an elegant mathematical form, discarding the complications that had made Boltzmann's work so difficult for his contemporaries to accept, while retaining the insights that had made it so effective.

Gibbs (1902) considered a general mechanical system comprising a very large number of particles, without any special assumptions about what state of matter it represented. The particles were assumed to obey the general laws of (frictionless) mechanics as formulated by Lagrange, Hamilton, and others. In this formulation the mechanical state of the system is specified by giving the positions and velocities of all the particles. If there are N particles, this specification requires $6N$ numbers, 3 for each particle (its x, y, and z coordinates) to specify its position in 3-dimensional space and 3 more to specify its velocity. He imagined these $6N$ numbers to be the coordinates of a point in a $6N$-dimensional space which he called *phase space*. As time progresses, this point moves around in the phase space in an extremely complicated way about which very little is known even now, but Gibbs, like Boltzmann before him, was able to overcome this ignorance by means of plausible probability assumptions. The difference from most of Boltzmann's work was that Gibbs made probability assumptions only about systems in equilibrium. It was this self-restraint that enabled him to make his theory so much more general.

Gibbs coined the term 'statistical mechanics' to describe his approach, which concentrated on equilibrium, but this term has since come to refer to any application of probability theory to a large mechanical system, whether or not it is in equilibrium. The term 'ensemble', which I have already used in this essay, also appears to be due to Gibbs. He used it to describe the probability models his work was based on. The best-known of these is the canonical ensemble, which he used to describe a system whose temperature T was prescribed. In the canonical ensemble, the probabilities of different parts of phase space are proportional to the so-called Boltzmann factor $\exp(-\mathbf{H}/kT)$ where $\mathbf{H}$ (not to be confused with Boltzmann's H) denotes the Hamiltonian of the mechanical system, that is, the energy written in terms of the $6N$ phase-space coordinates.

The best-known outcome of Gibbs' work is a set of formulas, which he called 'thermodynamic analogies', expressing various thermodynamic functions in terms of integrals over phase space. Gibbs used the cautious term 'analogies' because of the specific heat 'cloud' already mentioned; but the subsequent development of quantum statistical mechanics has dispelled the cloud and completely vindicated Gibbs' approach. Boltzmann's formula for the entropy, $S = k \log W$, can be regarded as the precursor of Gibbs' 'thermodynamic analogies'. The most important of these is his formula for the thermodynamic free energy F, defined as $E - TS$ where E is the thermodynamic energy and S is the entropy. The free energy of any object is a measure of the amount of useful work that can be obtained from the energy contained in that object. When journalists, politicians, and others talk about conserving energy (i.e. not wasting fuel), the more thermodynamically correct language for what they are saying would be conserving free energy.

The Gibbs formula for free energy is $F = -kT \log Z$ where Z is the so-called phase integral or partition function, defined by

$$Z = \text{const.} \int \exp(-\mathbf{H}/kT)\, d\omega$$

where $d\omega$ represents a volume element in the $6N$-dimensional phase space and the integration is taken over the whole of phase space. This formula, and its quantum-mechanical analogues, are used every day to calculate the equilibrium thermodynamic properties of various substances from molecular models.

FLUCTUATIONS

In statistical mechanics it frequently happens that we use the expectation of some dynamical variable as an estimate of the actual value of that variable. The question then arises how accurate this estimate is likely to be. The deviation of the actual value from the estimate is often referred to in statistical mechanics as a fluctuation, and so it is a question of estimating the likely size of the fluctuations. The calculation by Thomson in his 1874 'waterfall' paper, mentioned earlier, may have been the first fluctuation calculation, but he is only concerned to show that a certain very large fluctuation is fantastically unlikely; we are concerned now with fluctuations of a size that has a good chance of occurring, not ones that have virtually no chance of occurring.

The method used in statistical mechanics for estimating the likely size of fluctuations is exemplified by a calculation in Gibbs' book (1902). He considers the fluctuations of total kinetic energy in a large (classical) system. If the system comprises N particles, the expectation of their total kinetic energy at temperature T is easily calculated (for example from the Maxwell distribution function) to be $3/2\,NkT$. Using the notation for expectations introduced earlier, this fact can be written

$$\langle K \rangle = 3/2\,NkT$$

where K denotes the total kinetic energy of the system. But this is only an expectation. The actual energy of the system we are actually observing at a particular moment will be different from the ensemble average. We would like to know how far the actual energy is likely to deviate from the expectation value, that is to say, how large the deviation from $3/2\,NkT$ is likely to be. A conveniently calculated measure of the likely magnitude of these deviations is the expectation of its square, which is called in probability theory the *variance* of K and denoted **var** (K). In mathematical language, this definition is

$$\mathbf{var}\,(K) = \langle (K - \langle K \rangle)^2 \rangle$$

The variance, being the average of the squared deviation of K from its expectation, gives an estimate of the likely size of the square of the deviation. An estimate of the deviation itself is provided by the square root of the variance, known to statisticians as the standard deviation.

In the case of the kinetic energy, the variance can be calculated from the Maxwellian distribution; the result is

$$\mathbf{var}\,(K) = 3/2\,N(kT)^2$$

The most important part of this formula is the factor N; it tells us that the square of the deviation of the total kinetic energy from its mean is of order N, and hence that the deviation itself is of order $\sqrt{N}$ (i.e. the standard deviation is of order $\sqrt{N}$). This is much smaller (by a factor of order $1/\sqrt{N}$) than the mean itself, which we have just seen to be of order N. In general, such fluctuations are much too small to be observed, and in the applications of statistical mechanics to thermodynamics they can usually be neglected. For example, the expectation of the kinetic energy of the 3×10^{22} molecules in a litre of air is about 150 joules; the standard deviation of this estimate is smaller by a factor of around 10^{-11}, that is, its order of magnitude is a completely negligible 10^{-9} joules.

There are some situations, however, where fluctuations are not negligible. In 1828 the botanist Robert Brown observed that certain tiny particles in the pollen of plants were in continual motion. The motion was at first thought to be a manifestation of life, but it turned out that all sufficiently small objects, immersed in a liquid such as water, would behave in the same way. The random motion they execute came to be called *Brownian motion*. Obviously it is some kind of fluctuation phenomenon, but the theory about it was not understood until Albert Einstein's amazing year of 1905. Einstein (1905) obtained a probabilistic formula relating Δx, the distance travelled by the Brownian particle in a given time interval, to Δt, the duration of that time interval; the formula can be written

$$\langle (\Delta x)^2 \rangle = 2\,M\,kT\,\Delta t$$

where M denotes the mobility of the particle, that is, what (the expectation of) its speed would be if it were dragged through the fluid by a force of unit magnitude. The expression on the left is, of course, just the variance of Δx, since in this case $\langle \Delta x \rangle = 0$. Besides its intrinsic appeal, Einstein's result was important at the time because it provided a straightforward method of determining k experimentally, since M can be estimated from hydrodynamics and an estimate of the particle's radius. But today, when the value of k is known to at least six decimal places from other methods, one could use Einstein's formula in the opposite direction, as a method of determining the mobility of a given Brownian particle experimentally without having to estimate its radius.

Einstein's formula is an example of a *fluctuation-dissipation* relation: it relates the fluctuation variance $\langle (\Delta x)^2 \rangle$ to the mobility M which quantifies a dissipative effect. Later, other fluctuation-dissipation relations were discovered; one of them is the formula discovered experimentally by John Bertrand Johnson (1928) and explained theoretically by Harry Nyquist (1928) relating the fluctuations of electrical voltage across a resistor to the value of its resistance. Johnson noise can be thought of as arising from Brownian motion of the conduction electrons in the resistor. You may be able to hear this noise by turning up the volume on a battery-operated CD or audio-tape player when the CD or tape is not being played.

More recently a whole class of relations of fluctuation-dissipation type were discovered (Kubo 1957; Greenwood 1958). Just as the Gibbs phase integral is the basis of today's methods for evaluating equilibrium thermodynamic properties from molecular models, so these relations of fluctuation-dissipation type are the basis of today's methods for evaluating transport coefficients such as the coefficients of viscosity, heat conductivity etc. from these same models. The prototype of such methods would be the hypothetical determination of the mobility of a Brownian particle from Einstein's formula, mentioned above.

ONSAGER'S RECIPROCAL RELATIONS

For three-quarters of a century the questions raised by Thomson's formula $\Pi = T\sigma$ and the hypothesis he had used to derive it remained unanswered. However in 1931 Lars Onsager, a Norwegian scientist who had taken up residence in the USA, addressed the question, a piece of work for which he received the Nobel prize in Chemistry in 1968. He noted first that certain general relations which did not follow from the general principles of thermodynamics were used in other areas besides thermoelectricity. He considered the example of a chemical reaction system involving three different types of molecule A, B, C, each of which could transform into either of the others, so that two cyclic sequences of transformations are possible:

$$A \to B \to C \to A \quad \text{and} \quad A \to C \to B \to A.$$

At equilibrium the three transitions A $\rightarrow$ B, B $\rightarrow$ C, and C $\rightarrow$ A in the first cycle should all take place equally often, so that the amount of each type of molecule stays the same, and likewise the reactions A $\rightarrow$ C, C $\rightarrow$ B, and B $\rightarrow$ A in the second cycle should take place equally often; but equilibrium is maintained even if the two cycles go at different rates. However, as he noted, chemists are accustomed to impose an additional restriction, the detailed balance condition. The detailed balance condition has the effect that the rates in the two cycles are equal, that is, at equilibrium, the transitions A $\rightarrow$ B and B $\rightarrow$ A take place equally often. Guided by the efficacy of the chemists' detailed balance condition, Onsager formulated what he called the *principle of microscopic reversibility*; this principle says that at equilibrium every type of microscopic motion is just as likely to occur as its reverse.

The microscopic reversibility principle is more than a re-statement of the reversibility of the microscopic equations of motion, which Thomson had already drawn attention to in his 1874 'waterfall' paper; it is a probability assumption, saying that the equilibrium probability distribution is also reversible. This probability assumption was already implicit in the probability assumptions about equilibrium (the canonical ensemble, for example) that Gibbs had promulgated, but Onsager saw how to use it in a completely new way. His idea is related to the fluctuation-dissipation formulas already mentioned, but comes into play only when more than one dissipative process is going on at the same time—for example electric conduction and heat conduction, as in thermoelectricity.

To describe how microscopic reversibility affects fluctuations, we need some more definitions. Suppose X, Y are two dynamical variables. In any probability ensemble, the covariance of these two variables is defined by

$$\mathbf{covar}(\mathrm{X}, \mathrm{Y}) = \langle(\mathrm{X} - \langle X\rangle)(Y - \langle Y\rangle)\rangle$$

The covariance is a measure of the correlation between the two variables: if the two variables tend to vary in the same way, that is, if on most occasions either both are larger than their respective expectations or both smaller, then the covariance is positive, but if they vary in opposite ways, so that on most occasions when one is larger the other tends to be smaller, the covariance is negative.

In the above formula it is tacitly assumed that X and Y are measured at the same time, but they could instead be measured at two different times, say t_1 and t_2; the covariance of these two measurements would be $\mathbf{covar}(X(t_1), Y(t_2))$. Provided that the dynamical variables X and Y are time-reversal symmetric (i.e. symmetric under the reversal of all particle velocities), the microscopic reversibility principle implies

Fig. 15.2. Josiah Willard Gibbs. (Courtesy AIP Emilio Segre Visual Archives.)

that, for the equilibrium ensemble, this covariance is equal to **covar**$(X(-t_1), Y(-t_2))$ or indeed (adding $t_1 + t_2$ to both times, which does not affect the equilibrium properties) to **covar**$(X(t_2), Y(t_1))$, that is,

$$\mathbf{covar}(X(t_1), Y(t_2)) = \mathbf{covar}(X(t_2), Y(t_1))$$

From this it follows, subtracting **covar**$(X(t_1), Y(t_1))$ from both sides, that

$$\mathbf{covar}(X(t_1), [Y(t_2) - Y(t_1)]) = \mathbf{covar}([X(t_2) - X(t_1)], Y(t_1))$$

This is a general property of the equilibrium fluctuations of any time-reversal symmetric pair of dynamical variables. But now suppose that the non-equilibrium macroscopic behaviour of these dynamical variables is described by a self-contained pair of constitutive equations, expressing their (expectation) rates of change in terms of the (expectation) values of the dynamical variables themselves and nothing else. For the purposes of this discussion the constitutive equations will be assumed to express the rates of change of X and Y as linear combinations of the values of X and Y themselves so that, for small enough positive $t_2 - t_1$, the expectation values of $X(t_2)$ and $Y(t_2)$ (conditional on the given values of $X(t_1)$ and $Y(t_1)$) are related to those given values by

$$X(t_2) - X(t_1) = -[AX(t_1) + BY(t_1)]\,(t_2 - t_1)$$
$$Y(t_2) - Y(t_1) = -[CX(t_1) + DY(t_1)]\,(t_2 - t_1)$$

where A,B,C,D are constants.

The crucial point in Onsager's argument is what he calls a 'reasonable general assumption', which sets up the required connection between equilibrium fluctuations and macroscopic irreversible processes. The assumption is this: if at some time t_1 a set of dynamical variables such as X, Y, have certain particular values then it makes no difference to the subsequent average behaviour of X and Y whether they reached those particular values accidentally from fluctuations after the system has come to equilibrium, or whether they were put there purposely by external manipulations. Introducing this assumption Onsager says 'the principles involved are not new; they are classical theorems of statistical mechanics', but this is somewhat misleading, since the crucial idea at this stage is not a theorem at all but a probability assumption, on a par with (for example) Boltzmann's assumption about the number of collisions of different types in a gas.

Accepting Onsager's assumption, we can put the above constitutive relation into the covariance formula; then, after dividing by $t_2 - t_1$, we get

$$\mathbf{covar}(X(t_1), [CX(t_1) + DY(t_1)]) = \mathbf{covar}([AX(t_1) + BY(t_1)], Y(t_1))$$

Let us assume, for simplicity, that the variables $X(t_1)$ and $Y(t_1)$ are uncorrelated, so that $\mathbf{covar}(X(t_1),Y(t_1))$ is zero. Then the above formula simplifies (since, by definition, $\mathbf{covar}\,(X(t_1), X(t_1)) = \mathbf{var}\,(X)$, etc.) to

$$\mathrm{C}\,\mathbf{var}\,X = \mathrm{B}\,\mathbf{var}\,Y$$

The two variances can be evaluated (and also the covariance of X and Y, to check that it is zero) by methods based on the Gibbs canonical ensemble, which were set out by Einstein (1910) but will not be summarized here. We end up with a relation connecting two different coefficients B and C in the constitutive relation. Onsager's main result was a general method for obtaining relations of this kind, which he called *reciprocal relations*.

To illustrate how Onsager's method might be applied to thermoelectricity, consider a system consisting of two identical parallel metal plates, close together but

Fig. 15.3. Lars Onsager was awarded the 1968 Nobel Prize for Chemistry for the discovery of his reciprocal relations. (Courtesy Yale University and AIP Emilio Segre Visual Archives.)

separated by a thermal and electrical insulator, and joined by a wire. The entire system is isolated, and the only way that heat and electric charge can be transferred between the two plates is by flowing along the wire. Then we can take the two fluctuating variables, previously called X and Y, to be the electric charge Q on one of the plates (which is proportional to the voltage difference between the two plates) and the difference of the energies of the two plates (assumed proportional to the difference of their temperatures), which we may denote by ΔE. The electric charge on the second plate is assumed to be $-Q$, and the sum of the energies of the two plates is a constant. If the system is started off with given values for Q and ΔE, our assumption is that these variables will change according to linear constitutive relations like the ones written above; an alternative form for the constitutive relations in the present example would be

$$I = -AQ - B\,\Delta E$$
$$J = -CQ - D\,\Delta E$$

where I and J denote the electric current and the heat current in the wire (the rates of change of Q and ΔE), and the constants A, B, C, D depend on such things as the electrical resistance of the wire, the electrical capacity of the two-plate system, and so on. The coefficients B and C are[5] respectively proportional to the Seebeck coefficient σ and the Peltier coefficient Π and so Onsager's relation connecting B and C gives a relation connecting these two coefficients, which turns out to be the same as the formula $\Pi = T\sigma$ which Thomson obtained in 1854.

Onsager's theory is generally held to have superseded Thomson's original derivation of his formula $\Pi = T\sigma$, and Thomson's hypothesis is no longer used. Onsager's relations are more general and the circumstances where they apply are more clearly defined. Various people have worked out the implications of Onsager's theory, and the subject has come to be known as non-equilibrium thermodynamics[6]. Unfortunately there is no easy way of extending it to non-linear constitutive relations. A standard work on non-equilibrium thermodynamics is the book by de Groot and Mazur (1961).

Thomson's original argument was based on a formula for the rate of production of entropy, which he wrote in terms of the electric current and various temperatures. In the set-up described above, the analogue of Thomson's formula is a quadratic form that looks like this:

$$K_1(\Delta E)^2 + K_2 I^2 + K_3 (\Pi - T\sigma) I \Delta E$$

where K_1, K_2, K_3 are constants, of which the first two are positive. Thomson's hypothesis was that the heat conduction term $K_1(\Delta E)^2$ could be ignored; if this is accepted then, since the entropy production must be non-negative for all values of I, both positive and negative, the coefficient of $I \Delta E$ must be zero and hence $\Pi - T\sigma$ is zero, that is, $\Pi = T\sigma$. This result is perfectly correct, but if the entropy production had been written in terms of Q and ΔE, or the currents I and J, rather than in terms of ΔE and I the analogue of Thomson's method would have given a wrong answer. One would like to have a re-formulation of Thomson's method which makes it clear which variables should be used.

Onsager gave the necessary reformulation, replacing Thomson's hypothesis by a variational principle. One might say that Thomson's hypothesis was to Onsager's variational principle as the caterpillar is to the butterfly. In the particular thermoelectric example considered here, the variational principle says that for given temperature difference (i.e. given ΔE) the entropy production is bound to decrease if the circuit is broken: that is, the value of the quadratic form is a minimum (subject to the given value of ΔE) when $I = 0$. For the quadratic form to have this property, we must have $\Pi - T\sigma = 0$, the same result that Thomson obtained using his own hypothesis. Onsager's somewhat cryptic exposition of his variational principle has been elaborated by various authors, for example

Denbigh (1951), and is one of the main components of non-equilibrium thermodynamics.

ACKNOWLEDGEMENTS

In preparing this article I have benefited greatly from the books by Steven Brush (1976) and by Carlo Cercignani (1998).

16
Kelvin—The Legacy

Andrew Whitaker

KELVIN—AS VIEWED AT THE BEGINNING OF THE TWENTIETH CENTURY

When the Nobel Prize committees began their work of evaluating prospective Laureates at the beginning of the last century[1], one of their problems was to decide whether to award the prize to those scientists, still alive, who had played leading parts in establishing the structure of their disciplines, but who had done so several decades earlier. In his will, Alfred Nobel had referred to discoveries made 'during the preceding year', but the Nobel statutes recognized that earlier work should be considered if its significance had become apparent only recently.

Taken very seriously was the nomination of Lord Kelvin, then in his mid-70s for the Physics prize. He was nominated every year until his death in 1907, and his nominators included Wilhelm Röntgen, Lord Rayleigh, Hendrik Lorentz, and Pieter Zeeman, all winners of the prize in the first four years of its existence[1].

On one hand this might seem surprising as, at least from a modern perspective, his most important work had been carried out practically half a century before. Yet through that half-century, his perceived eminence as a physicist had steadily grown rather than the reverse. During the second half of the nineteenth century, his scientific work—on vortex atoms and the ether, on the age of the earth, on electromagnetism, and telegraphy—was treated with the utmost respect by the general scientific community. Only towards the very end of this period did some of the most progressive members draw attention to his failure to keep pace with modern approaches and modern successes.

For the more general public he remained until his death the embodiment of the immensely successful scientist. He would be judged probably the leading researcher of his period, but in addition he was the public face of science and engineering, acting as President of the Royal Society and the British Association, sitting on important committees and inquiries, chairing conferences and meetings, and giving innumerable public lectures and addresses. This was, of course, in addition to the technical achievements that made him most famous, his work on the Atlantic cable and the numerous other inventions and patents that had made him both respected and rich. When his peerage was awarded in 1891, it was in considerable part for his support of the Liberal Unionists in Lord Salisbury's coalition government, but it was generally assumed that it was for his work in science and engineering, and, as such, it was extremely widely welcomed[2].

KELVIN—AS VIEWED AT THE END OF THE TWENTIETH CENTURY

A century after his death, things seem very different. Even to the practising physicist, while Kelvin's name is attached to the unit of absolute temperature, the credit involved may easily be assumed to be moderate. It is often believed that all that was required was to adjust the ideal gas law $PV = nR(T + 273)$, with T in degrees Celsius, to $PV = nR\Theta$, with Θ in Kelvin, by use of the transformation $\Theta = T + 273$, a procedure which might seem relatively straightforward.

Kelvin's name is also associated with the Kelvin–Planck statement of the second law of thermodynamics, and several 'effects', in particular the Joule–Kelvin effect. However, contemporary scientists may be more influenced by whatever knowledge they may have of his later work, which, irrespective of its intrinsic merits, has made practically no contribution to the development of physics since his death, and indeed is very much removed from the style of work that has achieved spectacular progress in physics in the twentieth century.

The actual merit of this work should not be regarded as negligible. Harold Sharlin[3] has suggested a useful analogy between the scientific activities of the latter decades of Kelvin and Albert Einstein. Posterity has judged Einstein's years spent searching for a unified theory to be just as misguided as Kelvin's spent searching for a hydrodynamical model of the ether (though, as Sharlin admits, Einstein's reputation gained for his earlier work has certainly been more secure that that of Kelvin). Yet in neither case would one be justified in criticizing the decision to commence the particular type of research. When Einstein first considered a unified field theory in 1918, he was following in the footsteps of Hermann Weyl[4], whose reputation did not suffer for his having initiated this type of study, while Kelvin's work on vortex atoms[5] followed an 1858 paper of Hermann von Helmholtz, who has likewise received little if any criticism for working on this topic.

Where both Einstein and Kelvin might justify criticism is for persisting in their labours on these topics when it would have appeared more rational to have admitted lack of success and to have moved to other topics, or at least to other approaches to the same topics.

In Kelvin's case, his ideas were provisionally summed up in 1884 in his massive *Baltimore Lectures on Molecular Dynamics and the Wave Theory of Light*, eventually published with additional matter and several appendices two decades later[6]. This book is a *tour de force* providing a masterly account of Kelvin's elastic-solid theory of the ether. It contains extensive physical analysis, lengthy and detailed mathematical derivations, and much discussion of experimental evidence. The lectures clearly indicate the author to be a scientist of enormous ability, combining great imagination with immense mathematical skill and attention to detail, and also considerable stamina. They were received in Baltimore by their largely American audience of leading physicists as the work of an undisputed master. Indeed in these lectures, Kelvin appears to be practically as keen to demonstrate how a dynamical theory should be created as to present the actual theory.

Yet not just posterity, but leading physicists of the last 20 years of Kelvin's own life, have judged the Baltimore lectures much more harshly. Even Silvanus P. Thompson[7], in his immensely useful but scarcely critical biography of Kelvin, written in 1910, was moved to comment that certain passages 'led his audience to conclude that at that date the lecturer had never read Clerk Maxwell's book!' In fact Kelvin's elastic-solid model had become more and more complicated without making any genuine advance in describing physical behaviour.

The future was to lie with Maxwell's 1864 theory of electromagnetism, the novel element of which, the displacement current, was rejected by Kelvin because he believed it was drawn from purely mathematical rather than physical reasoning. By the 1880s, in fact, many contributions to and tests of Maxwell's theory had been made by such physicists as George Fitzgerald, Oliver Lodge, J. J. Thomson, John Poynting, and Oliver Heaviside, and in 1883 Heinrich Hertz demonstrated the existence of radio waves, a key prediction of Maxwell's theory. These successes of Maxwell's theory led Fitzgerald to launch a fierce attack on many aspects of the *Baltimore Lectures*.

In the last years of his life, Kelvin actually showed quite a high degree of flexibility by abandoning vortex atoms, but he retained the ether and a physical approach involving only matter, and he was never to accept Maxwell. Yet he had to admit that his own approach seemed to lead to a dead end; at his Jubilee of his Glasgow Chair he announced the result of his investigation over many decades of the ether in one word, 'failure'[8].

Another major theme of Kelvin's work was his attempt to limit the ages of the Earth and the Sun. By any standards this was in principle excellent and important work. As explained in particular by Smith and Wise[9], it was an aspect of one of the main themes of Kelvin's scientific programme, the necessity for change and decay in the universe as opposed to eternal stability. The most important application of this theme was, of course, the second law of thermodynamics, but it was also crucial in Kelvin's criticism of the uniformitarianism of Charles Lyell, which allowed an infinite timescale for geological and evolutionary processes.

Kelvin's work and its mathematical development were beyond criticism; indeed, as explained later in this chapter, one of his main physical assumptions was of great and lasting importance. However his extremely precise limits on the age of the Earth, tight enough to cause great difficulties for geologists and evolutionists, appeared to be a result of particular assumptions for parameters known only within fairly wide limits. In addition, his attitude, and even more so that of his friend and supporter Peter Tait, often seemed to be along the lines that the physicists and mathematicians were doing a great favour to the geologists and biologists by illuminating these lesser disciplines[10].

This attitude was necessarily irritating to supporters of evolution such as Thomas Huxley and John Tyndall, and behind them, Charles Darwin himself. Kelvin and Tait appeared unaware that geologists and evolutionists had good arguments for their long timescales for the age of the Earth, and progress would presumably be made by searching for weak points in the totality of the argument, rather than the mathematical arguments being assumed to be above suspicion, and the geologists being expected to perform whatever contortions might be required in order to accommodate them. In fact, as is well-known, the discovery of radioactivity greatly increased the timescales emerging from Kelvin's type of argument, and essentially saved the day for geology and evolution.

Kelvin himself was only a moderate opponent of evolution, relaxed enough about its general nature, though concerned about the emergence of life on Earth and also about natural selection; he preferred to retain some element of design. (His great scientific friend, George Stokes, was more conservative theologically and much cooler towards evolution[11].) Nevertheless, despite the correctness of Kelvin's actual arguments, posterity could hardly do other than to put him on the losing side in the great evolution debate, on the same side as Bishop Samuel Wilberforce, and the opposing one to Huxley.

Turning to Kelvin's entrepreneurialism, for most of the twentieth century such activities would have been at least unfashionable, and in fact broadly frowned upon in academic circles. Indeed even in Kelvin's lifetime, Huxley was persuaded to stand for the position of President of the Royal Society in 1883, because this would keep out Kelvin; Huxley commented that he did not wish 'to have the noble old Society exploited by enterprising commercial gents who make their profit out of the application of science'[12].

In fact, Huxley resigned as President due to ill-health in 1885, and, following a five-year Presidency of Stokes, Kelvin himself did become President in 1890. However, during this Presidency, Kelvin was subject to fierce criticism over a published advertisement for Kelvin's 'Patent Indestructible Water Taps', with the inventor specifically designated President of the Royal Society[13].

Through much of the following century, a common position in academic circles would be that it was acceptable to take out and exploit a patent if one's research work opened up this possibility. However putting considerable effort into the design of commercial products, setting up firms for the manufacture of these products, promoting them in an active way, or going to court in defence of patents, all of which absorbed much of Kelvin's energies, might be seen as breaching the dignity, perhaps the sanctity, of pure science, and took time which should be devoted to research 'for its own sake'.

It is relevant that Kelvin retained the Chair of Natural Philosophy in the University of Glasgow for 53 years. He had, of course, strong roots in both Glasgow and Cambridge. It is well-known that he was asked to take the Cavendish Chair of Experimental Physics at Cambridge three times, but each time refused. The first occasion was at the foundation of the chair in 1870, Maxwell instead serving from 1871, the second after Maxwell's death in 1879, when Rayleigh took the chair, and

Fig. 16.1. The man of Empire. Lord and Lady Kelvin dressed for the coronation of King Edward VII on August 9th 1902. (From A.G. King, *Kelvin the Man*, Hodder & Stoughton Ltd. 1925 op. p. 104.)

the third when Rayleigh resigned and J. J. Thomson was the eventual occupant. Kelvin was a fellow of Peterhouse from the time of his graduation until his marriage, and a special Life Fellow from 1872. He was nominated to be Master of the College in 1876, and declined the same position in 1900.

There is no doubt that Kelvin appreciated the academic tradition and the collegiate structure of Cambridge. He enjoyed the presence of a fair number of excellent mathematicians and scientists, and he remembered with fondness from his youth such activities as rowing and music. Yet his unwillingness to return permanently demonstrates clearly that his roots in Glasgow were much the stronger. He perhaps appreciated the fact that he was undisputed leader of his subject in Glasgow, and could and did forward his discipline as he thought best. Probably much more important, though, was the presence of students from a wider range of backgrounds than those at Cambridge, and in particular, outside the University, the opportunity to interact and form friendships with practical men such as factory owners and shipbuilders, and to understand their difficulties and to work towards realistic solutions.

Following Kelvin's retirement, the centre of gravity of British physics moved definitively and swiftly to Cambridge, with successive Cavendish Professors, J. J. Thomson, Ernest Rutherford, Lawrence Bragg, and Neville Mott, as well as several of Rutherford's protégés, all Nobel Prizewinners. Pure research came to be all-important, and Kelvin's brand of engineering physics appeared old-fashioned and a throwback to days of Victorian commercialism, even Victorian vulgarity. One could almost imagine his life caricatured alongside those of Cardinal Manning, General Gordon, and Thomas Arnold in Lytton Strachey's 'Eminent Victorians'[14].

The common twentieth-century impression of Kelvin as a Victorian scientist, obsessed with the importance of the models of the Baltimore lectures, was enhanced by a number of what turned out to be rather unwise views which Kelvin promulgated a little too freely towards the end of his life. The engineer may be shocked that, though a very keen supporter of the use of electric power, Kelvin believed until nearly the end of his life that the future lay with DC rather than AC. The scientist may be surprised that Kelvin was adamantly opposed to the use in mathematics, not only of quaternions, the brainchild of William Hamilton and the favourite of Tait, which are admittedly rarely seen today, but also of vectors, which are now used by even the most junior students of science and engineering. The general reader may be most shocked by statements as late as the 1890s about the impossibility of aeroplanes.

THE AIM OF THIS CHAPTER

Thus a host of factors have contributed to the comparatively low regard in which Kelvin is held by many today. Yet this view of Kelvin ignores or undervalues his many substantial and indeed crucial contributions to the development of physics during the nineteenth century, and his influence on the education of physicists, the application of physics, and the interactions between physicists and the political and service establishments. (Note that the terms 'physics' and 'physicists' will be used freely, though it is recognized that Kelvin preferred the Scottish terms of 'natural philosophy' and 'natural philosopher'; also note that, for convenience, he will always be referred to as 'Kelvin'.)

The remainder of this chapter will concentrate rather unashamedly on these contributions. As such, it will have a somewhat different ethos from most of the preceding chapters, which, being more strictly historical in nature, naturally make every effort to consider Kelvin's work in terms of the beliefs, ideas, and developments of the period, as a part of the scientific enterprise carried on by many. Here my aim is slightly different—to answer the questions, fair and common ones, 'What did Kelvin do to deserve the towering reputation he achieved in his lifetime?' and 'Does he deserve to be remembered in a similar way today?'

The approaches may not differ as much as one might expect. In the context of nineteenth-century physics, as much of what we now call classical physics was taking practically its final form, even more perhaps than in other periods, it is, for most aspects of this study, extremely difficult and probably damagingly misleading to attempt to assign particular 'discoveries' to individual scientists.

The prime example of this, very relevant to our discussion here, is the argument of Thomas Kuhn[15] regarding the 'simultaneous discovery' of energy conservation, not just by those usually mentioned, Julius Mayer, James Joule, Hermann von Helmholtz, and Ludwig Colding between 1842 and 1847, but by another eight scientists as well, including Sadi Carnot as early as 1832, Michael Faraday, William Grove, and Justus Liebig, in all cases except that of the late but independent work of Gustave-Adolphe Hirn, by 1845. Kuhn actually suggests that the number could be increased 'but not fruitfully'.

Kuhn thus differs from, for example, Yehuda Elkana[16], who writes of 'that concept-creating activity which finally resulted in the development of the concept of energy: the work of Hermann von Helmholtz'. He also differs from Clifford Truesdell, whose celebrated though controversial history of the origins of thermodynamics demands the achievement of total mathematical clarity before allowing any investigator to be regarded as a 'discoverer'. He writes of conservation of energy as the first law of thermodynamics: 'Who . . . is the discoverer of [this law]?', he asks[17], and replies that: 'I am not certain there was one, but I am certain that *before 1850 no "First Law" had been published by anyone* [italics in original]. If discoverer there were, certainly it was not Mayer, nor Helmholtz, not Joule'. Truesdell regards the law as only established in 1851 by the work of Rudolf Clausius and William Rankine, and in particular by that of Kelvin in 1850 and 1851.

Kuhn himself, incidentally, was rather dismissive of the contributions of Kelvin. He commented that, in Kelvin's famous paper of 1851[18], he had stated that Humphry Davy had established the triumph of the dynamical theory of heat over the caloric theory as early as 1799. Yet Kelvin himself, Kuhn remarked, had been using the caloric theory up till the previous year. I shall attempt to reconcile, as far as is possible, some of the contradictions of the last few paragraphs later in this chapter, and particularly to elucidate Kelvin's approach and to evaluate his contributions.

Just as in the development of thermodynamics in the nineteenth century, in that of electromagnetism, at least in the period when Kelvin was working, there are few cases where 'discoveries' may be ascribed to individuals; more often several workers came up with broadly similar results or ideas. Yet there are two major exceptions to that rule. First, Faraday would be recognized for his idea of introducing lines of force and the concept of the electric and the magnetic field. Secondly, Maxwell was certainly independently responsible for his set of equations for electromagnetism that predicted the existence of electromagnetic waves. Actually, in both these cases, Kelvin did play a considerable role in the creation and presentation of their work, and this again will be discussed below.

The general stance of this chapter, though, will be not so much to pick out particular contributions of Kelvin, though that will not be neglected, as to show where his work and his more general views were decisive in moving forward our general understanding of the universe, and how his opinions and leadership on the education, organization, and application of what we will call physics (though he would not) became models which were to be taken up, to a greater or lesser extent, not only in Britain but through the world.

HEAT, ELECTRICITY AND MAGNETISM—EARLY WORK AND ITS INFLUENCE

Kelvin matriculated at the University of Glasgow in 1834 when he was ten years old, and studied there until he went up to Cambridge in 1841. By that time he had waiting for publication in the *Cambridge Mathematical Journal* two papers which were to be published, under the pseudonym P. Q. R., later that year. By the time he took his Cambridge examinations in 1845, he had published 12 papers, all but one in the same journal, the exception being a note sent to *Liouville's Journal* in France[19].

Even by 1846, when Kelvin stood for the Glasgow Chair, the Rev. Henry Cookson, his tutor at Peterhouse, was able to write of him that he was 'regarded here, by the most competent judges, as the first man of science, of the rising generation, in the country', and Robert Ellis, Senior Examiner for the examinations that Kelvin had taken just the year before, expressed the belief that 'Mr Thomson will hereafter occupy a very distinguished place among the scientific men of Europe'[20].

By the end of his decade, and following his appointment to the Glasgow Chair, Kelvin's number of publications had risen to around 50. The general range of topics of most of these papers was indicated by Augustus de Morgan's testimonial for the Glasgow position:

> [H]is mathematical speculations have turned very much in that line which is now so essential to a teacher of Physics—I mean the analysis by which the mathematical theories of electricity, heat and light have been brought to their present form'[21].

Kelvin's research career started with an intense study of the now universal methods of Jean-Baptiste Fourier; indeed his first paper was a defence of Fourier against an ill-advised attack by Philip Kelland, Professor of Mathematics at Edinburgh University. Fourier had died in 1830; his famous book, which had been published in 1822, was titled *Théorie Analytique de la Chaleur*, but in fact the general approach was as relevant to aspects of electricity, magnetism, and gravitation as to heat.

By the end of the eighteenth century, it may be said that the physical basis of many areas of physics had been set up. For example Henry Cavendish in 1771 and Auguste Coulomb around 1786 had established with great accuracy the force between electrically charged spheres (today's 'Coulomb's law'). Similarly basic ideas concerning the conduction of heat had been established since the days of Newton. However, further physical questions naturally presented themselves; in particular what heat actually was, and whether electricity should be considered from a one-fluid or a two-fluid point of view.

The approach of Fourier, and following him Kelvin and many other leading mathematically minded scientists of this period, was to leave aside these physical questions, and concentrate on the further mathematical analysis of the central physical results. Fourier himself established fundamental mathematical methods for studying the movement of heat in bodies and between bodies. The basic differential equations were built up and solved, and it was in the course of this work that Fourier developed his famous Fourier series, central to all aspects of mathematical physics today. It was on the question of the probity of these series that Kelvin defended Fourier against Kelland.

It may be said that Kelvin worked on many areas of physics during these early years of his research—electricity, gravitation, hydrodynamics, and heat in particular. However it would be more meaningful to say that he was successful in establishing the analogies between the mathematical treatments of these different areas, so that results in one could be translated into the other areas. In his third paper, written in 1841[22], Kelvin showed how Fourier's treatment of heat could be applied to electrical attraction; in this aspect of the work his approach was unique, and one he was to develop fruitfully over the coming years. As will be seen shortly, the approach was also the base of Maxwell's theory.

The results obtained by Kelvin and the other workers in this period in the areas of gravitation and electrostatics may best be described in today's terms as the rudiments of field theory and potential theory. They established much of the formal relationship between mass and electric charge as the source of gravitational and electrostatic potentials respectively, the potentials themselves, and the forces resulting from these potentials. They also built up a powerful mathematical structure for evaluating the distributions of charges and potentials in different geometries. The concept of the field was, of course, a central part of this construction, but we shall consider this separately when we discuss the relationship between the work of Faraday and that of Kelvin in the following section.

One aspect of this work should be credited to Kelvin independently of others. This is the simple but brilliant idea of electrical images[23]. The simplest example this is for a point charge outside a plane conducting surface. The potential distribution in space may be obtained by replacing the charge distribution over the surface by an equal but opposite charge to the original point charge positioned as an *image*, with the conducting surface playing the role of the mirror. Many more complicated problems may be solved by the same technique. Again we see Kelvin's genius in translating problems from one area of physics to another. The method of images is now standard in any elementary treatment of electrostatics.

In most aspects of these developments, though, it is probably fruitless to attempt to allocate credit to individuals. A couple of anecdotes may help to make this point. The first relates to Kelvin's third paper mentioned above. While, as has been said, the analogy between heat and electrostatics was discovered by Kelvin alone, he soon discovered that some of his actual results had been anticipated by the famous French mathematicians Michel Chasles and Jacques Charles Sturm; later he found that the even more famous German mathematician Karl Gauss had also found some of the results.

This however was far from the end of the matter. More than a decade before, in 1828, George Green, a mathematically (highly)-inclined miller of Nottingham, had published by private subscription a work of 72 pages titled *An Essay on the Application of Mathematical Analysis to the Theories of Electricity and Magnetism*. This work established Green as a considerable mathematical genius, and he was to become first a student and then a researcher at Cambridge between 1833 and 1839 or 1840, before dying in Nottingham in 1840. Fewer than 100 copies of his celebrated essay were printed, and it soon became extremely difficult to get hold of.

In 1845, having taken his degree at Cambridge, Kelvin prepared to travel to Paris to spend some time with Joseph Liouville and other French mathematicians. He confided to William Hopkins, the famous mathematical coach who had played a large part in Kelvin's studies, his sadness in never having had access to Green's essay. Hopkins announced that he actually had three copies, and gave Kelvin a copy for himself and one to take to Liouville. On reading Green's work, Kelvin discovered that Green had anticipated nearly all the general theorems of Gauss, Chasles, Sturm, and Kelvin himself.

Some weeks later Sturm, who had been informed by Liouville that Kelvin had a copy of the elusive memoir, turned up unannounced at Kelvin's Parisian lodgings to see for himself. Scanning its contents at great speed, he quickly found, with perhaps mingled horror and delight, all his own work. 'Ah! Voilà mon affaire ... Mon Dieu, oui', he cried[24].

The theorems these mathematical scientists discovered, written in vector form, are fundamental in any modern treatment of electromagnetism, where they are attributed to such names as Gauss and Green. That such attribution should not always be taken too seriously is made clear by our second anecdote. Another of these central theorems in vector calculus is invariably known as Stokes' Theorem. In fact this theorem was discovered by Kelvin in 1842. Kelvin informed Stokes of it in a letter, but did not publish it. A few years later, Stokes asked Cambridge students to prove the theorem in their degree examinations, and his own name somehow became attached to it[25].

While we have picked out a few important aspects of the work where Kelvin should be acknowledged as unique discoverer, just as significant a truth that may be gleaned from this section is that, at a remarkably young age—at the time of his French trip he was still only twenty, Kelvin was able to perform independent work of the standard of mathematicians whose names have come down to us as among the greatest in the history of mathematics; Sturm, Gauss, Green. He was able to meet in France on equal terms with Liouville, Chasles, and Sturm, and also Augustin Cauchy, another French mathematician of immense achievement. And Sylvanus Thompson points out that, even where Kelvin repeated the results of previous workers, his methods were different from, and often more powerful than, those of the earlier work.

This mathematical approach to physical laws perhaps suited Kelvin extremely well in these years when his work centred on and then followed from the Cambridge mathematical tripos, where concern with the physical universe was expressed in fairly strictly mathematical terms. However in these same years, Kelvin's interaction with Faraday propelled him towards a more directly physical approach to scientific study.

KELVIN AND FARADAY

Students of electromagnetism today are taught that Faraday, with his limited education and practically non-existent formal mathematics, was able to produce methods centred around the field concept which were more physically meaningful and, as it was to turn out, far more fruitful in application than what might be called the 'sterile' mathematical methods of the academics. One would not question the significance of the field concept in the least, or Faraday's point that the medium between the various charges plays a crucial role in determining the forces between them. It must be said, though, that the battle between the two approaches was rather less one-sided than usually thought, and that Kelvin was largely responsible for bringing together the two points of view.

Indeed at first Kelvin was by no means a supporter of Faraday. An entry from Kelvin's diary while he was at Cambridge in 1843 reads:

> I found Gregory [David Gregory, the exceptionally promising young Cambridge mathematician, who sadly was to die the following year]... I asked him about where I could see anything on electricity, and we had a long conversation in which Faraday ... got abused to the *n*th power.

As a proud Cambridge mathematician, Kelvin felt 'disgust' for Faraday's theory, which he regarded as no more than 'speaking of the phenomena'[26].

The problem was not just one of conceit on the academic side. Faraday himself had, at least initially, profound theoretical misconceptions. He believed that Coulomb's law entailed that what Faraday himself would call lines of force must be straight, whereas he had shown they were usually curved. On this basis Faraday built up a whole series of 'objections' to the standard mathematical theory.

David Thomson, a cousin of Faraday, had been standing in for the ill Professor of Physics at Glasgow since 1840, in 1845 becoming Professor himself at Aberdeen. Kelvin had assisted him in the earlier years, and reported that David Thomson gradually 'inoculated [him] with Faraday fire'[27]. In 1843 he found the French mathematicians somewhat concerned about Faraday's 'objections' and he promised to write a paper to show that these arguments, far from being objections to Coulomb's law, were verifications of it.

Yet as he went through four successive drafts of this promised paper, Kelvin's attitude changed from one of barely concealed scorn for Faraday to a much greater understanding and indeed appreciation of his methods. The two men met in 1845, and from then on their relationship became increasingly positive, from both the personal and the scientific viewpoint.

Misconceptions on Faraday's part were ironed out. Faraday described Kelvin as having come 'the nearest to understanding what I meant' by the lines of force.

Kelvin was able indeed to put Faradays' lines of force in more mathematical form; the lines of force, he showed, must be perpendicular to the equipotentials, while the force on a charge particle was along the tangent to the line of force. This is of course, the account given to every student of elementary physics to this very day. Smith and Wise[28] report that: 'From 1845 Thomson's analysis would proceed step for step with Faraday's investigations. The result was field theory'.

Through the second half of the 1850s, Faraday and Kelvin interacted exceedingly fruitfully over the nature of the magnetic field and magnetic materials. The stimulus for this work was Faraday's discovery in 1845 of magneto-optic rotation—the plane of polarization of light rotates when the light is transmitted along the lines of force in particular materials, an investigation directly suggested by Kelvin earlier that year. During this period the two reached an understanding of the distinction between ferromagnets and diamagnets, and an understanding of the behaviour of each in a magnetic field.

This sharing of ideas was so intense that there has been considerable discussion over which of the two should be credited as the actual inventor of the magnetic field. Barbara Doran[29] has suggested that Kelvin should be given the credit, David Gooding[30], though, argues that it was Faraday who invented the concept, but that Kelvin was instrumental in enabling Faraday to draw out the idea from his own understanding of the phenomena. It is certainly true that, as Gooding says, '[Kelvin] made a greater contribution to the early development of field theory than is usually recognized'.

KELVIN AND MAXWELL

In emphasizing Kelvin's contributions to field theory, this section continues the theme of the previous one. I must start, though, by saying that the comparative status of Maxwell and Kelvin has changed dramatically over the last century. While 100 years ago it would have been taken for granted by all except, perhaps, the most scientifically knowledgeable that Kelvin's position was considerably higher than that of Maxwell, today the reverse is the case. In a poll to discover views on the leading physicist of all time carried out by *Physics World* in 1999[31], Maxwell was in third place behind Einstein and Newton; Kelvin did not obtain a single vote.

Indeed the most common view of the relationship between these two men and Faraday (who, incidentally came eleventh equal in the poll) would be that Faraday and Maxwell had interacted positively, Maxwell putting Faraday's physical ideas into mathematical form to very great effect, while the interaction between Kelvin and Maxwell was broadly negative, Kelvin failing to appreciate the most crucial components of Maxwell's theory.

That such a view is, at best, an extremely limited one, may be seen from the fact that, by Maxwell's own admission, most of what he knew of electricity and magnetism he owed to Kelvin, and much of the content of his own theories was based solidly on earlier work of Kelvin. It may be said that Maxwell encountered Faraday through the illuminating lens of Kelvin. In fact when Maxwell had graduated from Cambridge in 1854, he wrote to Kelvin asking for help in understanding electricity since he wished to prepare himself for research in that area. In their subsequent correspondence, their roles were very much Kelvin and Maxwell as master and novice respectively.

By the summer of 1855, when Maxwell felt confident enough to attempt to make his own contribution to the area, he listed his debts to Kelvin; he had learned of Faraday's lines of force, Ampère's current laws together with Faraday's understanding of them, the work of Green and the idea of the potential, and Kelvin's own research on the analogy of electricity with incompressible elastic solids, and on the application of Ampère's theories of circulating current to solenoidal distributions of magnetism[32].

Maxwell's main worry in preparing to spend considerable efforts in developing his own ideas on electricity and magnetism was that he rather naturally assumed that Kelvin himself already had the whole subject at his fingertips. He clearly, Maxwell said, had the mathematical part available. Probably he had the specific application to electrical phenomena in rough form, and could put everything together once he had 'worked out Heat or got a little spare time'[32].

Once reassured that this was not the case, Maxwell commenced work. Moyer[33] has assembled a list of the places where Maxwell acknowledged Kelvin's major contributions to his own work. He said that Kelvin's paper on Faraday's methods, written for Liouville and mentioned above, was the 'germ' of his own speculation by which he gradually developed the material significance of Faraday's idea.

In particular Maxwell was highly influenced by Kelvin's demonstrations that solutions from one area of science could be carried over to another, and his early work was based on Kelvin's arguments concerning analogies between electricity, heat conduction, and incompressible elastic solids. Also central to Maxwell's work was the suggestion of Kelvin in 1856 that electromagnetic phenomena should be considered in terms of the inertia and pressure of the basic elements of the medium, and one should not speculate on the nature of these elements.

Maxwell's work, in fact, started from Kelvin's demonstration of the mathematical analogy between heat conduction and electrical and gravitational attraction. He recognized the appearance in problems of heat conduction of two directly related types of quantity—a 'flux' or flow of heat, and an 'intensity' or temperature gradient. Using Kelvin's analogy, Maxwell classified electromagnetic variables as related 'fluxes' and 'intensities'. Current density, $\mathbf{k}$, a flux, was related to $\mathbf{E}$, electric field, an intensity; **D**, electric displacement, and **B**, magnetic induction, were fluxes, related to intensities, **E** and magnetic field, **H**, respectively.

Maxwell then adapted solutions of problems in different branches of mechanics to the language of electromagnetism. He thus obtained partial differential equations which related the values of electromagnetic variables to those at immediately adjacent positions, exactly as required for a field theory. It would be natural to describe this work of Maxwell as using the methods of Kelvin to produce the results of Faraday. Sharlin[34] says that: 'This landmark of nineteenth-century science was the capstone of the mathematical study of electricity which [Kelvin] had begun in 1842'.

Yet what could have been a jewel in Kelvin's reputation was to become almost the complete opposite. As Norton Wise[35] writes:

> Only a small step separated [Kelvin's] mathematics from Maxwell's theory, yet a chasm lay between. The 'new set of electrical notions' was clearly Maxwell's and not [Kelvin's].

Kelvin did not accept Maxwell's formalism and results, and, almost paradoxically, as the theory gained advocates and achieved a measure of experimental confirmation, Kelvin's opposition became more pronounced. As late as 1896 he referred to Maxwell's theory as 'nihilism'[36]. Posterity was to judge him harshly for what is generally considered a major error of judgement.

There were technical reasons for Kelvin's position[35]. For all that he was a strong advocate of Faraday, he still felt that the ultimate reality was action at a distance in the sense of Coulomb's law. He had, after all, shown at the beginning of his career that Faraday's results could be obtained from Coulomb's law rather than being opposed to it. Action at a distance implied a simple relationship between electrostatic or magnetic force and matter and thus no requirement for a duality of vectors: **E** *and* **D**, **H** *and* **B**.

Although Maxwell had taken his idea on analogy between different branches of science from Kelvin, he was willing to use it in a more thorough way. While Kelvin was prepared to make *mathematical* use of equivalent formalisms in different subjects, he could not accept that use of an analogy from one area to another could provide any information of *physical* significance in the second area. Maxwell, in contrast, considered field theory as a starting-point for further analysis, rather than merely as a deduction from any more primitive concepts, and this further analysis led to his own theory including the duality of vectors.

More generally this is an example of an important strand of Kelvin's approach to scientific theory. He would include in his analysis only terms that were empirically based or which appeared in more general theories; duality of vectors constituted, to Kelvin's mind, a duality in the nature of force; it was this that he considered 'nihilism'. His criterion became sharper as science developed during the middle portion of the nineteenth century, and became a requirement for visualization. In his *Baltimore Lectures*[6] he wrote, specifically of Maxwell's theory:

> I never satisfy myself until I can make a mechanical model of a thing. If I can make a mechanical model I can understand it. As long as I cannot make a mechanical model all the way through I cannot understand; and that is why I cannot get the electromagnetic theory.

With the coming of relativity and, in particular, quantum theory in the twentieth century, this statement of Kelvin's has frequently been subject to ridicule. Whatever the justice of this, and however unfortunate his inability to accept Maxwell's theory may be, it is clear that of immensely more importance is the positive fact that Kelvin's work was central, perhaps indispensable, for the actual creation of the theory.

KELVIN AND THERMODYNAMICS

Kelvin is today probably best known for his work on thermodynamics. Several scientists and engineers played central roles in the creation of thermodynamics: Carnot, Joule, Rudolf Clausius, and William Rankine as well as Kelvin. Indeed the first statement of what would shortly be known as the first and second laws of thermodynamics was given by Clausius in his seminal paper of 1850[37].

Once that is admitted, it is possible to make several points serving to make clearer the significance of Kelvin's own contributions. First, as Cardwell[38] says, it is possible for Clausius to be described as a 'disciple, at one remove, of Kelvin'. Clausius had not actually read the crucial memoir of Carnot, of which much more is said below, being familiar only with the discussions of its ideas by Emile Clapeyron and Kelvin himself. Clausius was able to take advantage of Kelvin's intense and long-lasting study of the nature of heat over many years, culminating in the stated dilemma in Kelvin's paper of 1849[39]—how to reconcile the work of Carnot with that of Joule. Unfortunately for Kelvin, Clausius succeeded in achieving this reconciliation first, but he had certainly built on foundations provided by Kelvin.

Cardwell speaks of Kelvin being 'overtaken' by both Clausius and Rankine in the later half of 1849. The latter, who would, in 1855, become a colleague of Kelvin as Professor of Engineering at Glasgow, was interested in the same questions as Kelvin and Clausius, but, rather than using macroscopic methods of some generality as they did, he used a detailed and rather unconvincing model of an atomic nucleus surrounded by an elastic atmosphere, the oscillations and revolutions of which constituted heat—a model of so-called molecular vortices. His ideas were rather vague and unconvincing in detail, but helped to convince Kelvin that heat and work were mutually convertible.

In contrast, though Kelvin had noticed Clausius's paper of 1850 before he published his own paper of 1851, he claimed that his own work was completely independent of that of Clausius. Indeed he always remained somewhat critical of Clausius's argument, arguing that Clausius had used simplifying assumptions, in particular that of an ideal gas, while part of the reason for Kelvin's delay had been his search for detailed experimental evidence on, in particular, the physical properties of steam. (See the full titles of Refs. 18 and 39.)

Perhaps a more convincing reason for Clausius's speed and Kelvin's tardiness was that, as Crosbie Smith[40] points out, Clausius's interests were much narrower than those of Kelvin, being restricted to the theory of heat engines and the nature of heat. Kelvin, in contrast, had concerns in many general aspects of physical science including conservation, irreversibility, and the nascent quantity of energy, as well as some quite strong and relevant theological beliefs.

Smith quotes Joseph Larmor[41] from an obituary of Kelvin:

> In their parallel developments of the subject, while Clausius kept mainly to the theory of heat engines, applications over the whole domain of physical science crowded on Kelvin.

It may thus not have been too surprising that Clausius came fairly quickly to a solution of a rather restricted problem, though one, of course, of immense importance, while Kelvin took a little longer to reach a much broader and subtler synthesis.

Let us now turn to the set of opposing beliefs that caused Kelvin to think unceasingly for many years, to debate to great effect with his brother James, and finally to reach this broad synthesis. As Smith and Wise[42] have discussed in detail, from the time of his student days in Glasgow, he was taught and came to believe strongly that God was the eternal creator, that it is unthinkable that human beings could themselves create or destroy. Clearly this is a call for conservation and stability in the physical universe and likewise in physical theory. From the scientific point of view these beliefs matched the system of Pierre Laplace, though, in his case, there was no theological backing. Kelvin and Joule were united in believing strongly that the power to annihilate mechanical effort was the privilege of God alone[43].

However, this set of beliefs was challenged by the discovery in 1833 that Encke's comet was slowing down[44]. This indicated the presence of a resistive medium in the universe, which would ultimately destroy the motion of all celestial objects, and thus bring the universe together, of course, with all life on it, to a halt. To Kelvin this too came to seem an obvious and accepted truth. As early as 1841, for example, he discussed how the production of tides would retard the Moon. He came to see such decay, which he was to call 'irreversibility' or 'dissipation' as a central and universal feature of physical systems: mountains are eroded, humans and animals die.

It is important to recognize that this deep belief in a decaying creation as opposed to a timeless creator was as much theological as his belief in conservation and stability. In a draft of his 1851 paper, he included the words: 'The earth shall wax old &c', referring to Isaiah 51:6: 'for the heavens shall vanish away like smoke, and the earth shall wax old like a garment' (or possibly to a similar verse from the Psalms). The two beliefs, equally strongly based, seemed in conflict, and hence his intense inner struggles, before he eventually came to an original and creative combination of the two in the laws of thermodynamics.

Before studying the physical arguments leading to this resolution we must discuss briefly views of the nature of heat in the 1840s, in particular Kuhn's rather dismissive remark that Kelvin at this stage was a believer in caloric. A belief in caloric is often thought to imply a detailed and, to modern tastes, unnecessary and unrealistic description of heat as a physical substance.

However, as Truesdell[45] says, all that was stated by Laplace, who originated the caloric idea, is that heat is never created or destroyed. Laplace put the idea a little more clearly by saying that the heat in a body is a function of pressure, density, and temperature. In more modern terminology, heat is a *state function*; from the state of a particular system, we may say how much heat it contains. Though Thomson used the term 'materiality of heat' to express the caloric theory, he also used the term 'permanence of heat', and there is no need to assume that his belief in the caloric theory in the late 1840s meant any more than his belief that heat was a state function. Such a belief is by no means foolish. Indeed it is probably held to this day by practically everybody who has not had the advantage of an education in basic thermodynamics.

It was implicit in Carnot's theory of the reversible engine, produced in 1824. Carnot died prematurely in 1832, and his work was popularized by Clapeyron in 1834, but Kelvin's first major task in the study of heat, after he had with great difficulty tracked down a copy of Carnot's famous memoir, was to present in 1849 an 'Account' of Carnot's theory[39], which was actually Kelvin's first important paper in this area of research.

Carnot had argued that the heat-engine that was most efficient in obtaining mechanical effort from a supply of fuel was a reversible one, in which there were no losses due, for example, to friction. He also stressed that, as well as a source of heat at a high temperature, a heat-engine must have a sink of heat at a lower temperature, and the amount of mechanical effort produced in a reversible cycle depended only on these two temperatures.

The model followed the analysis of the water wheel. Cardwell[46] has pointed out that the heat-engine was taking over from the supremely successful water wheel as the main source of power for industry, and it is scarcely surprising that the theoretical analysis of the water wheel was taken over as well. In the water wheel, the mechanical effort is obtained from the fall of the water, but, of course, the amount of water is unchanged. The water is merely altered from one state to another; in no way is it transformed into the work produced by the wheel.

By analogy, it was generally assumed that in a heat-engine the heat *fell* from high to low temperature, but there was no change in the actual amount of heat; in a cyclic process, of course, there could not be if one held to the idea that heat was a state function, because at the end of each cycle, the state of the system is the same. Work was performed by the engine, but it seemed impossible to imagine that heat had been transformed into work.

For Kelvin and his brother James, who was very much involved in the analysis of the various problems, the main difficulty with this picture lay in comparison with the related situation where heat passed from a high temperature source to a low temperature sink in a more simple way—merely by conduction. The heat was transformed in exactly the same way, but now there was no production of work. It seemed as though work which *might* have been produced was, in effect, lost, but the brothers found that concept unacceptable because of their belief in conservation: God could enable work to 'disappear', but man could not.

From 1847 Kelvin became increasingly interested in the experiments and ideas of Joule, as did his brother and also Stokes; initially they were in a very small minority of scientists concerned with Joule's work. Joule was convinced that, quite contrary to the caloric theory and the idea of heat as a state function, heat *could* be transformed into work and vice versa. Thus in the steam engine, part of the heat emerging from the high temperature source is transformed to become the mechanical effort, the remainder being deposited as heat at the lower temperature.

Kelvin was very willing to agree that work could be transformed into heat, as in friction, but much less happy to accept the possibility of (low temperature) heat being transformed to work. The latter would imply that there were no genuine losses even in the case of conduction or friction; the heat produced could in principle be transformed back to work. But Kelvin was convinced of the central importance of dissipation and irreversibility. In 1849 the clash between conservation and irreversibility seemed impossible to resolve.

By 1851, he had the answer. His first proposition (soon to be christened the first law of thermodynamics), which Kelvin attributed to Joule, was very much about conservation. Heat and work were both forms of energy and thus each could be transformed into the other. Thus heat could not, of course, be a state function. Heat was energy in flow from one system to another; Kelvin's term was the 'dynamical theory' of heat.

However there was a function, the internal energy of the system, which was the sum of energies of each particle, and therefore *was* a state function. The value of this function could be changed by transferring heat to or from the system, *or* by the system doing work or work being done on the system.

In accordance with conservation, the energy that is seemingly lost, in conduction for example, is not lost 'in the material world'. However dissipation comes in through the idea that this energy—we may call it low temperature heat—is 'lost to man irrecoverably'. Kelvin discussed this in terms of what became famous as Kelvin's statement of the second law of thermodynamics: 'It is impossible ... to derive mechanical effect from any portion of matter by cooling it below the temperature of the coldest of the surrounding objects'. Some heat must be deposited in a low temperature reservoir or sink, and it is then inaccessible to us, except if the engine is worked in reverse mode. (The statement is often called the Kelvin–Planck statement, as Max Planck stressed much later in the century that there was a restriction to cyclic processes.)

If the engine is reversible, all *may* then be returned to its original state, but of course the work produced by the engine will have been returned to the high temperature source as heat. If, however, the engine is irreversible, there must be dissipation; heat is produced which can never be made use of. Kelvin's second proposition, which he attributed to Carnot and Clausius, was indeed expressed in terms of reversibility and irreversibility. From Kelvin's point of view, his conundrum between conservation and dissipation had been solved in the most beautiful and subtle way.

Clausius's own solution had come before that of Kelvin, but had, in a sense, been a technical solution to a technical problem. Smith[40] stresses that Clausius had not been interested in general arguments concerning conservation or irreversibility. Only in the 1860s did he start using Kelvin's terminology of 'energy', and only then did he make fundamental contributions to the study of irreversibility, introducing the crucial idea of 'entropy'. The most powerful statement of the second law is that entropy remains constant in reversible processes, but increases in irreversible processes. The quantity of entropy is extremely useful, not only in thermodynamics, but also in statistical mechanics, where, as opposed to thermodynamics, the atomic nature of matter is taken into account.

From the late 1840s, Kelvin's concerns, in contrast to those of Clausius, had been both broad and deep, encompassing the future of the universe, the so-called 'heat death' when irreversibility will have led to a sameness in the properties of the universe, inimical to the presence of living creatures; the directional aspect of physical process; and fundamental beliefs and recent developments in theology. The development of thermodynamics owed an immense amount to both men.

THEORY AND EXPERIMENT IN THERMODYNAMICS

In the previous section, there was a general account of Kelvin's work on thermodynamics. Here we look at a number of special issues, some of which played important parts in the development of his ideas.

The first, the important discovery of the effect of pressure in lowering the freezing point of water[47], was a joint study of Kelvin and his brother James. It was carried out between 1847 and very early in 1850, that is to say following the brothers' discovery and acceptance of the Carnot memoir, but before Kelvin's solution of the problem of thermodynamics. William and James imagined a Carnot engine that froze water; the engine would perform work because water expands when it freezes, but since both reservoirs, source and sink, of the engine will be at the same temperature, the freezing point of water, no input of energy would seem to be required. It would appear that what is described is a machine producing perpetual motion, and that would be quite unacceptable to any scientist.

It was almost certainly James who realized that the only possible solution to the apparent paradox must be that the freezing point of the water became lower as the pressure that the water was subject to was increased, and it was also James who did the detailed calculations of the magnitude of effect expected. It was William who then performed the extremely sensitive experiment needed to confirm the results in practice; the most delicate thermometer yet built was required to measure a change of freezing point of around a quarter of a degree Fahrenheit produced by a pressure of 18 atmospheres.

It may be said that this work was the first ever utilizing and confirming the results of thermodynamics; even though it pre-dated the work of Clausius, as well as that of Kelvin which established the theory, it tested ideas of Carnot which are fully part of modern thermodynamics.

The second piece of work dates from the same period, again using the result of the Carnot theory. Up to this point, there had been no universal scale of temperature. Any temperature scale depended on the properties of a specific substance; an air thermometer was usually used as a standard for convenience, but this was equivalent to assuming that the air was acting as an ideal gas.

Kelvin[48] suggested that the Carnot engine could be used as the basis of a universal temperature scale, in principle independent of the properties of any individual substance. A unit of heat dropping from temperature T_0 to temperature $T_0 - 1$ would produce the same amount of work whatever the value of T_0, and this enabled a universal scale to be set up, or in fact an air thermometer to be calibrated in terms of this scale. Kelvin was able to show in practice that the temperature given by the air thermometer differed only slightly from that given by the new scale. Again it will be noted that the theory was set up in terms of caloric, but could be immediately translated to modern terms after the work of Clausius and Kelvin in 1850 and 1851.

Next we turn to the so-called Joule–Kelvin (or Joule–Thomson) effect. One of Kelvin's criticisms of Clausius had been that Clausius assumed the ideal gas law while Kelvin considered his own work to be more generally based. Following the establishment of modern thermodynamics, Joule and Kelvin collaborated to test this law in great detail. The test[49] was to drive a gas under high pressure through a porous plug. If it behaved exactly as an ideal gas, its temperature would not change. In practice though, most gases show a slight cooling, the Joule–Kelvin effect, and this property may be used for the liquefaction of gases.

We now turn to a much broader topic, perhaps the broadest in the whole of physics, that of the concept of energy and its conservation. As was said earlier, many workers developed components of this idea with various degrees of breadth and depth. Helmholtz' famous paper[50] of 1847 'On the conservation of force' gave a quite general treatment of various types of energy: mechanical, thermal, electrostatic, and magnetic, and the transformations between them. He is often credited with being the first to grasp the full significance of the idea, though, as mentioned earlier, Truesdell felt that it was impossible to believe that a proper appreciation of energy could have existed before the coming of modern thermodynamics due to Clausius and Kelvin.

It is certainly true that it was Kelvin who seized on the idea of energy. He gave the modern scientific quantity its name in the draft of his 1851 paper, a fact of which he remained very proud. Until this time, terms such as work, effort, and *vis viva* represented different aspects of what Kelvin now called energy, and he was responsible for ensuring that energy, in Smith's term[40], attained 'primary status' in science and theology as well, its conservation being practically an *a priori* requirement. Of course, the word 'energy' had been used in many ways, some more technical than others, for millennia; Elkana[51] discusses its use by Aristotle, Thomas Young, and Humphry Davy. But it was Kelvin who established its modern usage. As said before, Clausius, took up the idea of energy seriously only as late as the 1860s[40].

One of the main vehicles for promotion of the central nature of energy was the famous *Treatise on Natural Philosophy*[52], usually referred to as 'T and T′' after its authors, William Thomson and Peter Tait, the latter having moved from Queen's College Belfast to the Chair of Natural Philosophy at Edinburgh in 1860. The book was intended to provide a popular text for the relatively unsophisticated students at Glasgow and Edinburgh, thus to increase their numbers, and so to augment the incomes of the authors. The mathematical content was to be limited, the main thrust being towards experimental description and demonstration.

From the start, though, a subsidiary aim was to centre the text on energy conservation and modern thermodynamics. In Tait's plan[53], the proposed second volume would conclude with 'a great section on the *one* law of the Universe, the Conservation of Energy'. During the long struggle between the two men in which the book was written, this aim rather took over. The book that emerged was celebrated, but it was mathematically and conceptually sophisticated. Smith and Wise[53] describe it as 'replacing Newton's *Principia* of force with a new *Principia* of energy and extrema' (the latter term referring to Kelvin's favoured method of stating mechanical principles). The authors stressed the universality of kinetic energy, as contrasted to the dependence of gravitational force on locality on the Earth's surface.

Rather than appear to be attempting to replace Newton, Kelvin and Tait opted to claim that Newton had in all essence discovered the importance of energy himself. His third law, they said, stated that the real significance of the concept of force was that its action represented an exchange of energy. Elkana[54], though, speaks of the 'absurdity' of attempting to read the conservation of energy into Newton's work.

Indeed, in any assessment of Kelvin's legacy it is ironic to observe the extreme success of the project to install energy as the central aspect of physical theory, and realize that it is the very success of this project that has effectively prevented Kelvin from retaining much credit for the achievement. Physicists of today take the primacy of energy as natural, perhaps obvious, and are unaware of the struggles, of Kelvin in particular, first to formulate the belief for himself and then to persuade the scientific community of its truth. Attempting to claim that the idea actually came from Newton could only serve to reduce even further any appreciation of Kelvin's role.

We now turn to Kelvin's discussion of the age of the Sun. We saw earlier that Kelvin's arguments with geologists and evolutionists over several decades on the ages of the Earth and Sun have done his reputation little but harm. In fact, though, the *principle* behind his arguments—that the universe and all its constituents have a finite lifetime, that there is an irreversibility in its development, is correct and extremely important. Incidentally it should not be felt that this was a deduction, on Kelvin's part, from the second law of thermodynamics. Rather the idea came earlier, from the same general belief in irreversibility that helped his passage to the second law.

Here we stress his development of the idea, originated by Helmholtz, of the nature of the source of the Sun's energy. The Helmholtz-Kelvin theory[55] was that the source was the original gravitational potential energy of the universe. An early suggestion was that meteors played a large part in bringing potential energy to the Sun, but the more mature position was that the loss of potential energy of the Sun as it cools provides that heat radiated from the Sun. In detail, of course, the theory needs to include radioactivity, but the basic idea has been central in astrophysics ever since, being particularly picked up by Arthur Eddington in the 1930s.

Much of Kelvin's work from 1851 on could be described as studying aspects of the conservation and transformation of energy. To conclude this section we give a brief account of an important example—that of thermoelectricity[56]. Already two thermoelectric effects had been discovered. In 1822 Thomas Seebeck had discovered the Seebeck effect; an electric current is created in a circuit if the junctions between different metals are maintained at different temperatures. In 1834 Jean Peltier discovered the opposite effect, the Peltier effect; when an electric current passes through a junction, heat is evolved or absorbed.

Kelvin now applied to thermoelectricity the recently discovered laws of thermodynamics, and realized that there must be a third effect, usually called the Thomson effect; if an electric current flows through a conductor, the ends of which are maintained at different temperatures, heat must be evolved at a rate approximately proportional to the product of the current and the temperature gradient. Thermoelectricity is, of course, an excellent example of transformations of energy, but it has also been of use, for example, in the liquefaction of gases.

KELVIN: ENGINEERING AND INVENTION

Much more could be said about Kelvin's scientific work, but we will move on to discuss briefly his practical work on the Atlantic telegraph and other later cables, the magnetic compass, his sounding apparatus and in many other applications of science. A full account is given in Bernard Crossland's article in this book, and here only a few general points will be made.

Kelvin, of course, wished his work on the Atlantic cable and his many inventions to be successful from an engineering and from a marketing point of view, and he took it for granted that he himself would draw a considerable income from such work. Yet this was certainly not the only motivation, not the original motivation, often not even the main motivation for this work.

His first ideas on the possibilities of the Atlantic cable were stimulated by a theoretical inquiry from Stokes in 1854, and these ideas led to a sustained controversy between Kelvin, the expert on theory as applied to practical matters, and the retired medical man and self-appointed expert on practical cabling, Wildman Whitehouse. Whitehouse dismissed Kelvin's suggestion that retardation of the signal in a long cable could only be avoided by a substantial increase in the thickness of the copper conductor.

When The Atlantic Telegraph Company was founded in 1856, Whitehouse was appointed electrician; Kelvin was merely one of 18 directors elected by the subscribers[57]. Nevertheless when cable-laying began in August 1857, Whitehouse declined to board the vessel on the grounds of ill-health, and, on the request of the directors, it was Kelvin who, without salary and without position other than that of director, joined the expedition. This attempt soon failed, but a second attempt the following year was successful in establishing a connection across the Atlantic, though unfortunately a connection that worked only intermittently and after a short period failed altogether.

Whitehouse disobeyed the strict orders of the Board and again refused to sail, and for a second time it was Kelvin who did took part in the expedition[58]. Indeed it was only when a type of galvanometer designed and built by Kelvin replaced that of Whitehouse that the cable transmitted messages at all.

The official letter of the Board remarked that Whitehouse's investigations had cost £12000 to the company, yet would have rendered it a laughing-stock had not the Board been:

> fortunate enough to have an illustrious colleague who had devoted his mind to this subject, and whose inventions produced *in his own* study—*at small expense*—and from his own resources are available to supersede the useless portions of apparatus prepared at great labour and enormous cost for this special occasion[59].

Whitehouse was dismissed, and when further attempts at laying the cable were made in 1865, and finally and successfully in 1866, it was Kelvin on whom the Board relied almost entirely for technical management of the project. It was also Kelvin who gained prestige and indeed fame from the achievement; he became Sir William, and was able to go on to make a great deal of money from advice, assistance, and production of scientific instruments for future attempts at cable-laying.

Nevertheless it is important to remember that his work on the Atlantic cable was carried out as a result of scientific interest, and, as he saw it, duty to the Board. As Sylvanus Thompson says: 'The work which he undertook for it was enormous; the sacrifices he made for it were great. The pecuniary reward was ridiculously small[57]'. In addition he accepted the far from negligible dangers of carrying out the difficult task of laying heavy cables in weather liable to be stormy, dangers with which Whitehouse declined to engage. (As a tragic example of these dangers, it may be mentioned that Kelvin's nephew, David Thomson King, who was a skilled electrical worker and had made his career in the cabling industry, was drowned when the *La Plata* foundered while carrying cable to South America in November 1874, 60 lives being lost in all[60].)

Kelvin was himself a keen sailor, and the purpose of many of his important inventions, in particular the magnetic compass, was to make seamanship safer. While again there is no question that he wanted these inventions to be successful financially, again there can also be no question that he was genuinely concerned with the safety of those travelling to sea. Indeed he also[61] took part in many Admiralty committees on the design and safety of ships, and also drew up tables for finding the position of a ship at sea.

Sylvanus Thompson[62] was assured by 'the present highest authority in the British Navy' that he considered Kelvin to be 'the man who had done by far the most for

Fig. 16.2. Thomson's nephew, David Thomson King, who followed his uncle into telegraph cable laying work, was one of 60 men drowned when the *La Plata* foundered while carrying cable to South America on 29 November 1874. (From A.G. King, Kelvin the Man, Hodder & Stoughton Ltd. 1925 opposite p. 64.)

the advancement of navigation' in their time. He quoted a sailor in the distant parts of the East as saying that: 'I don't know who this Thomson may be, but every sailor ought to pray for him every night'.

An aspect of Kelvin's work connected with his interest in cables and other practical applications of electricity, and also with his academic interests, was his influence on the development of electrical measurement and units. Though the general principles of electromagnetism emerged mainly during the course of the nineteenth century, use for technical or scholarly purposes required establishment of a foundation of standards and units[63].

Kelvin was one of those mainly responsible for a number of reports and then a series of International Congresses through the second half of the nineteenth century, which established fully the practice of electrical technology for the twentieth century. He was at his best working with the world's leading scientists, and persuading them to agree to suitable compromises on nomenclature and procedures.

Successful as Kelvin was as an engineer and inventor, there must remain a question as to whether this practical work limited his ultimate scientific achievement by restricting the time available for serious thought; one might say that there remains a question as to whether his example should be looked on as a wholly profitable one for today's professor of science.

While paying Kelvin a visit in 1884, Helmholtz[64] wrote to his wife that:

> I have an impression that Sir William might do better than apply his eminent sagacity to industrial undertakings ... He is simultaneously revolving deep theoretical projects in his mind, but has no leisure to work them out quietly.

However he immediately continued that:

> I did Thomson an injustice in supposing him to be wholly immersed in technical work; he was full of speculations as to the original properties of bodies ..., and as you know he will not stop for meals or any other considerations.

Yet one may still wonder whether, for all the effort, the lack of much genuine achievement in the latter part of his career may have been, at least in part, due to outside pressures. As had been said, for most of the twentieth century, his position would have been open to criticism by those establishing the ethos of scientific work. However in more recent decades, university managements around the world have been much more enthusiastic about entrepreneurial activities of their staff; Kelvin may perhaps appear to be ahead of, rather than behind his times.

THE FIRST STUDENT PHYSICAL LABORATORY

Kelvin's own scientific education had been spent studying for the mathematical tripos at Cambridge, with its emphasis on the theory of natural phenomena, but certainly not on experiment. (The Cavendish laboratory was not founded until 1871.) At Glasgow there had been a strong tradition of lecture demonstrations, and, while Kelvin was a candidate for the Glasgow chair, his father was insistent that he should take every opportunity of getting experience in this area. However there was no expectation that the students themselves would take part in any laboratory work.

However Kelvin found his hand forced. He became aware of the desperate need for accurate data on which to base his theoretical investigations. He was able gradually[65] to extend, by legitimate and illegitimate means, the space available to him for experimental research, and to recruit voluntary helpers from among his students, stimulated by his enthusiasm. Work was performed initially on thermoelectric and elastic properties of materials, and also on heat and magnetism[11]. This constituted the first physical laboratory to be put at the disposal of students in any university.

Forty years later, when he opened the Physics Laboratories at Bangor in North Wales, Kelvin[66] spoke of his pride on this achievement, and of how much he felt the students had gained from this exposure to the demand for accurate measurement and perseverance. In the intervening decades, undergraduate laboratories had evolved into something like their modern pattern with specialized apparatus, prescribed experiments, and specially recruited demonstrators. Nevertheless Kelvin deserves immense credit for being the first to allow students to be more than passive watchers of experiments demonstrated by the lecturer.

CONCLUSIONS

Enough has been said of Kelvin's scientific work to make it clear that he should be regarded as one of the leading physicists of the nineteenth century. From the time of his entry into research in 1840, he played a large part in all the major developments, in particular those in potential theory, field theory, and thermodynamics. In other words he was a central contributor to the creation of classical physics,

Fig. 16.3. The statue of Lord Kelvin in Belfast's Botanic Gardens being unveiled by Sir Joseph Larmor in June 1913. (Courtesy of Queen's University, Belfast.)

and became a great supporter of its ideas and methods. It is scarcely surprising that his reputation suffered with the coming of relativity and quantum theory at the beginning of the twentieth century. Yet it should be remembered that classical physics underpins modern physics through, for example, the correspondence principle, so Kelvin should be remembered as one of the most important founders of today's physics.

When one includes his work on the cable projects and his multitude of other inventions, his innovative teaching at Glasgow for over 50 years, his tireless work for scientific societies and conferences, and his important role in numerous national and international commissions aimed at advancing science and technology, it is clear that he was a hugely important figure in the development of the applications and structure of science, as well as its actual content. Far from sinking into obscurity, Kelvin deserves to be remembered by all those interested in science and its history.

NOTES ON CONTRIBUTORS

Peter Bowler FBA is Professor of History of Science at Queen's University, Belfast.

Alex Craik is Emeritus Professor at the University of St Andrews.

Sir Bernard Crossland FRS is Emeritus Professor of Mechanical Engineering at Queen's University, Belfast.

C. W. Francis Everitt is Research Professor at the W. W. Hansen Experimental Physics Laboratory, Stanford University.

Raymond Flood is University Lecturer in Computing and Mathematics at the Department for Continuing Education, Oxford University.

Elizabeth Garber is Professor Emerita in the Department of History, State University of New York at Stony Brook.

Ivor Grattan-Guinness is a Professor at Middlesex University at Enfield, Middlesex.

Patrick N. Wyse Jackson is a Lecturer in the Department of Geology, and a Fellow of Trinity College, Dublin.

Colin Latimer is Professor of Physics at Queen's University, Belfast.

Mark McCartney is a Lecturer in Mathematics at the University of Ulster.

Iwan Rhys Morus is a Senior Lecturer in the Department of History and Welsh History at the University of Wales, Aberystwyth.

Oliver Penrose FRS is Professor Emeritus at Heriot-Watt University.

Sir Brian Pippard FRS is Emeritus Professor of Physics at the University of Cambridge.

John Roche is a Historian of Science at Linacre College, Oxford.

Denis Weaire FRS is Erasmus Smith's Professor of Natural and Experimental Philosophy at Trinity College, Dublin.

Andrew Whitaker is Professor of Physics at Queen's University, Belfast.

Alastair Wood is Emeritus Professor in the School of Mathematical Sciences, Dublin City University.

NOTES, REFERENCES, AND FURTHER READING

CHAPTER 1

References

1. Elizabeth Thomson King (ed.) *Lord Kelvin's Early Home*, Macmillan &Co. Ltd., London (1909) [ETK], p. 126.
2. ETK, p. 41.
3. ETK, p. 147.
4. Silvanus P. Thompson, *The Life of William Thomson Baron Kelvin of Largs*, 2 volumes, Macmillan & Co., London, 1910 [SPT], p. 17.
5. SPT, p. 32.
6. SPT, p. 37.
7. EKT, p. 204.
8. Agnes Gardiner King, *Kelvin the Man*, Hodder & Stoughton Ltd., London (1925) [AGK], p. 13.
9. SPT, p. 87.
10. ETK, p. 230.
11. SPT, p. 98. The tale is also related by Andrew Gray in *Lord Kelvin: An Account of His Scientific Life and Work*, J. M. Dent & Co., London (1908), p. 27.
12. Leonard Roth, 'Old Cambridge Days', in Donald M. Campbell & John Higgins, *Mathematics: People, Problems, Results*, volume 1, Wadsworth International Belmont, California, 1984, p. 98.
13. SPT, p. 48.
14. SPT, p. 1154.
15. Harold I Sharlin, *Lord Kelvin: The Dynamic Victorian*, The Pennsylvania State University Press, University Park & London, 1979, [HIS] p. 73.
16. SPT, p. 185.
17. David B. Wilson (ed.), *The Correspondence between Sir George Gabriel Stokes and Sir William Thomson, Baron Kelvin of Largs*, 2 volumes, Cambridge University Press, 1990.
18. ETK, p. 236.
19. Crosbie Smith and M. Norton Wise, *Energy and Empire, A Biographical Study of Lord Kelvin*, Cambridge University Press, (1989) [SW], p. 123.
20. David Murray, *Lord Kelvin as Professor in the Old College of Glasgow*, Maclehose Jackson & Co., Glasgow (1924) [DM] p. 1 & p. 6.
21. SW p. 124, footnote.
22. DM, p. 9.

23. SPT, p. 193, footnote.
24. SPT, p. 787.
25. SPT, p. 302.
26. Andrew Gray, *Lord Kelvin: An Account of His Scientific Life and Work*, J. M. Dent & Co., London (1908), [AG] pp. 67–68.
27. SPT, pp. 194–5.
28. AG, pp. 71–72.
29. SPT, p. 265.
30. Donald S. L. Cardwell, *James Joule: A Biography*, Manchester University Press, Manchester & New York (1989), p. 85.
31. William Thomson and Peter Guthrie Tait, *Treatise on Natural Philosophy*, Oxford (1967), expanded to two volumes, Cambridge (1879, 1883)
32. SPT, pp. 46, 47, 62, 66, 81, 129, 189, 225, 232.
33. SW, pp. 141–146.
34. SW, p. 142.
35. SPT, Appendix B pp. 1223–1274 gives what SW (p. 815) describe as a 'virtually complete list of Thomson's published papers'.
36. SPT, pp. 111–2.
37. Charles Darwin, *On The Origin of Species*, Murray, London (1859), p. 282. Quoted in Joe. D. Burchfield, *Lord Kelvin and the Age of the Earth*, Science History Publications, New York (1975), p. 70.
38. SPT, p. 546.
39. SPT, pp. 456–7.
40. HIS, p. 182.
41. SPT, pp. 612–3.
42. AGK, pp. 119–121.
43. SPT, p. 616.
44. R. J. Strutt, *John William Strutt, 3rd Baron Rayleigh*, Edward Arnold, London (1924), p. 145
45. William Thomson, *Baltimore Lectures on Molecular Dynamics and the Wave Theory of Light*, Cambridge University Press (1904). This 1904 version was an extensively reworked version of the original notes taken at the time by A. S. Hathaway. Hathaway's version can be found in Robert Kargon and Peter Achinstein (editors) *Kelvin's Baltimore Lectures and Modern Theoretical Physics: Historical and Philosophical Perspectives*, MIT Press, Cambridge MA (1987).
46. AG, p. 261.
47. SPT, p. 906.
48. SW, pp. 807–8.
49. For example, the Keith Medal of the Royal Society of Edinburgh (1864), Royal Medal (1856), Copley Medal (1883) of the Royal Society of London.
50. This number swelled more in the Jubilee year itself, and by the time of his death Thomson had been elected to over 100 learned societies across the world.
51. Lindley, David, *Degrees Kelvin: The Genius and Tragedy of William Thomson*, Aurum Press, London (2004).

CHAPTER 2

References

1. See e.g. SW, pp. 51, 135–39; ETK, pp. *vii*, 138, 219, 239.
2. ETK, pp. 73–81.
3. SW, p. 32; ETK , pp. 94–97.
4. SW, p. 15.
5. ETK, pp. 87–88.
6. ETK, p. 91.
7. ETK, p. 108.
8. ETK, p. 109.
9. William won prizes in Latin (Humanity) in 1836 and 1841, in logic in 1838, and in astronomy in 1840. William and James (always in that order) took first and second prizes in mathematics in 1837 and 1838, and in natural philosophy in 1839. See SPT, pp. 9–10.
10. SPT, p. 20.
11. ETK, p. 121.
12. SW, Ch.2.
13. SW, p. 75, quoting a letter from Anna to William Thomson, 8th Nov. 1842. However, the able and resourceful Nichol eventually recovered his equilibrium by writing several successful books on astronomy and by giving popular lectures that attracted huge crowds.
14. DBW, 1985, 26–29.
15. SW, p. 167.
16. SPT, pp. 19–20; DBW, 1985, pp. 29–30; SW, p. 213. Thomson and Faraday finally met in 1847 when William visited the Royal Institution in London.
17. ETK, pp. 146–55, 168–69.
18. SW, p. 51.
19. ETK, p. 191.
20. See ADDC, Ch.11. Also Pattenden, Philip 2002. 'The noblest prospect: Scotland in Peterhouse.' *Peterhouse Annual Record, 2001/2002*, pp. 97–106. Peterhouse, Cambridge.
21. ETK, pp. 196–7, 199.
22. SPT, p. 25.
23. ETK, pp. 198, 204.
24. SPT, p. 29.
25. SPT, pp. 29, 32.
26. SPT, p. 32.
27. SPT, p. 32.
28. For more about William's musical activities, see SPT, pp. 69–76.
29. See ADDC, Ch.2.
30. Peacock, George 1820. *A Collection of Examples of the Application of the Differential and Integral Calculus.* Cambridge: Smith, Deighton.
31. SPT, p. 35.
32. SPT, pp. 36–37.

33. SPT, p. 87.
34. SPT, pp. 109–10.
35. SPT, pp. 36, 59. One of the group was George Wirgman Hemming, the Senior Wrangler of 1844 and later an eminent lawyer.
36. The Colquhoun Silver Sculls was a race for individuals, for which William borrowed a boat superior to his own. He won a cup worth about 15 guineas and had custody of a silver trophy for a year. See SPT, pp. 60–61, 64; SW, p. 78.
37. See SW, p. 78. Present-day students will surely be amazed that, during the early summer of 1844, Thomson had classes with Hopkins at 7a.m. every other day, and attended college chapel at that hour on the remaining days. What's more, he reported that, before doing so, 'I always bathe [in the river Cam], before I go, with Blackburn, . . . so that I am up every day at six': quoted in SPT, p. 77.
38. Quoted in SW, p. 76.
39. ADDC, Ch. 5(*b*).
40. Letters in Special Collections, University of St Andrews Library (msdep7); quoted in ADDC, Ch. 5(*b*).
41. See Crosbie Smith, 'Geologists and mathematicians: the rise of physical geology.' In Peter M. Harman, (ed.) *Wranglers and Physicists; Studies on Cambridge Physics in the Nineteenth century*, pp. 12–48. Manchester: Manchester Univ. Press. Also ADDC, Chapters 5(*d*) and 12(*b*).
42. Quotation from Rawnsley, (Canon) Hardwicke D. 1896. *Harvey Goodwin, Bishop of Carlisle, a Biographical Memoir.* London: John Murray. p. 41.
43. The Thomson and Crum families lived in Glasgow and Largs, and visited each other frequently. Margaret was a young woman of delicate health and poetic disposition. Sadly, she became a long-term invalid not long after marriage; and, despite William's best efforts, she died in 1870. Though William was a devoted husband, Margaret Crum had not been his only love: during 1850–52, he three times proposed to Sabina Smith, sister of his friend Archibald Smith, and was each time refused. In 1874, William Thomson (by then Lord Kelvin) was married again, to Fanny Blandy of Madeira. See SPT, p. 66; AGK, pp. 32–35; SW, pp. 141–146.
44. SPT, p. 67.
45. Quoted in SW, p. 79.
46. SW, p. 80.
47. ETK, p. 229.
48. Quoted by SW, p. 80.
49. SPT, pp. 156, 158.
50. Glasgow was then staffed by a superabundance of Thomsons! David Thomson's situation was resolved when he was appointed Professor of Natural Philosophy at King's College, Aberdeen in 1845. When Aberdeen's two colleges, King's and Marischal, merged in 1860, it was he as the senior professor, and not the young James Clerk Maxwell, who retained his post.
51. SW, p. 102.
52. SW, p. 103.

53. SW, p. 105.
54. Chalmers retained an interest in science, particularly astronomy, and early in his career had been an assistant in mathematics at St Andrews. He had been friendly with the Thomson family since their days in Belfast. In 1843, his popular fame (or, in some quarters, notoriety) grew when he led the Disruption of the established Church of Scotland and helped to found the rival Free Church.
55. SW, p. 106. Blackburn was later to be William's colleague at Glasgow, as Professor of Mathematics.
56. SW, p. 107–8.
57. SPT, pp. 113–33; AG, pp. 29–31; SW, p. 216.
58. See DBW, 1987, DBW, 1990; ADDC, Ch. 8(*c*). All of Thomson's early papers in the *Cambridge Mathematical Journal* are reprinted in volume 1 of his collected *Mathematical and Physical Papers*: Thomson (1882–1911).
59. SW, pp. 113–14.
60. The full list of referees and the texts of many of their testimonials are in SPT, pp. 167–182.
61. Quoted in SPT, pp. 171.
62. Though William strenuously denied the rumour, subsequent accounts of his ill-organized lecturing at Glasgow lend support to it. Andrew Gray, Thomson's successor and former assistant at Glasgow, had no doubts. After describing Thomson's unplanned and digressive lecturing style, Gray concluded that: 'Professor Thomson was, it must be admitted, too discursive for the ordinary student, and perhaps did not study the art of boiling down physical theories to the form most easily digestible. His eagerness of mind and width of mental outlook gave his lectures a special value to the advanced student, so that there was a compensating advantage': AG, pp. 90–91.
63. The other candidates, though reputable enough, were not formidable opposition. They were: David Gray (Professor of Natural Philosophy at Marischal College, Aberdeen), Thomas Miller (Rector of Perth Academy), Thomas Aitken (formerly lecturer in Natural Philosophy at the Royal Institution, Liverpool), Andrew Bell (mathematical master at Dollar Institution), William Brydone Jack (Professor of Mathematics and Natural Philosophy at King's College, Fredericton, New Brunswick). List given in SPT, p. 167.
64. ETK, pp. 231–32.
65. ETK, pp. 232–33.
66. SW, p. 137; ETK, p. 239.

Further Reading

Craik, Alex D. D. 2007. *Mr Hopkins' Men: Cambridge Reform and British Mathematics in the Nineteenth Century*. New York etc.: Springer. [ADDC]

Gray, Andrew 1908. *Lord Kelvin; An Account of his Scientific Life and Work*. London: J. M. Dent; New York: E.P. Dutton. [AG]

King, Agnes Gardner 1925. *Kelvin the Man*. London: Hodder & Stoughton. [AGK]

King, Elizabeth (née Thomson) and King, Elizabeth Thomson 1910. *Lord Kelvin's Early Home*. London: Macmillan & Co. [ETK]

Smith, Crosbie and Wise, M. Norton 1989. *Energy and Empire. A Biographical Study of Lord Kelvin*. Cambridge: Cambridge Univ. Press. [SW]

Smith, Crosbie 2002a. 'Thomson, James (1786–1849) mathematician'. In *Oxford Dictionary of National Biography*. Oxford: Oxford Univ. Press.

Smith, Crosbie 2002b. Thomson, James (1822–1892) mechanical engineer. In *Oxford Dictionary of National Biography*. Oxford: Oxford Univ. Press.

Smith, Crosbie 2002c. Thomson, William, Baron Kelvin (1824–1907), mathematician and physicist. In *Oxford Dictionary of National Biography*. Oxford: Oxford Univ. Press.

Thompson, Silvanus P. 1910. *The Life of William Thomson Baron Kelvin of Largs*. 2 vols. London: Macmillan Volume 1. [SPT]

Thomson, William 1882–1911. *Mathematical and Physical Papers*, 6 vols. Cambridge: Cambridge University Press.

Wilson, David B. 1985. The educational matrix: physics education at early Victorian Cambridge, Edinburgh and Glasgow Universities. In Peter M. Harman, (ed.) *Wranglers and Physicists; Studies on Cambridge Physics in the Nineteenth Century*, pp. 12–48. Manchester: Manchester Univ. Press. [DBW1985]

Wilson, David B. 1987. *Kelvin & Stokes, a Comparative Study in Victorian Physics*. Bristol: Adam Hilger. [DBW1987]

Wilson, David B. (ed.) 1990. *The Correspondence between Sir George Gabriel Stokes and Sir William Thomson Baron Kelvin of Largs*. 2 vols. Cambridge, Cambridge Univ. Press. [DBW1990]

Notes

The main modern biography of William Thomson is Smith and Wise (1989); and the publications of Wilson (1985; 1987; 1990) are also important. Other relevant material is in Craik (2007). Good brief biographies of William Thomson, his father James and his brother James (Jr.) are Smith (2002a, b, c) in the *Oxford Dictionary of National Biography*.

Works by Thomson's own contemporaries are numerous. The most valuable are Silvanus P. Thompson's two-volume *Life* (Thompson 1910), and Andrew Gray's shorter account of Thomson's life and work (Gray 1908). Recollections of family life are those of William Thomson's sister Elizabeth, as edited by her daughter (King 1910), and a later volume (King 1925) prepared by another niece.

Many letters and diary extracts are reproduced in the above-mentioned works, but more remain unpublished. The full scientific correspondence between Thomson and G. G. Stokes is in Wilson (1990). Throughout this chapter, references are given only to published works (which in turn give full references to quoted archival sources). Much of the archival material is housed in Cambridge University Library and in Glasgow University Library (a fuller list is given in Smith 2002c).

CHAPTER 3

References

1. See S. E. Despaux, Launching mathematical research without a formal mandate: the role of university-related journals in Britain, 1837–1870, *Historia Mathematica,* **34** (2007), 89–106.
2. W. Thomson, 1872 *Reprint of Papers in Electrostatics and Magnetism,* 1st ed., Cambridge, Cambridge University Press.
3. W. Thomson, 1882. *Mathematical and Physical Papers,* vol. 1, Cambridge, Cambridge University Press.
4. See I. Grattan-Guinness,1990 *Convolutions in French Mathematics, 1800–1840. From the Calculus and Mechanics to Mathematical Analysis and Mathematical Physics,* 3 vols., Basel (Birkhäuser) and Berlin Deutscher Verlag der Wissenschaften, , esp. ch. 9 on heat, ch. 13 on electricity and magnetism, and chs. 16–17 on the new generation of French mathematicians. Space prohibits citation of the massive original or historical literature in this chapter.
5. See Thompson, 1910, 111–112.
6. See G. Bachelard, 1928 (repr. 1973). *Etude sur l'Évolution d'un Problème de Physique. La Propagation Thermique dans les Solides,* Paris, Vrin.
7. For a survey of Laplace's book see I. Grattan-Guinness (ed.), 2005 *Landmark Writings in Western Mathematics 1640–1940,* Amsterdam, Elsevier ch. 18.
8. On the physical aspects of this story see R. Fox,1973 The rise and fall of Laplacian physics, *Historical Studies in the Physical Sciences, 4,* 81–136; on the mathematical ones, see Grattan-Guinness (note 4), ch. 7.
9. For accounts of Thomson's progress through a conflicting and complicated sequence of theories, with somewhat different emphases, see J. Z. Buchwald, William Thomson and the mathematization of Faraday's electrostatics, *Historical Studies in the Physical Sciences,* **8** (1977), 102–136; and M. N. Wise, William Thomson's mathematical route to energy conservation, *ibidem,* **10** (1979), 49–83 and The flow analogy in electricity and magnetism. Part 1: William Thomson's reformulation of action at a distance, *Archive for History of Exact Sciences,* **25** (1981), 19–70. Much further background is provided in R. Reiff and W. Sommerfeld, Standpunkt der Fernwirkung. Die Elementargesetze, in *Encyklopädie der Mathematischen Wissenschaften,* vol. 5, sec. 2, 3–62 (1912, article V 12); R. Gans, Elektrostatik und Magnetostatik, in *ibidem,* 289–349 (1906, article V 15); E. T. Whittaker, *History of the Theories of Aether and Electricity. The Classical Theories,*1951 rev. ed., London, Nelson, esp. chs. 5–6; G. N. Cantor and M. J. S. Hodge (eds.), *Conceptions of Ether,* 1981 Cambridge, Cambridge University Press, chs. 6–8; F. Bevilacqua, *The Principle of the Conservation of Energy and the History of Classical Electromagnetic Theory,*1983 Pavia, La Goliardica Pavese, chs. 3–4; P. Harman (ed.), *Wranglers and physicists,*1985 Manchester, Manchester University Press, chs. 4–6; O. Darrigol, 2000 *Electrodynamics from Ampère to Einstein,* Oxford, Oxford University Press, ch. 3; K.-H. Schlote, 2004 *Zu den Wechselbeziehungen zwischen Mathematik und*

Physik an der Universität Leipzig in der Zeit von 1830 bis 1904/05, Stuttgart and Leipzig (Verlag der Sächsischen Akademie der Wissenschaften zu Leipzig in Kommission bei S. Hirzel: *Abhandlungen der Sächsischen Akademie der Wissenschaften zu Leipzig, mathematisch-naturwissenschaftliche Klasse*, vol. 63, no. 1), ch. 6; and G. Wolfschmidt (ed.), 2005 *Vom Magnetismus zur Elektrodynamik anlässlich des 200. Geburtstages von Wilhelm Weber (1804–1891) und des 150. Todestages von Carl Friedrich Gauss (1777–1855)*, Hamburg, Fachbereich Mathematik, Universität esp. ch. 5.

10. Thomson (1843: *Elec Mag*, **136**); see also (1853: *Elec Mag*, **92**) and the 1854 note on pp. 1–2.
11. See S. Earnshaw, On the nature of the molecular forces which regulate the constitution of the luminiferous ether, *Transactions of the Cambridge Philosophical Society*, **7** (1842), 97–112.
12. For a survey of Green's book see Grattan-Guinness (note 7), ch. 30. An edition of all his publications is *Mathematical papers* (ed. N.M. Ferrers), 1871 (repr. New York (Chelsea), 1970); London (MacMillan), the book is on pp. 1–115.
13. Green *ibidem*, 23–26.
14. *Ibidem*, **12**.
15. *Ibidem*, **7**, 83.
16. Thomson (1845: *Elec Mag*, 104–107); compare Green (note 12), 100–106.
17. See C. J. de la Vallée Poussin, Gauss et la théorie du potentiel, *Revue des questions scientifiques*, **133** (1962), 314–330; and A. F. Monna, 1975 *Dirirchlet's Principle*, Utrecht (Oosthoek, Scheltema & Holkema). A very large story is at hand here, involving many figures up to the end of the century, though seemingly not Thomson. Some of the issues interacted with the truthhood of Gauss's theorem mentioned in section 5.
18. See D. B. Wilson (ed.), 1990 *The Correspondence Between Sir George Gabriel Stokes and Sir William Thomson, Baron Kelvin of Largs*, 2 vols., Cambridge, Cambridge University Press 96–97.
19. See D. B. Wilson, 1987 *Kelvin and Stokes*, Bristol, Adam Hilger.
20. See J. D. Burchfield, *Lord Kelvin and the Age of the Earth*, 1975 New York, Science History Publications; and Smith and Wise, *Energy and Empire, A biographical study of Lord Kelvin*, Cambridge University Press (1989), chs. 10, 16 and 17.
21. See G. Bacharach, *Abriss der Geschichte der Potentialtheorie*, 1883 Würzburg, Thein.
22. J. B. J. Fourier, *The Analytical Theory of Heat* (ed. A. Freeman), 1878 (repr. New York (Dover), 1955) Cambridge (Cambridge University Press).
23. W. Thomson, 'Heat', in *Encyclopaedia Britannica*, 9th ed., vol. 11 (1880), 554–589; this part also in *Mathematical and Physical Papers*, 1884 vol. 2, Cambridge, Cambridge University Press, 41–60.
24. See R. Olson, *Scottish Philosophy and British Physics, 1750–1880. A Study in the Foundations of the Victorian Scientific Style*, 1975 Princeton, Princeton University Press; he consciously omits Thomson from his account of the influence because of Thomson's silence about it.
25. See W. Thomson, Address on installation as Chancellor of the University of Glasgow (1904), in *Mathematical and Physical Papers*, vol. 6 (ed. J. J. Larmor),1911 Cambridge, Cambridge University Press 370–376 (p. 373).

Further reading

There are some general remarks on the work reviewed here in S. P. Thompson, *The life of William Thomson, Baron Kelvin of Largs,* 2 vols., London (MacMillan), 1910, chs. 1–4 *passim.* Its general context is rather disappointingly treated in ch. 6 of the fine recent biography C. Smith and M. N. Wise, *Energy and Empire. A Biographical Study of Lord Kelvin,* Cambridge (Cambridge University Press), 1989; however, the account of Thomson himself on electricity and magnetism in chs. 7–8 draws valuably on manuscript sources.

CHAPTER 4

References

1. See for instance Peter J. Bowler, *The Fontana History of the Environmental Sciences* (London: Fontana, 1992) which provides background on some of the issues discussed below.
2. See Crosbie Smith and M. Norton Wise, *Energy and Empire: A Biographical Study of Lord Kelvin* (Cambridge: Cambridge University Press, 1989). This book also contains a wealth of information on James Thomson.
3. In addition to the discussions in Smith and Wise, *Energy and Empire*, see also Joe D. Burchfield, *Lord Kelvin and the Age of the Earth* (New York: Science History Publications, 1975).
4. James Thomson to William Thomson, 7 April 1862, quoted in Sir Joseph Larmor and James Thomson, eds., *Collected Papers on Physics and Engineering by James Thomson* (Cambridge: Cambridge University Press, 1912) pp. lv–lvii, and William's reply of 21 April 1862, *ibid.*, pp. lvii–lviii.
5. On Thomson's career see the 'Biographical Sketch' in his *Collected Papers*, pp. xiii–xci and the 'Obituary Notice' by J. T. Bottomley reprinted from the *Proceedings of the Royal Society* (1893) *ibid.*, pp. xcii–cii. I am grateful to Crosbie Smith for sending me a copy of his entry on Thomson forthcoming in *The New Dictionary of National Biography*. See also Sir Bernard Crossland, 'James Thomson' in Crossland and John S. Moore, eds., *The Lives of the Great Engineers of Ulster, Vol. 1* (Belfast: NE Consultancy for Belfast Industrial Heritage Ltd., 2003), pp. 149–156. I am grateful to Sir Bernard for sending me a copy of the proof of this article in advance of publication.
6. See the drawings of September 1866 held in Queen's University Library, Belfast, Thomson Papers MS13, section L, item 5.
7. See for instance James Thomson to William Thomson, 21 September 1891, Thomson Papers, section H.
8. James Thomson, 'On public parks in connection with large towns, with a suggestion for the formation of a park in Belfast', reprinted in *Collected Papers*, pp. 464–472.
9. See Thomson's account to the British Association for the Advancement of Science, 1852, 'On the vortex water wheel', reprinted in *Collected Papers*, pp. 2–16.

10. Thomson, 'On a centrifugal pump with exterior whirlpool, constructed for draining land', reprinted from *Proceedings of the Engineers of Scotland*, 1858, in *Collected Papers*, pp. 16–24.
11. Thomson, 'On a jet pump, an apparatus for drawing up water by the power of a jet', reprinted from the report of the British Association for the Advancement of Science, 1852, *Collected Papers*, pp. 26–27.
12. James Thomson, 'Theoretical considerations on the effect of pressure in lowering the freezing point of water', reprinted from *The Cambridge and Dublin Mathematical Journal*, 1850, in *Collected Papers*, pp. 196–203 and William Thomson, 'The effect of pressure in lowering the freezing point of water experimentally demonstrated', reprinted from *Proceedings of the Royal Society of Edinburgh*, 1850, *ibid.*, pp. 204–208.
13. There is a copy of the report of the paper to the Belfast Natural History and Philosophical Society, with a later note dated 1866 in Queen's University Library, Thomson Papers, section I, item 1. See also Thomson, 'On the plasticity of ice, as manifested in glaciers', reprinted from *Proceedings of the Royal Society*, 1856–57, *Collected Papers*, pp. 208–211, and the letters between Thomson and Faraday, *ibid.*, pp. 212–219.
14. A number of papers on the topic are reprinted in *Collected Papers*, pp. 276–333. There are also many items on the topic in Queen's University Library, Thomson Papers, section J. See J. S. Rowlinson, 'The Work of Thomas Andrews and James Thomson on the liquefaction of gases', *Notes and Records of the Royal Society of London*, **57** (2003): 143–159. I am grateful to Professor Rowlinson for supplying me with a copy of his paper.
15. See James Thomson's letter of 9 July 1862 to William, quoted in Smith and Wise, *Energy and Empire*, p. 285 (original in the Kelvin Collection, University of Glasgow).
16. James Thomson, 'Experimental demonstration in respect to the origin of the windings of rivers in alluvial plains', reprinted from Proceedings of the Royal Society, 1873, in *Collected Papers*, pp. 100–101, and numerous other papers on this topic reprinted in the same volume.
17. See the notes for papers dated 1887–89, Queen's University Library, Thomson Papers, section E.
18. Thomson, 'On the parallel roads of Lochaber', reprinted from the *Edinburgh New Philosophical Journal*, 1848, in *Collected Papers*, pp. 407–420. See Martin J. S. Rudwick, 'Darwin and Glen Roy: A "great failure" in scientific method?' *Studies in the History and Philosophy of Science*, **5** (1974): 97–185.
19. James Thomson, 'Bakerian Lecture—On the great currents of atmospheric circulation', reprinted from *Philosophical Transactions of the Royal Society*, 1892, in *Collected Papers*, pp. 153–195; on the 1857 address see p. 182.
20. Queen's University Library, Thomson Papers, section H.
21. James Thomson, 'On the jointed prismatic structure in basaltic rocks', reprinted from *Transactions of the Geological Society of Glasgow*, 1877, in *Collected Papers*, pp. 422–440.

Note

This article will also appear in Alvin Jackson and David N. Livingstone, eds., *Queen's Thinkers: Essays on the Intellectual Heritage of a University* (Blackstaff, Belfast, to be published).

CHAPTER 5

References

1. Wilson, David B., editor, 1990, *The Correspondence between Sir George Gabriel Stokes and Sir William Thomson Baron Kelvin of Largs in 2 volumes*. Cambridge, Cambridge University Press. [DBW] Many extracts from the correspondence, along with their dates, are quoted in this chapter. Because the correspondence has been arranged in chronological order by Wilson, individual letters are easily located and the page number of each extract is not given.
2. Larmor, Sir Joseph, editor, 1907, *Memoir and Scientific Correspondence of the late Sir George Gabriel Stokes*. Cambridge, Cambridge University Press. [JL] p. 3.
3. JL, p. 35.
4. JL, p. 6.
5. DBW, Preface to Volume 1, p. xli.
6. Stokes, Sir G.G., 1880–1905, *Mathematical and Physical Papers Volumes I-V*. Cambridge, Cambridge University Press.
7. JL.
8. Knox, Kevin C. and Noakes, Richard, editors, 2003, *From Newton to Hawking: a History of Cambridge University's Lucasian Professors of Mathematics*. Cambridge, Cambridge University Press. [KN] p. 297.
9. KN, Timeline in Preface.
10. KN, p. 256.
11. Wilson, David B., 1987, *Kelvin and Stokes, a Comparative Study in Victorian Physics*. Adam Hilger, Bristol.
12. JL, pp. 141–4.
13. Smith, Crosbie, and Wise, M. Norton, 1989, *Energy and Empire: A Biographical Study of Lord Kelvin*. Cambridge, Cambridge University Press. p. 804.
14. JL, p. 18.

CHAPTER 6

References

1. M. Purser, *Jellett, O'Brien, Purser and Stokes*, Prejmer Verlag, Dublin, 2004.
2. C. Smith and M. Norton Wise, *Energy and Empire: A Biographical Study of Lord Kelvin*, Cambridge University Press, Cambridge, 1989.
3. T. West, *The Bold Collegians*, Lilliput Press, Dublin, 1991.
4. A. Warwick, *Masters of Theory*, University of Chicago Press, Chicago, 2003.
5. J. Larmor, ed., *The Scientific Writings of the late George Francis Fitzgerald*, Hodges, Figgis, Dublin, 1902.
6. B. Hunt, *The Maxwellians*, Cornell University Press, Ithaca, 1991.

CHAPTER 7

References

1. Thompson S. P. 1910 *The Life of William Thomson* 2 vols (London: Macmillan) [SPT] **1**, pp. 212, 430, 444.
2. SPT, pp. 819, 1070, 1138.
3. Thomson W. 1882–1911 *Mathematical and Physical Papers* 6 vols (Cambridge: Cambridge University Press) [MPP] **3**, p. 457.
4. Thomson W. 1872 *Reprint of Papers on Electrostatics and Magnetism* (London: Macmillan) [EM]; MPP; Thomson W. 1882–1911 *Mathematical and Physical papers* 6 vols (Cambridge: Cambridge University Press); Thomson W. 1884 *Notes of Lectures on Molecular Dynamics and the Wave Theory of Light* (Baltimore: Johns Hopkins University); [Thomson W] (1904) Lord Kelvin *Baltimore Lectures* (C. J. Clay and Sons); Thomson W. 1889–94 *Popular Lectures and Addresses* 3 vols (London); Thomson W. and Tait P. Guthrie 1872 *A Treatise of Natural Philosophy: Part 1* (Cambridge: Cambridge University Press); Whittaker E. T. (1961–2) *A History of the Theories of Aether and Electricity* 2 vols (Thomas Nelson, 1910, 1953) This text still has enormous authority because of its grasp of electromagnetism and it scholarship. Nevertheless, Whittaker is rarely faithful to the mathematical notation of his authors, and this sometimes alters the meaning of the original text.
5. SPT; Harman P. M. ed. 1985 *Wranglers and Physicists* (Manchester: Manchester University Press); Crosbie Smith W. and Norton Wise M 1989 *Energy and Empire: A Biographical Study of Lord Kelvin* (Cambridge: Cambridge University Press). [SW] Although an extraordinary achievement, the latter work is sometimes over interpreted.
6. Heilbron J. L. 1979 *Electricity in the 17th and 18th Centuries* (U California Press) chs 8, 12, 14, 18, 19; L. Pearce Williams *1965 Michael Faraday: A Biography* (London: Chapman and Hall) chs 4–6.

Gooding, D. and James, F. A. J. L ed. and intro 1985 *Faraday Rediscovered: Essays on the Life and Work of Michael Faraday, 1791–1867* (Basingstoke: Macmillan) chs 6, 7, 10, 11.

7. EM, p. 346.
8. SW, ch 4.
9. Harman P. M. ed. 1985 *Wranglers and Physicists* (Manchester: Manchester University Press) ch 4.
10. SPT, chs 1–2; SW, ch 2–5; Wilson D. W. 1987 *Kelvin and Stokes* (Bristol: Adam Hilger) ch 2.
11. Thomson W. 1867 On vortex atoms, *Philosophical Magazine* **34** 15–24; SPT, 84, 114–6, 265–76, **2** 1116–7.
12. SPT, pp. 1–16, 49, 51, 64, 176, 186, 291, 1116.
13. SPT, pp. 171, 178, 187, 192, 250, 253, 299, 801.
14. SPT, pp. 1123.
15. SPT, pp. 821, 11178–; Knudson O 1985 Mathematics and physical reality in William Thomson's electromagnetic theory, in *Wranglers and Physicists* Harman P M ed (Manchester: Manchester University Press) ch 4.
16. SPT, pp. 822, 821, 827, 830, 835.
17. SPT, p. 835.

18. SPT, pp. 1116–21.
19. SPT, p. 239.
20. Maxwell J. C. 1954 *A Treatise on Electricity and Magnetism* 2 vols (New York: Dover: Original edition Oxford, Clarendon Press 1873) **2** 25; SPT, **2** 1118; EM § 30.
21. MPP, **1** p. 521.
22. Roche J. 2003 What is potential energy? *Eur. J. Phys.* **24** 185–196, 186–8, 192.
23. MPP, **1** pp. 235–6.
24. Thomson W. and Tait P. G. 1867 *Treatise of Natural Philosophy* (Cambridge: Cambridge University Press) p. 36.
25. EM, p. 484.
26. EM, p. 472.
27. SPT, pp. 831–2.
28. MPP, **3** p. 442.
29. MPP, **4** p. 51.
30. MPP, **4** p. 126.
31. MPP, **4** p. 149.
32. SPT, pp. 535, 1097, 1104; SW, pp. 641–5.
33. EM, p. 340.
34. SW, p. 151.
35. MPP, **4** pp. 125–128.
36. SW, chs 6, 7, 8.
37. EM, §20, 23; MPP **2** § 91; **3** appendix a, d.
38. http://www.physics.gla.ac.uk/Physics3/Kelvin_online/Patents.htm; SW, ch 22.
39. SPT, p. 1065.
40. SPT, p. 1064.
41. SPT, p. 1117.
42. SW, ch 6–8, 11–13, 19–20, 22; Darrigol O. 2000 *Electrodynamics from Ampère to Einstein* (Oxford: Oxford University Press) chs 1–3
43. MPP, **1** p. 76.
44. MPP, **1** p. 98.
45. MPP, **1** p. 99.
46. EM, p. 33.
47. SPT, ch 20, *830*.
48. SW, p 151.
49. SPT, pp. 13–19.
50. MPP, **3** p. 296.
51. Fourier J. (1955) *The Analytical Theory of Heat,* trans A. Freeman (Dover; 1st French ed 1822).
52. Newton I. 1999 *The Principia,* transl. and guide by I. B. Cohen and A. Whitman (Berkeley: University of California Press) 767–8; Poisson S. D. 1821–2 Mémoire sur la théorie du magnétisme, *Mem Acad R Sci Inst France* **5** 247-338, *302;* Thomson W. 1843 On the motion of heat in solid bodies and its connexion with the mathematical theory of electricity, *Cambridge Mathematical Journal* **3** 71–84.

53. Thomson W. 1843 On the motion of heat in solid bodies and its connexion with the mathematical theory of electricity, *Cambridge Mathematical Journal* **3** 71–84.
54. Thomson W. 1846 On the elementary laws of static electricity, *Cambridge and Dublin Mathematical Journal* **1** 74–96, 83, 92; SW, pp. 219–29.
55. Thomson W. 1845 Extrait d'une letter de W Thomson, *Journal de mathematiques* **10** 365–367; Thomson W. 1872 Note of induced magnetism in a plate, *Reprint of Papers on Electrostatics and Magnetism* (London: Macmillan) 105–7; Thomson W. 1847 On electrical images, *British Associate for the Advancement of Science* (BAAS) **17** 6–7; Thomson W. 1858 On the mathematical theory of electricity in equilibrium: iii Geometric investigations with respect to the distribution of electricity on spherical surfaces, *Cambridge and Dublin Mathematical Journal* **3** 141–8; 206–7; 4 (1849), 276–84; **5** (1850) 1–9.
56. SPT, pp. 114–120; SW, pp. 215–8.
57. Thomson W. 1845 Note sur les loix élementaire d'électricité statique, *Journal de mathematiques* **10** 209–221.
58. Faraday M. 1839, 1844, 1855 *Experimental Researches in Electricity* 3 vols (London: Taylor and Francis) **1**, 361–2 (§ 1162), 363 (§ 1166), 380 (§1215).
59. SPT, p. 1057.
60. Thomson W. 1845, Note sur les loix élementaire d'électricité statique, *Journal de mathematiques,* **10**, 217–8.
61. Poisson S. D. (18212–a) Mémoire sur la théorie du magnétisme *Mem Acad R Sci Inst France,* **5**, 247–338; Faraday M. 1839, 1844, 1855 *Experimental Researches in* 3 vols (London: Taylor and Francis) § 1165, §1679.
62. Thomson W. 1845 Note sur les loix élementaire d'électricité statique, *Journal de Mathematiques,* **10**, 216–221.
63. Thomson W. 1845 Note sur les loix élementaire d'électricité statique, *Journal de Mathematiques,* **10**, 218.
64. Faraday M. 1971 Pearce Williams L. ed. *The Selected Correspondence of Michael Faraday* 3 vols (Cambridge: Cambdge University Press), **1**, 458–60, 534.
65. Poisson S. D. anneé 1821–2 Mémoire sur la théorie du magnétisme' *Mem Acad R Sci Inst France* (Paris 1826) **5**, 247-338, *271–293;* 488–533, *459*; Poisson S. D. anneé 1823 Mémoire sur la théorie du magnétisme en mouvement, *Mem Acad R Sci Inst France* (Paris 1827), **6**, 441–570; Mossotti O. 1847 Récherches théorique sur l'induction électrostatique envisagé d'apres les ideés de Faraday, *Archives des Sciences Physiques et Naturelles* (Genéve) **6** 193–6; Mossotti O. 1850 Discussione analitica sull' influenza che l'azione di un mezzo dielettrico la sull distribuzioe dell'elettricà alla superfice di pì u corpi ellectrici disseminati in esso, *Memorie di Mathemici e di Fisica dellà Societa Italiana delle Scienze* **24** (Modena), 49–74; Mossotti O. 1981 *Dictionary of Scientific Biography (DSB)* C. Gillispie ed. (New York: Charles Scribner's Sons), **9**, 547–549.
66. Thomson W. 1843, On the motion of heat in solid bodies and its connexion with the mathematical theory of electricity, *Cambridge Mathematical Journal,* **3**, 71–84, *72*; Thomson W. 1846 On the elementary laws of static electricity, *Cambridge and Dublin Mathematical Journal,* **1**, 74–96, 83.

67. Thomson W. 1846, On the elementary laws of static electricity, *Cambridge and Dublin Mathematical Journal,* **1**, 74–96, *84–6.*
68. Thomson W. 1846, On the elementary laws of static electricity, *Cambridge and Dublin Mathematical Journal,* **1**, 74–96; Thomson W. 1854, On the uniform motion of heat in homogeneous solid bodies, and its connexion with the mathematical theory of electricity, *Phil Mag,* **7**, 502–515; Buchwald J. 1977 William Thomson and the mathematization of Faraday's electrostatics, *Hist Studies Phy Sci,* **8**, 101–136.
69. Poisson S. D. 1821–2, Mémoire sur la théorie du magnétisme, *Mem Acad R Sci Inst France,* **5**, 247–338, *302;* Fourier J. (1822) *Théorie Analytique de la Chaleur* (Paris); Thomson W. 1846 On the elementary laws of static electricity, *Cambridge and Dublin Mathematical Journal* **1** 74–96, *84.*
70. Faraday M. 1839, 1844, 1855 *Experimental researches in Electricity* 3 vols (London: Taylor and Francis), **1**, §1228.
71. EM, p. 32.
72. Thomson W. 1846, On the elementary laws of static electricity, *Cambridge and Dublin Mathematical Journal,* **1**, 74–96, *85.*
73. Poisson S. D. anneé 18212– Mémoire sur la théorie du magnétisme *Mem Acad R Sci Inst France* (Paris 1826) **5**, 296–299; Thomson W. 1846 On the elementary laws of static electricity, *Cambridge and Dublin Mathematical Journal,* **1**, 74–96, *88–90.*
74. Thomson W. 1845, On the elementary laws of static electricity, *Cambridge and Dublin Mathematical Journal,* **1**, 33–5; also EM, p. 33.
75. Poisson S. D. anneé 1821–2, Mémoire sur la théorie du magnétisme *Mem Acad R Sci Inst France* (Paris 1826), **5**, 296–299; EM, p. 33.
76. Poisson S. D. Anneé 1823, Mémoire sur la théorie du magnétisme en mouvement, *Mem Acad R Sci Inst France* (Paris 1827), **6**, 441–570; 464–6.
77. Thomson W. 1846, On the elementary laws of static electricity, *Cambridge and Dublin Mathematical Journal,* **1**, 74–96, 83, 92.
78. Hertz H. 1993, *Electric Waves* trans D. E. Jones, with a preface by Lord Kelvin (London: Macmillan) preface; SPT, **2**, 1056.
79. Thomson W. 1845, On a mechanical representation of electric, magnetic and galvanic forces, *Cambridge and Dublin Mathematical Journal* **2** 61–4; MPP, **1**, 76–80; SW, pp. 256–259.
80. MPP, **1**, 75, 78–80; SW, p. 257.
81. Faraday M. 1971, Pearce Williams L. ed. *The Selected Correspondence of Michael Faraday* 3 vols (Cambridge: Cambridge University Press), **1**, 458–60.
82. Weyl H. 1922, *Space-Time-Matter* (London: Methuen. 1st German ed 1918) pp 46–47; Roche J. 2001, Axial vectors, skew-symmetric tensors and the nature of the magnetic field, *European Journal of Physics,* **22**, 193–201, *193–4.*
83. MPP, **1**, 78–80; SW, pp. 258–9; Roche J. 1990 A critical study of the vector potential, in *Physicists Look Back* (Bristol: Adam Hilger) 150–151.
84. SPT, pp. 203–4, 1016.
85. SPT, pp. 203–4; SW, p. 260.
86. Gauss C. F. 1833 *Intensitas Magneticis Terrestris ad Mensuram Absolutem Revocata* (Göttingen: Sumtibus Dieterichianis) 44.

87. Gauss C. F. and Weber W. (1841), Results of the observations made by the Magnetic association in the year 1836, in *Scientific Memoirs* ed R. Taylor, *2* (London), 20–97, *71–3;* See also EM pp. 344–5, 360.
88. EM, ch 24; SPT, pp. 211–218; SW, pp. 261–263.
89. EM, pp. 340, 466.
90. Thomson W. 1872 On the electric currents by which the phenomena of terrestrial magnetism may be predicted EM, p. 463; EM, pp. 102, 341, 419; SPT 1017; Buchwald J. 1977 William Thomson and the mathematization of Faraday's electrostatics *Historical Studies in the Physical Sciences,* **8**, 101–136, *102, 118–9.*
91. Thomson W. 1847, On the electric currents by which the phenonema of terrestrial magnetism may be produced, EM, pp. 409–410, 463–4.
92. EM, pp. 372–5.
93. Poisson S. D. 18212– Mémoire sur la théorie du magnétisme, *Mem Acad R Sci Inst France,* **5**, 271–2, 294; EM, p. 351.
94. EM, pp. 19, 26, 30–37, 504–8, 528–534, 575–7; SPT, 19.
95. Thomson W. 1851, On the theory of magnetic induction in crystalline and non crystalline substances, in EM, p. 467.
96. Sears F., Zemasky M., and Young H. 1987, *University Physics* (Reading, Mass: Addison Wesley) 546.
97. EM, pp. 535–6.
98. Poisson S. D. 1821–2 Mémoire sur la théorie du magnétisme, *Mem Acad R Sci Inst France,* **5**, 294–303.
99. EM, pp. 346, 351–5, 357, 362–3, 424–5.
100. *Transactions of the Royal Society* June 1848; EM, pp. 340–378, 424.
101. EM, § 463–475.
102. EM, p. 424.
103. EM, p. 424.
104. EM, pp. 363–409.
105. EM, p. 33.
106. Thomson, W. 1851, A mathematical theory of magnetism, *Phil. Trans.* **141**, 243–268; 269–285; EM, pp. 360–365, 383, 387, 392–8.
107. Poisson S. D. 1821–22 Mémoire sur la théorie du magnétisme, *Mem. Acad. R. Sci. Inst. France,* **5**, 272, 276, 292, 298, 301; Poisson S. D. 1823; Mémoire sur la théorie du magnétisme en mouvement, *Mem. Acad. R. Sci. Inst. France,* **6**, 441–570, 453–461; Green G. 1828 *An Essay on the Application of Mathematical Analysis to the Theory of Electricity and Magnetism* (Privately published, Nottingham) 53.
108. EM, p. 362.
109. EM, p. 361.
110. EM, pp. 361, 383, 482.
111. EM, p. 362 n†.
112. EM, pp. 363–4.
113. EM, p. 362.
114. EM, pp. 361–2, 365.

115. EM, pp. 393–406, 424–5.
116. EM, pp. 424–5.
117. Thomson W. 1851, A mathematical theory of magnetism, *Philosophical Transactions,* **149**, 243–85; EM, pp. 378–405.
118. EM, pp. 379–80; Letter to Stokes 25 January 1850, Cambridge Mss add. 7656/K38.
119. EM, pp. 363 (footnote), 389–90.
120. EM, pp. 390–393.
121. EM, pp. 342, 360–371.
122. EM, p. 397.
123. EM, pp. 412–7 art 28 446–7.
124. EM, pp. 392–3, 421–2, 425; art 25; 402–5, 446–7.
125. EM, pp. 389–90, 393, 396–7.
126. EM, pp. 395–396, 545–546..
127. EM, pp. 421–2.
128. John J. Roche, 2000, B and H, the intensity vectors of magnetism: A new approach to resolving a century-old controversy *Am J Phys,* **68**, 438–449, *440–442.*
129. EM, pp. 421–2.
130. Poisson S. D. 1823, Mémoire sur la théorie du magnétisme en mouvement, *Mem Acad R Sci Inst France* (Paris 1827), **6**, 441–570; Neumann F. E. 1848 Ueber die magnetisirung eines rotations-ellipsoids *Journal fur die Reine und Angewandte mathimatik (Crelle's Journal)*, **37**, 21–50, *39*. Neumann's term χ is now the standard term for susceptibility.
131. EM, pp. 367–370, 484–5.
132. EM, p. 472.
133. EM, pp. 5, 484–5.
134. Thomson, W. 1855, Observations on the 'magnetic medium' on the effects of compression, *Phil Mag,* **9**, 290.
135. EM, pp. 485–6.
136. EM, pp. 396.
137. EM, p. 33.
138. Lorentz H. A. 1935– 1939 *Collected Papers* 9 Vols (The Hague: Nijhoff), **1**, 227.
139. Roche J. 2003, What is potential energy? *E J Phys* **24** 185–196, *186–7.*
140. MPP, **1**, 174–332, *175, 222–223, 502.*
141. Gauss C. F. 1843, General propositions relating to attractive and repulsive forces in *Scientific Memoirs,* ed R Taylor (London: Taylor) **3** 153–196,*181.*
142. Thomson, W. 1872, Propositions in the theory of attraction *in* EM, pp. 132–138, *136.*
143. SW, pp. 240–5.
144. SW, p. 245; see also EM, p. 92.
145. SW, p. 245.
146. SW, p. 245.
147. EM, pp. 136–7; SW, p. 253.
148. SW, p. 255.
149. Thomson W. 1849, Notes on hydrodynamics. On the vis-viva of a liquid in motion *Cambridge and Dublin Mathematical Journal* **4**, 90–94; MPP, **1**, 107–112, *110, 112n.*

150. MPP, **1**, pp. 107–112, *112*.
151. EM, pp. 447.
152. Helmholtz H. 1853, On the conservation of force, *Scientific Memoirs* (London: Taylor and Francis), **2**, 114–162. 1st published in 1847.
153. MPP, **1**, pp. 91–2.
154. MPP, **1**, pp. 472–502.
155. EM, p. 93.
156. On transient electric currents, *Phil Mag* June 1853; MPP, **1**, 534–553.
157. Thomson, W. 1872, On the mechanical values of distributions of electricity, magnetism and galvanism in MPP, **1**, 525–530.
158. MPP, **1**, p. 530.
159. Maxwell J. C. 1873, *A Treatise on Electricity and Magnetism* 2 vols (New York: Dover; original ed Oxford, Clarendon Press), **2**, 273–4.
160. MPP, **1**, p. 529–30.
161. MPP, **1**, pp 530–1.
162. MPP, **1**, p. 541.
163. MPP, **1**, p. 542.
164. Maxwell J. C. 1954, *A Treatise on Electricity and Magnetism* 2 vols (New York: Dover: Original edition Oxford, Clarendon Press 1873), **2**, 215.
165. MPP, **1**, pp 531, 551.
166. MPP, **1**, p. 532.
167. MPP, **1**, p. 532; EM, p. 434.
168. Maxwell J. C. 1873, *A Treatise on Electricity and Magnetism* 2 vols (New York: Dover; original ed Oxford, Clarendon Press), **2**, 273–4.
169. SW, pp. 317–27.
170. MPP, **1**, p. 175.
171. MPP, **1**, p. 175; SW, pp. 334, 337.
172. MPP, **4**, p. 2.
173. EM, p. 419.
174. Thomson W. 1849, Notes on hydrodynamics. On the vis-viva of a liquid in motion, *Cambridge and Dublin Mathematical Journal 1847,* **2** *282–286,1848* **3** *89–93* **3** *121–127,* **3**, *209–219; 1849* **4** (5) 90–94; **4** 219–240.
175. SW, p. 396; see also p. 345.
176. Thomson W. 1856, Dynamical illustrations of the magnetic and the helicoidal rotatory effects of transparent bodies on polarised light, *Proc Roy Soc,* **8**, 150–158.
177. Maxwell J. C. 1954, *A Treatise on Electricity and Magnetism* 2 vols (New York: Dover: Original edition Oxford: Clarendon Press. 1st ed 1873), **2**, 462–468; Mansuripur M. 1999 *The Faraday Effect* (Optics & Photonics News: http://www.mmresearch.com/articles/article3/accessed 21 Feb 2007).
178. Helmholtz H. 1858, Ueber Integrale der hydrodynamischen Gleichungen, welche den Wirbelbewegungen entsprechen *Wissenschtliche Abhandlungen* I (Leipzig 1882), **1**, 101–134; Helmholtz H. 1867 On integrals of the hydrodynamic equations which express vortex motion, *Philosophical Magazine,* **33**, 485–512.

179. Thomson W. 1867, On vortex atoms, *Philosophical Magazine,* **34**, 15–24, *15*; SPT, 510–11, 517–9.
180. MPP, **4**, 13.
181. Thomson W. 1867, On vortex atoms, *Philosophical Magazine,* **34**, 17–18.
182. Thomson W. 1867, On vortex atoms, *Philosophical Magazine,* **34**, 21, 23–24; Thomson W. 1882–1911, On vortex motion, MPP, **4**, 30–32.
183. Thomson W. 1867, On vortex atoms, *Philosophical Magazine,* 34, 16.
184. Thomson W. 1889–94 *Popular Lectures and Addresses* 3 vols (London), 1, 145; SPT, **2**, 744.
185. MPP, **4**, pp 94–96; 151.
186. MPP, **4**, pp 135–6.
187. Thomson W. 1882–1911, On vortex motion, *Mathematical and Physical Papers* 6 vols (Cambridge: Cambridge University Press), **4**, 31–2, 95; Helmholtz H. 1867, On integrals of the hydrodynamic equations which express vortex motion, *Philosophical Magazine,* **33**, 485–512; Thomson W. 1867, On vortex atoms, *Philosophical Magazine,* **34**, 21–24.
188. MPP, **4**, p. 116.
189. MPP, **4**, p. 125.
190. MPP, **4**, p. 122.
191. MPP, **4**, p. 477.
192. MPP, **3**, p. 495.
193. MPP, **3**, pp 497–8.
194. MPP, **3**, p. 500.
195. MPP, **3**, p. 500.
196. MPP, **3**, p. 498.
197. MPP, **3**, p. 498.
198. MPP, **3**, p. 498.
199. MPP, **3**, p. 501.
200. MPP, **4**, p. 124.
201. SPT, p. 1047.
202. SPT, p. 421.
203. Whittaker E. T. (1961–2), *A History of the Theories of Aether and Electricity* 2 vols (Thomas Nelson, 1910, 1953) chs 4, 9; SW, pp. 482–8.
204. Young Thomas, 1855, Experiments and calculations relative to physical optice, *Transaction of the Royal Society in Miscellaneous Works of Thomas Young,* ed. G. Peacock (London: John Murray) 188; Swenson, Lloyd S. 1972 *The Ethereal Ether* (Austin: University of Texas Press) 15–19.
205. Swenson, Lloyd S. 1972, *The Ethereal Ether* (Austin: University of Texas Press) 20.
206. MPP, **3** 503; Thomson W. 1901 Nineteenth century clouds of the dynamical theory of heat and light, *Phil Mag* **2** 1–40, *2–3*; SPT, 817–8.
207. SPT, p. 819.
208. SPT, p. 833.
209. Thomson W. 1888, On reflection and refraction of light, *Phil Mag,* **26**, 415.

210. Thomson W. 1888, On reflection and refraction of light, *Phil Mag*, **26**, 414–5, 420.
211. Faraday M. 1839, 1844, 1855, *Experimental Researches in Electricity* 3 vols (London: Taylor and Francis), **3**, § 2702–7, pp 450–452; Faraday M. 1971 Pearce Williams L. ed. *The Selected Correspondence of Michael Faraday* 3 vols (Cambridge: Cambridge University Press), **1**, 458–60.
212. SPT, p. 819.
213. MPP, **3**, p. 484.
214. MPP, **3**, p. 490.
215. MPP, **3**, p. 427.
216. Thomson W. 1882–1911, On a Gyrostaic adynamic constitution for the ether, in MPP, **3**, 471.
217. Thomson W. 1882–1911, On a Gyrostaic adynamic constitution for the ether, in MPP, **3**, 467–8.
218. MPP, **3** pp 505–8.
219. Thomson W. 1882–1911, On a Gyrostaic adynamic constitution for the ether, in MPP, **3**, 472.
220. Thomson W. 1882–1911, On a Gyrostaic adynamic constitution for the ether, in MPP, **3**, 472.
221. MPP, **3**, p. 462.
222. MPP, **3**, p. 503.
223. MPP, **3**, p. 503.
224. MPP, **3**, p. 502.
225. MPP, **3**, pp 442, 462.
226. MPP, **3**, pp 509–510.
227. MPP, **3**, pp 509–19.
228. MPP, **3**, p. 510.
229. MPP, **3**, pp. 446, 449.
230. MPP, **3**, p. 463.
231. MPP, **3**, pp. 450–462.
232. MPP, **3**, p. 450.
233. MPP, **3**, 504; Maxwell J. C. 1954, *A Treatise on Electricity and Magnetism* 2 vols (New York: Dover: Original edition Oxford, Clarendon Press 1873) **2** 278–281; SPT, **2**, 1048.
234. Maxwell J. C. 1954, *A Treatise on Electricity and Magnetism* 2 vols (New York: Dover: Original edition Oxford, Clarendon Press 1873), **1**, 157–161; SPT, 1064–5.
235. MPP, **3**, p. 465.
236. MPP, **3**, p. 465.
237. SPT, pp. 1048, 1058.
238. SPT, p. 1083.
239. Thomson W. 1884 *Notes of lectures on Molecular Dynamics and the Wave theory of Light* (Baltimore: Johns Hopkins University) pp. 5–6, 271; SPT, pp. 819, 836, 1024.
240. SPT, p. 1058.
241. MPP, **3**, p. 427, 490.
242. SPT, pp. 1065; 819, 1072.

243. Maxwell J. C. 1873, *A Treatise on Electricity and Magnetism* 2 vols (New York: Dover; original ed Oxford, Clarendon Press), **1**, 12.
244. Maxwell J. C. 1873, *A Treatise on Electricity and Magnetism* 2 vols (New York: Dover; original ed Oxford, Clarendon Press), **2**, 273–4.
245. Roche J. 2000, B and H, the intensity vectors of magnetism. A new approach to resolving a century-old controversy, *American Journal of Physics,* **68**, 440–442.
246. Thomson W. 1843, On the motion of heat in solid bodies and its connexion with the mathematical theory of electricity *Cambridge Mathematical Journal,* **3**, 71–84, *74*; Thomson W. 1846, Note on hydrodynamics II. *Cambridge and Dublin Mathematical Journal,* **1**, 282–6, *284*; SPT, **2**, 1071.
247. Thomson W. 1842, Propositions in the theory of attraction *The Cambridge Mathematical Journal*, **3**, 201–6, *203;* EM, pp. 445, 484, 581–6; SPT, **2**, 1071.
248. SPT, p. 1028.
249. SPT, p. 1025.
250. SPT, p. 1058.
251. MPP, **3** § 99, § 100, § 102.
252. SPT, p. 1063.
253. SPT, p. 1066.
254. SPT, p. 1063.
255. SPT, p. 1064.
256. SPT, p. 1072.
257. SPT, p. 1083.
258. Fitzgerald; SPT, p. 1065.
259. SPT, p. 984.
260. Sommerfeld A. 1910, Zur Relativittstheorie. I. Vierdimensionale vektor algebra' *Annalen der Physik* **32** 749–776 (p 750).
261. Weyl H. 1922, *Space-Time-Matter* (London: Methuen. 1st German ed 1918), 46–47; Roche J. 2001, Axial vectors, skew-symmetric tensors and the nature of the magnetic field, *European Journal of Physics,* **22**, 193–201.
262. MPP, **6**, Section 1; SPT, p. 954, 1083, 1078–81.
263. SPT, p. 1048.

CHAPTER 8

References

1. James Prescott Joule, 'On the Mechanical Equivalent of Heat', *Reports of the British Association for the Advancement of Science*, 1847, **17:** 55.
2. The story is related in a number of biographies. Most recently see Donald Cardwell, *James Joule: A Biography* (Manchester: Manchester University Press, 1989).

3. For an overview of different eighteenth and nineteenth-century views on the nature of heat and its relationship to mechanical work see Donald Cardwell, *From Watt to Clausius: The Rise of Thermodynamics in the early Industrial Age* (London: Heinemann, 1971) [DC] and Robert Fox, *The Caloric Theory of Gases: From Lavoisier to Regnault* (Oxford: Clarendon Press, 1971).
4. DC, note 1.
5. For an overview of nineteenth-century physics putting thermodynamics in broader context see Iwan Rhys Morus, *When Physics became King* (Chicago IL: University of Chicago Press, 2005).
6. E. Mendoza (ed.), *Reflections on the Motive Power of Fire by Sadi Carnot, and Other Papers on the Second Law of Thermodynamics by É. Clapeyron and R. Clausius* (New York NY: Dover Publications, 1960), [EM] p. 4.
7. EM, pp. 6–7.
8. EM, p. 7.
9. The classic study of Laplacian physics is Robert Fox, 'The Rise and Fall of Laplacian Physics', *Historical Studies in the Physical Sciences*, 1974, **4:** 89–136.
10. Clapeyron's essay was first translated into English as Émile Clapeyron, 'Memoir on the Motive Power of Heat', Richard Taylor (ed.), *Scientific Memoirs* vol.1 (London, 1837), pp. 347–76.
11. For an account of Manchester's scientific community at the beginning of the Victorian period see Robert Kargon, *Science in Victorian Manchester: Enterprise and Expertise* (Manchester: Manchester University Press, 1977). For the Royal Victoria Gallery's metropolitan equivalents see Richard Altick, *The Shows of London* (Cambridge MA: Belknap Press, 1978), Iwan Rhys Morus, 'Manufacturing Nature: Science, Technology and Victorian Consumer Culture', *British Journal for the History of Science*, 1996, **29:** 403–34.
12. James Prescott Joule, 'Description of an Electro-magnetic Engine', *Annals of Electricity*, 1838, **2:** 122–23.
13. Nathan Reingold (ed.), *The Papers of Joseph Henry* 9 vols published (Washington DC: Smithsonian Institution Press, 1972 –), vol.4, p,152, Joseph Henry to Asa Gray, 1 November 1838.
14. For the London Electrical Society see Iwan Rhys Morus, 'Currents from the Underworld: Electricity and the Technology of Display in early Victorian England', *Isis*, 1993. **84:** 50–69; Iwan Rhys Morus, *Frankenstein's Children: Electricity, Exhibition and Experiment in early Nineteenth-century London* (Princeton NJ: Princeton University Press, 1998).
15. 'Davenport's Electro-magnetic Engine', *Mechanics' Magazine*, 1837, **27:** 404–05, on p. 405.
16. James Prescott Joule, 'Investigations in Magnetism and Electro-magnetism', *Annals of Electricity*, 1839–40, **4:** 131–37, on p. 134.
17. For discussions of early Victorian views on waste and efficiency see Norton Wise & Crosbie Smith, 'Measurement, Work and Industry in Lord Kelvin's Britain', *Historical Studies in the Physical Sciences*, 1986, **17:** 147–73; Norton Wise, 'Mediating Machines', *Science in Context*, 1988, **2:** 77–113; Norton Wise (with the collaboration of Crosbie Smith), 'Work and Waste: Political Economy and Natural Philosophy in Nineteenth-century Britain', *History of Science*, 1989, **27:** 263–301; 391–449; 1990, **28:** 221–61.

18. James Prescott Joule, 'On the Calorific Effects of Electricity, and on the Mechanical Equivalent of Heat', *Philosophical Magazine*, 1843, **23:** 263–76; 347–55; 435–43, on p. 263.
19. Otto Sibum, 'Reworking the Mechanical Value of Heat: Instruments of Precision and Gestures of Accuracy in early Victorian England', *Studies in History and Philosophy of Science*, 1994, **26:** 73–106.
20. James Prescott Joule, 'On the Changes in Temperature Produced by the Rarefaction and Condensation of Air', *Philosophical Magazine*, 1845, **26:** 369–83, on p. 381–82.
21. Quoted in Crosbie Smith, *The Science of Energy: A Cultural History of Energy Physics in Victorian Britain* (London: Athlone Press, 1998), [CS] on p. 72.
22. The classic overview is Susan Cannon, *Science in Culture: The Early Victorian Period* (New York NY: Science History Publications, 1978). See also Jack Morrell and Arnold Thackray, *Gentlemen of Science: Early Years of the British Association for the Advancement of Science* (Oxford: Oxford University Press, 1981).
23. William Thomson, 'Notice of Stirling's Air-engine', *Proceedings of the Glasgow Philosophical Society*, 1847, **2:** 169–70.
24. For 'north British' physics see Crosbie Smith, op. cit. note 20. For reformed Cambridge mathematics see Andrew Warwick, *Masters of Theory: Cambridge and the Rise of Mathematical Physics* (Chicago IL: University of Chicago Press, 2003).
25. On Charles Babbage and the Analytic Society see Harvey Becher, 'Radicals, Whigs and Conservatives: The Middle and Lower Classes in the Analytic Revolution at Cambridge in the Age of Aristocracy', *British Journal for the History of Science*, 1995, **28:** 405–26.
26. Quoted in Crosbie Smith and Norton Wise, *Energy and Empire: A Biographical Study of Lord Kelvin* (Cambridge: Cambridge University Press, 1989), [SW] on p. 56.
27. For an overview of Cambridge mathematics and British physics see Peter Harman (ed.), *Wranglers and Physicists: Studies on Cambridge Physics in the Nineteenth Century* (Manchester: Manchester University Press, 1985).
28. SW, note 16, p. 289.
29. SW, note 26, pp. 292–93.
30. William Thomson, 'On an Absolute Thermometric Scale, founded on Carnot's Theory of the Motive Power of Heat, and Calculated from the Results of Regnault's Experiments on the Pressure and Latent Heat of Steam', *Philosophical Magazine*, 1848, **33:** 313–17.
31. For other contemporary views regarding just what the paddle wheel experiment could be taken to demonstrate see Crosbie Smith, 'Faraday as a Referee of Joule's Paper "On the Mechanical Equivalent of Heat"', *Isis*, 1976, **67:** 444–49.
32. SW, note 62, p. 310.
33. William Thomson, 'An Account of Carnot's Theory of the Motive Power of Heat, with Numerical Results Deduced from Regnault's Experiments on Steam', *Transactions of the Royal Society of Edinburgh*, 1849, **16:** 541–74.
34. William Thomson, 'On the Dynamical Theory of Heat; with Numerical Results Deduced from Mr. Joule's "Equivalent of a Thermal Unit" and M. Regnault's "Observations on steam"', *Transactions of the Royal Society of Edinburgh*, 1853, **20:** 261–98; 475–82, on p. 264.
35. William Thomson, 'On the Dynamical Theory of Heat; with Numerical Results Deduced from Mr. Joule's "Equivalent of a Thermal Unit" and M. Regnault's "Observations on

Steam"', *Transactions of the Royal Society of Edinburgh*, 1853, **20:** 261–98; 475–82, on p. 264.

36. William John Macquorn Rankine, *Manual of the Steam Engine and Other Prime Movers* (Glasgow: Griffin, 1859).
37. William Thomson, 'On the Age of the Sun's Heat', *Macmillan's Magazine*, 1862, **5:** 288–93, on p. 290.
38. William Thomson, 'On the Secular Cooling of the Earth', *Philosophical Magazine*, 1863, **25:** 1–14.
39. For a comprehensive account of nineteenth-century controversies surrounding the age of the Earth and Sun, and William Thomson's role in them, see Joseph Burchfield, *Lord Kelvin and the Age of the Earth* (Chicago IL: University of Chicago Press, 1975).
40. CS, note 20, p. 172.
41. William Thomson and Peter Guthrie Tait, *Treatise on Natural Philosophy* (Oxford: Clarendon Press, 1867), p. *vi*.

CHAPTER 9

References

1. Silvanus P. Thompson, *The Life of Lord Kelvin* (Chelsea Publishing Company, New York, 2 vols, 1974), p. 1297.
2. Gillian Cookson, The cable – the wire that changed the world, *Tempus*, p. 160 (2003).
3. W. Thomson, On the theory of the electric telegraph, *Proc. Roy. Soc.* **7,** 382–99 (1855).
4. W. Thomson, W. J. M'Q. Rankine and J. Thomson, Improvements in electrical conductors for telegraphic communications, UK Patent, no. 2547, p. 3 (1854).
5. W. Thomson, On practical methods for rapid signalling by the electric telegraph, *Proc. Roy. Soc.* **8,** 299–307 (1856).
6. W. Thomson, On the electrical conductivity of copper of various kinds, *Proc. Roy. Soc.* **8,** 550–5 (1857).
7. W. Thomson, On machinery for laying submarine telegraph cables, *Engineer* **4,** 185 and 280 (1857).
8. W. Thomson, Apparatus for applying and measuring resistance to motion of rotating wheels, shafts, etc. UK Patent, no. 437, p. 6 (1858).
9. W. Thomson, Improvements in testing and working electric telegraphs, UK Patent, no. 329, p. 29 (1858).
10. W. Thomson, Electric telegraph, UK Patent, no. 329, p. 29 (1858).
11. W. Thomson, Receiving or recording instruments for electric telegraphs, UK Patent, no. 2147, p. 10 (1867).
12. W. Thomson, The mariner's compass, *Good Works* 69–72 (Jan 1874).
13. W. Thomson, Improvements in the mariner's compass and in the means of ascertaining and correcting its errors, UK Patent no. 1339, p. 7 (1876).

14. W. Thomson, Improvements in the mariner's compass and in appliances for ascertaining and correcting its errors, UK Patent no. 4876, p. 17 (1876).
15. W. Thomson, On deep-sea sounding by pianoforte wire, *Soc. Telegr. Engrs. Journ.* **3,** 206–219, 220–221, and 222–224 (1874).
16. W. Thomson, Deep sea sounding apparatus, UK Patent no. 3452, p. 8 (1876).
17. W. Thomson, On his new navigational sounding machine and depth gauge, *United Services Inst. Journal* **3,** 374–381 (1881).
18. W. Thomson, The tide-gauge, tidal harmonic analyser and tide-predictor, *Inst. Civil Engrs. Proc.*, **65,** 2–25 (1889).
19. W. Thomson, On the economy of the heating or cooling of buildings by means of currents of air, *Glasgow Phil. Soc. Proc.* **3,** 269–72, (1852).
20. W. Thomson, The power required for the thermodynamic of buildings, *Cambridge and Dublin Mathematical Journal*, **8,** 25–34 (1853).
21. T. R. Morley, The reversed heat engine as a means of heating buildings, *The Engineer* **133,** 145–6,(1922).

CHAPTER 10

References

1. S. G. Brush, 'Finding the age of the earth by physics or by faith', *Journal of Geological Education*, vol. **30** (1982), 34–58.
2. W. Thomson, 'On the mechanical energies of the Solar System', *Philosophical Magazine*, Series 4, vol. **8** (1854), 409–430 [MPP 2 (1884), 1–27, quotation from p. 8].
3. W. Thomson, 'On the mechanical energies of the Solar System', *Philosophical Magazine*, Series 4, vol. **8** (1854), 409–430 [MPP 2 (1884), 1–27, p. 11].
4. W. Thomson, 'On the age of the Sun's heat', *Macmillan's Magazine*, vol. **5** (1862), 288–293 [PLA 1 (1889), 349–368, quotation from p. 368].
5. W. Thomson, 'On the reduction of observations of underground temperature; with application to Professor Forbes' Edinburgh observations, and the continued Calton Hill series. *Transactions of the Royal Society of Edinburgh,* vol. **22** (1861), 405–427 [MPP **3** (1890), 261–294].
6. Gunter Buntebarth, 'Temperature measurements below the Earth's surface: a history of records', *Earth Sciences History*, vol. **21** (2002), 190–198.
7. W. Thomson, 'On the secular cooling of the Earth', *Transactions of the Royal Society of Edinburgh*, vol. **23** (1864), 157–170 [MPP **3** (1890), 295–311, quotation from p. 295].
8. W. Thomson, 'On Geological Time', *Transactions of the Geological Society of Glasgow*, vol. **3**, part 1 (1868), 1–28, quotation from p. 25.
9. This conversation is given in W. Thomson [Lord Kelvin] 'The age of the earth as an abode fitted for life', *Science* vol. **9**, (1899), p. 668.

10. T. Mellard Reade, ' The age of the World as viewed by the geologist and the mathematician', *Geological Magazine*, n.s. Decade 2, vol. **5** (1878), p. 154.
11. T.H. Huxley, Anniversary Address of the President', *Proceedings of the Geological Society*, volume **25** (1869), p. xlviii.
12. Osmund Fisher, 'On the age of the World, as depending on the condition of the interior', *Geological Magazine*, n.s. Decade 4, vol. **2** (1895), p. 246.
13. B. Shipley 'Had Lord Kelvin a right?': John Perry, natural selection and the age of the Earth, 1894–1895', C.L.E. Lewis and S.J. Knell (eds), *The Age of the Earth: from 4004 BC to AD 2002* (Geological Society of London Special Publication 190, London, 2001), pp. 91–105, quotation from p. 94.
14. George F. Becker. 'Age of a cooling Globe in which the initial temperature increases directly as the distance from the surface', Science, n.s., vol. 27, number 684 (1908), p. 233.
15. The quotation from Mark Twain appears in J. D. Burchfield, *Lord Kelvin and the Age of the Earth*, 2nd edition (University of Chicago Press, 1990), p. ix.
16. Archibald Geikie, 'Address to the Geological Section', *Report of the 69th meeting of the British Association for the Advancement of Science, Dover, 1899* (1900), p. 6.
17. Archibald Geikie, 'Address to the Geological Section', *Report of the 69th meeting of the British Association for the Advancement of Science, Dover, 1899* (1900), p. 8.
18. Archibald Geikie, *A long life's work*. Macmillan and Co., London (1924), p. 350.
19. W. Thomson 'The age of the earth as an abode fitted for life', *Science* vol. **9**, (1899), p. 666.
20. W. Thomson, 'On the age of the Sun's heat', *Macmillan's Magazine*, vol. 5 (1862), 288–293 [PLA 1 (1889), 349–368, p. 368].
21. Rutherford's recollection of this encounter with Kelvin is documented by Arthur S. Eve, *Rutherford*, Macmillan, New York (1939), p. 107, and also in Burchfield, *Lord Kelvin*, p. 164 where it is discussed more fully.

Further Reading

The most comprehensive accounts of William Thomson's geochronological research are J. D. Burchfield, *Lord Kelvin and the Age of the Earth*, 2nd edition (University of Chicago Press, 1990) and S. G. Brush, 'Finding the age of the earth by physics or by faith', *Journal of Geological Education*, vol. **30** (1982), 34–58. Other useful treatments are A. Stinner, 'Calculating the age of the Earth and the Sun', *Physics Education*, vol. **37**, part 4 (2002), 296–305, and F. M. Richter, 'Kelvin and the Age of the Earth', *Journal of Geology*, vol. **95** (1986), 395–401.

Thomson's publications may be read either in their original place of publication or in compilations that appeared towards the end of his life. A number of the papers listed following appeared in *Popular Lectures and Addresses* [PLA], Macmillan & Co. vol. **1** (1889), vol. **2** (1894), or in *Mathematical and Physical Papers* [MPP], Cambridge University Press, vol. **2** (1884), vol. **3** (1890). His most important papers on geochronology were:

(1) Relating to the age of the Sun:

'On the mechanical energies of the Solar System', *Philosophical Magazine*, Series 4, vol. **8** (1854), 409–430 [MPP 2 (1884), 1–27].

'On the age of the Sun's heat', *Macmillan's Magazine*, vol. **5** (1862), 288–293 [PLA 1 (1889), 349–368].

'On the Sun's heat', *Proceedings of the Royal Institution*, vol. **12** (1889), 1–12 [PLA 1 (1889), 369–422].

(2) On the cooling rate of the Earth:

'On the secular cooling of the Earth', *Transactions of the Royal Society of Edinburgh*, vol. **23** (1864), 157–170 [MPP 3 (1890), 295–311].

'On Geological Time', *Transactions of the Geological Society of Glasgow*, vol. **3**, part 1 (1868), 1–28 [PLA 2 (1894), 10–64].

'The age of the Earth', *Nature*, vol. **51** (1895), 438–440.

(3) On tidal friction and the Earth's rotation and shape:

'On Geological Time', *Transactions of the Geological Society of Glasgow*, vol. **3**, part 1 (1868), 1–28 [PLA 2 (1894), 10–64].

Thomson's final paper contribution to the geochronological debate was 'The age of the earth as an abode fitted for life', *Annual Report of the Smithsonian Institution* for 1897, (1898), 337–357. Also published in *Journal of the Transactions of the Victoria Institution*, vol. **31** (1899), 11–34, the *London, Edinburgh, and Dublin Philosophical Magazine and Journal of Science*, Series 5, vol. **57** (1899), 66–90, and in *Science* vol. **9**, (1899), 665–674; 704–711.

John Perry's objections are discussed in two papers: B. Shipley 'Had Lord Kelvin a right?': John Perry, natural selection and the age of the Earth, 1894–1895', C. L. E. Lewis and S. J. Knell (eds), *The Age of the Earth: from 4004 BC to AD 2002* (Geological Society of London Special Publication 190, London, 2001), pp. 91–105; and P. England, P. Molnar and Richter, F. 'John Perry's neglected critique of Kelvin's age for the Earth: a missed opportunity in geodynamics', *GSA Today*, vol. **17**, part 1 (2007), pp. 4–9.

CHAPTER 11

References

1. C. G. Knott, *Life and Scientific Work of Peter Guthrie Tait* (Cambridge, 1911). [CGK], p. 50.
2. CGK, p. 50.
3. Crosbie Smith and M. Norton Wise, *Energy and Empire: A Biographical Study of Lord Kelvin* (Cambridge University Press, Cambridge, 1989) [SW] p. 348.
4. Silvanus P. Thompson, *The Life of William Thomson: Baron Kelvin of Largs* (Macmillan, London, 1910) [SPT] p. 450.
5. University of Edinburgh Journal **15** (1949–1951), p. 164.
6. J. M. Barrie, *An Edinburgh Eleven* (Hodder & Stoughton, 1889), p. 46–47.
7. SPT, pp. 453–454.
8. SW, p. 352.
9. W. Thomson and P. G. Tait, *Treatise on Natural Philosophy* (Cambridge University Press, Cambridge, 1879). [T&T′] preface p. v.

10. SPT, p. 447.
11. T&T′ preface p. viii.
12. SW, p. 352.
13. SW, p. 351.
14. SW, p. 351
15. George Chrystal 'Professor Tait', *Nature*, **64**, 25 July 1901, p. 305–307 (f/n 306).
16. SW, p. 356.
17. SW, p. 353.
18. SW, p. 359.
19. SW, p. 359.
20. SPT, p. 473.
21. SPT, p. 476.
22. SPT, p. 477.
23. T&T′ Preface p v.
24. T&T′, p. 341.
25. CGK, p. 183.
26. T&T′, p. 229.
27. T&T′, p. 239.
28. SW, p. 354.
29. SW, p. 355.
30. SW, p. 355.
31. T&T′, p. 312.
32. T&T′, p. 479.
33. T&T′, p. 1.
34. T&T′, p. 54.
35. T&T′, p. 54.
36. T&T′, p. 219.
37. T&T′, preface, p. vi.
38. T&T′, p. 220.
39. T&T′, p. 265.
40. T&T′, p. 246.
41. T&T′, p. 265.
42. T&T′, p. 265.
43. SW, p. 390.
44. T&T′, p. 338.
45. SW, p. 392 and 395.
46. SPT, p. 468.
47. SPT, p. 468.
48. CGK, p. 201.
49. *Handbuch der theoretischen Physik* / von W. Thomson und P. G. Tait ; autorisirte deutsche Übersetzung von H. Helmholtz und G. Wertheim. 1. Bd, Braunschweig : F. Vieweg, 1871–1874.

CHAPTER 12

References

1. Kelvin did not differentiate between atoms and molecules until the 1880s. Earlier, the terms seem interchangeable in his papers.
2. William Thomson and Peter Guthrie Tait, *Treatise on Natural Philosophy*, Oxford University Press, 1867, second edition Vol. I (1879), vol. II (1883).
3. For Thomson's research in the 1840's see, Crosbie Smith and M. Norton Wise, *Energy and Empire: A Biographical Study of Lord Kelvin* (Cambridge University Press, 1989)[SW] chs. 7–8, especially pps. 263–75. See also, Ole Knudsen, 'From Lord Kelvin's Notebook: Ether Speculations,' *Centaurus*, **16** (1971): 41–53, and Harold Sharlin, 'William Thomson's Dynamical Thinking: An Insight into a Scientist's Thinking,' *Annals of Science*, **32** (1975): 133–148.
4. There are references to this throughout Kelvin's career. For one important example of this see, Bruce J. Hunt, 'How my Model was Right: G. F. FitzGerald and the Reform of Maxwell's Theory,' 299–321, p. 310, in Robert Kargon and Peter Achinstein eds., *Kelvin's Baltimore Lectures and Modern Theoretical Physics*, (1884) (Cambridge MA: MIT Press, 1987), [KA] pps. 3, 207.
5. William Thomson, 'On a mechanical representation of electric, magnetic and galvanic forces,' *Cambridge and Dublin Mathematical Journal*, **2** (1847): 61–64.
6. SW, 446–463, and, 'The practical imperative: Kelvin challenges the Maxwellians,' in KA, 323–348.
7. KA, Lecture XIV,135–136, and 139–141.
8. KA, Lecture XX, on refraction, which was expanded in *Baltimore Lectures*, 2nd edition (London: C. J. Clay and Sons, 1904) [WT1904], Lecture XX.
9. William Thomson, 'Nineteenth Century Clouds over the Dynamical Theory of Heat and Light,' *Proceedings of the Royal Institution*, **16** (1902) : 363–397, reprinted in WT1904, Appendix A, 486–527.
10. David Knight, *Atoms and Element* (London: Hutchinson, 1967) and William H. Brock and David Knight, 'The Atomic Debates: "Memorable and Interesting Evenings" at the Chemistry Society,' *Isis*, **56** (1965): 5–25.
11. Elizabeth Garber, Stephen G. Brush, and C. W. F. Everitt eds., *Maxwell on Molecules and Gases*, (Cambridge MA: MIT Press, 1986) 38–39, 278–291, 483–484.
12. William Thomson 'On the Size of Molecules,' *Nature*, **2** (1870): 56–57, p. 56.
13. William Thomson, 'The Size of Atoms,' *Nature*, **1** (1870) : 551–553. Kelvin, 'The Size of Atoms,' *Proceedings of the Royal Institution* **10** (1883) : 80–85. These estimates were repeated and added to in the Baltimore Lectures, his estimates getting more confident over broader range of phenomena. WT1904, Lecture XVII: 279–323.
14. See Robert Silliman, 'William Thomson: Smoke Rings and Nineteenth Century Atomism,' *Isis* **54** (1969): 461–474, and on vortex atoms in general, see, Helge Krage, 'The Vortex Atom: A Victorian Theory of Everything,' *Centaurus* 44 (2002) : 32–114.

15. Glasgow University Library, Kelvin Papers, T-89, T-93, 94, T-96, and, Silvanus P. Thompson,*The Life of William Thomson, Baron Kelvin of Largs* 2 vols. (London: Macmillan, 1910) [SPT], 527–528.
16. Keith Hutchinson, 'Der Ursprung der Entropiefunktion bei Rankine und Clausius,' *Annals of Science*, **30** (1973) :341–364, and 'W. J. M. Rankine and the Rise of Thermodynamics,' *British Journal for the History of Science*, **10** (1981) : 1–26. See also SW, 318–321.
17. William Thomson, 'Dynamical Illustration of the Magnetic and Helcoildal Rotary Affects of Transparent Bodies in Polarised Light,' *Proceedings, Royal Society of London* **8** (1856) : 150–158, reprinted in, WT1904, Appendix F, 569–577, p. 571, and, ' On Vortex Motion,' *Proceedings, Edinburgh Royal Society*, 6 (1867) : 1–12, p. 1, and *Transactions, Edinburgh Royal Society*, 25 (1869) : 217–260, reprinted in WT1904, Appendix F, 569–577.
18. Hermann von Helmholtz, 'Über die Integrale der hydrodamischen Gleichungen welche den Wirbelbewegung entsprechen,' *Journal für Reine und Angewandete Mathematik* **55** (1858) : 25–55, translated as, 'On the Hydrodynamical Equations of Vortices,' *Philosophical Magazine, Supplement* **33** (1867): 458–510, and, 'Über discontinuirliche Flussigkeitsbewegungen,' *Monats Bericht, Berlin*, (1868): 215–228, translated as 'Discontinuous Motions in Fluids,' *Philosophical Magazine* 36 (1868): 337–346.
19. Kelvin described this apparatus and the many attributes of 'Wirbelbewegung' including the stability of knotted vortices, in a letter to Helmholtz in January 1867. See, SPT, 513–516.
20. Frederick Guthrie, 'On Approach caused by Vibration,' *Proceedings, Royal Society of London* **18** (1870): 93–94., and *Philosophical Magazine* **39** (1870): 309, **40:** 345–354. Kelvin, On Approach caused by Vibration,' *Proceedings Royal Society London*, **19** (1871): 271–273.
21. William Thomson, 'On the Forces experienced by Solids Immersed in a Moving Liquid,' *Proceedings Royal Society Edinburgh* **7** (1872): 60–63: p. 60, and, 'On the Ultramondane Corpuscles of Le Sage,' *Proceedings Royal Society of Edinburgh* 7 (1872): 577–589.
22. William Thomson, 'On the Forces experienced by Solids Immersed in a Moving Liquid,' *Proceedings Royal Society Edinburgh* **7** (1872): 60–63: p. 61.
23. J. J. Thomson, *A Treatise on the Motion of Vortex Rings*, (London, Macmillan, 1883, reprinted London, Dawsons, 1968) on the apparent attraction of vortices p. 70, the interaction of two vortices, Part II, 37–67, and linked vortices, Part II, 75–92.
24. J. J. Thomson, *A Treatise on the Motion of Vortex Rings*, London, Macmillan, 1883, reprinted London, Dawsons, 1968, p. 107.
25. J. J. Thomson, *A Treatise on the Motion of Vortex Rings*, London, Macmillan, 1883, reprinted London, Dawsons, 1968, 120–124.
26. James Clerk Maxwell, 'Atom,' *Encyclopedia Brittanica* 9th edition, vol. 3, 36–49, 43–47. Elizabeth Garber, 'Molecular Science in late Nineteenth-Century Britain,' *Historical Studies in the Physical Sciences* **9** (1978): 265–297.
27. See SW, 436–438.
28. William Thomson, 'Presidential Address to the British Association meeting at Edinburgh,' *Report of the Forty-First Meeting of the BAAS* (1871)lxxxiv–cv.

29. Rayleigh, 'Remarks on Maxwell's Investigation respecting Boltzmann's Theorem,' *Philosophical Magazine* 33 (1892): 356–359 reprinted in, *Scientific Papers,* ,New York: Dover Reprint, 1964, vol III, 554–557, and, 'The Law of Partition of Kinetic Energy,' *Philosophical Magazine,* **49** (1900): 98–118, *Papers,* vol. IV, 433–451.
30. William Thomson, ' On the Instability of Fluid Motion,' *Philosophical Magazine* 23(1887): 287–292.
31. WT1904, p. 527.
32. William Thomson, 'Vibrations and Waves in a Stretched Uniform Chain of Symmetrical Gyrostats,' *Proceedings, London Mathematical Society* **6** (1875); 190–194, p. 190.
33. K A, Lecture XVIII, p. 187.
34. KA, Appendix, 'Improved Gyrostatic Model,' 218–255, p. 248.
35. See Bruce Hunt, *The Maxwellians* Ithaca NY: Cornell University Press, 1991. On the early teaching of Maxwell's theory in Cambridge, see, Andrew Warwick, *Masters of Theory: Cambridge and the Rise of Mathematical Physics* Chicago Il: University of Chicago Press, 2003, ch. 6.
36. For more details see, KA, 'Introduction,' p. 1.
37. The Lectures were initially published as they had been recorded stenographically. This edition already contained changes, and additions. The 1904 edition is much longer and different yet again. We have the progression of Kelvin's ideas on the subjects covered over a period of 20 years. The foundations of physics that he presented was still mechanics.
38. WT1904, 106.
39. WT1904, 258–259. By this date Kelvin had published a number of articles on crystal structure. The most detailed account was Kelvin, 'The Molecular Tactics of a Crystal,' *Baltimore Lectures* (1904), Appendix H, 602–642. This was the Robert Boyle Lecture delivered at Oxford University, 1893.
40. Kelvin introduced the linear vibration system in Lecture II and the 'wave machine' in Lecture XIV of the Baltimore Lectures. He claimed that he had used the latter for years in his lectures at Glasgow University. See also SW, 463–469.
41. WT1904, Lecture XII., 135–162, 156.
42. Roger Joseph Boscovich, *De viribus vivis*, Rome, 1745, A sketch of the theory appeared in *Theoria Philosophiae Naturalis*, Venice (1758), translated as *A Theory of Natural Philosophy* (Cambridge MA: MIT Press, 1966) trans. J. M. Child, p. 40.
43. William Thomson, 'Molecular Constitution of Matter,' *Proceedings, Royal Society of Edinburgh,* **24** (1889): 693–724.
44. William Thomson, 'Elasticity of Crystals according to Boscovich,' *Proceedings, Royal Society of London* **54** (1894): 59–75, reprinted in *Baltimore Lectures* (1904) Appendix I, 643–66, 'Piezo-electric Properties of Quartz,' *Philosophical Magazine* **36** (1893): 331–342, 'On Translational and Vibrational Energies of Vibrators after Impacts on Fixed Walls,' *Nature* **52** (1895): 532.
45. William Thomson,' Molecular Dynamics of a Crystal,' *Proceedings, Royal Society of Edinburgh* 24 (1902): 205–224, and, *Baltimore Lectures,* (1904) Appendix J.
46. Joseph Larmor, *Ether and Matter* Cambridge: Cambridge University Press, 1900. Kelvin discussed magnetism and molecular rotation, yet again in, 'Magnetism and molecular rotation,' *Proceeding of the, Royal Society of Edinburgh* 22 (1899): 631–635.

47. William Thomson, 'The Duties of the Ether for Electricity and Magnetism,' *Philosophical Magazine* **50** (1900): 305–307.
48. William Thomson, 'Aepinus Atomized,' *Philosophical Magazine* **3** (1902): 257–283, reprinted in *Baltimore Lectures* (1904), Appendix E, 541–568
49. WT1904, Lecture XX. This Lecture was rewritten between 1902 and 1904.
50. William Thomson, 'On the Motions of the Ether produced by Collisions of Atoms or Molecules, containing or not containing Electrions,' *Philosophical Magazine* **14** (1907): 317–324.
51. William Thomson, 'Plan of a Combination of Atoms having the Properties of Polonium or Radium,' *Philosophical Magazine* 8 (1904): 528–534, 'Plan of an Atom to be Capable of Storing an Electrion with enormous Energy for Radio-activity,' *Philosophical Magazine* 10 (1905): 695–698, 'An Attempt to Explain the Radioactivity of Radium,' *Philosophical Magazine,* **13** (1907): 313–316.

CHAPTER 13

References

1. Fukasaku, Yukiko (1992) *Technology and Industrial Development in Pre-war Japan*. Routledge, London and New York.
2. Checkland, Olive (1989) *Britain's Encounter with Meiji Japan, 1868–1912*. Macmillan, Basingstoke & London.
3. Sandersin, M. *The Universities in the Nineteenth Century*, quoted in Reference [2]
4. *Kelvin Centenary Oration and Addresses Commemorative* (1924). Percy Lund Humphries, London.
5. Dyer, H. (1893) *Technical Education in Glasgow and the West of Scotland*, Glasgow University.
6. Constable, T. *Memoir of Lewis D B Gordon*, quoted in Reference 9.
7. Fox, Grace (1969) *Britain and Japan 1858–1883*, Oxford University Press, Oxford.
8. Murray, David (1924) *Lord Kelvin as Professor in the Old College Glasgow*, MacLehose and Sons, Publishers to the University, Glasgow.
9. Brock, W. H. (1981) 'The Japanese connexion: engineering in Tokyo, London and Glasgow at the end of the nineteenth century'. *British Journal for the History of Science*, Vol **14**, No 48, p. 227.
10. *Oxford Dictionary of National Biography* (2004) Ayrton, William Edward (G. J. N. Gooday).
11. Gooday Graeme J. N. and Low, Morris F. (1999) *Technology Transfer and Cultural Exchange – Western Scientists and Engineers Encounter Late Tokugawa and Meiji Japan*. Osaris, 1998, 13:99–128.
12. Perry, John (1909) *Death of Professor Aryton*. JIEE, 42:3–6.
13. Davidson, Charles (1927) *The Founders of Seismology*. Cambridge University Press, Cambridge.
14. Tawara, Hiro (1991) *Pioneers of Physics in the Early Days of Japan*. North Holland, Amsterdam.

15. Hudson, Donald E. (1992) *A History of Earthquake Engineering*. Proc IDNDR International Symposium on Earthquake Reduction Technology Tsukuba, Japan: Building Research Institute. Quoted in Ref 18.
16. Lord Rayleigh (1936) 'Some reminiscences of scientific workers of the past generation, and their surroundings', *Proc Phys Soc,* Vol 48, No 265, p15.
17. Glazebrook, R. T. (1935) 'James Alfrew Ewing 1855–1935', *Obit Not Fellows Roy Soc* Vol 1.
18. Reitherman, Robert *The Importance of the Development of Instruments in the History of Earthquake Engineering*, 2004 CUREE Calendar, Consortium of Universities for Research in Earthquake Engineering, Richmond.
19. Kelvin Papers, K2, University Library Cambridge, From T Kato 27 January 1883.
20. Lewis, John (2002) *The Emergence of Physics* in Promoting Physics and Supporting Physicists, Ed. John Hewis, IoP Publishing, Bristol.
21. Obit Notices of Fellows Deceased, *Proc Roy Soc* CVL i-vii.
22. Craik, Alex D. D. (2007) 'Science and technology in 19th century Japan: The Scottish connection'. *Fluid Dynamics and Research*, 39, pp. 24–48.
23. Moss, Michael; Lindsay, Alison, and Russell, Iain (1992) An Honourable Tradition, Avenue, No 11 Glasgow University.
24. Wilson, David B. (1987) *Kelvin and Stokes – A Comparative Study in Victorian Physics*, Adam Hilger, Bristol.
25. King, Agnes Gardner (1925) *Kelvin the Man*. Hodder and Stoughton, London
26. *An Account of the Celebrations on the Occasion of Lord Kelvin's Jubilee as Professor*, MacLehose and Sons, Publishers to the University, Glasgow (1899).
27. Glazebrook, Sir Richard, in Reference 4.
28. Yoshida, Haruyo and Sugiyama, Shigeo (1997) 'Aikitu Tanakadate and the beginning of the physical researches in Japan'. *Historia Scientiarum* Vol 7–2 p 93.
29. Koizumi Kenkichiro (1975) 'The emergence of Japan's first physicists: 1868–1900'. *Historical Studies in the Physical Sciences* **6**, 3–108. Princeton, U.S.A.
30. Wakai, N (1995) *100 Years of Raido,* IEE Conf Pub Vol **4**,11, p 51.
31. *Japan Weekly Mail* (1878) 2, 52.
32. Kelvin Papers, T1 and T2, 1906, University of Cambridge Library.

CHAPTER 15

Notes

1. By 'thermal effect' he means what is now called entropy production.
2. This is the so-called 'scientific' notation, as used in calculators, which enables very large and very small numbers to be written down in a way that is easily read. The notation 10^{25} stands for 1 followed by 25 zeros, i.e. 10,000,000,000,000,000,000,000,000, and 10^{-23}

stands for 1 preceded by 23 zeros, of which the first is at the left of the decimal point, i.e. 0.000,000,000,000,000,000,000,01.

3. A nanometre is 10^{-9} metres.
4. As if the colliding molecules had been taken from two beams which came from completely different places, not from parts of the same gas.
5. The Peltier coefficient of a metal is the value of J/I when there is no temperature gradient, and is therefore equal to C/A. The Seebeck coefficient is the ratio of (minus) the voltage to the temperature difference when $I = 0$ and is therefore proportional to B/A.
6. Some work in non-equilibrium thermodynamics appears to have been done with insufficient care and has attracted vigorous criticism from Truesdell's school of 'rational thermodynamics' (Coleman and Truesdell 1960; lecture 7 of Truesdell 1984; for more information see Ball and James 2002). Onsager's own paper, however, is written with the greatest care.

References

Bernoulli, D (1738), 'De affectionibus atque motibus fluidorum elasticorum, praecipue autem aeris', section X in his *Hydrodynamica, sive de viribus et motibus fluidorum commentarii* (Argentorati); Dover reprint 1968. English translation in Brush (1965).

Brown, Robert (1828) 'A brief account of microscopical observations made in the months of June, July and August, 1827, on the particles contained in the pollen of plants; and on the general existence of active molecules in organic and inorganic bodies'. *Phil. Mag.* **4**, 161–173; cited by Brush (1976).

Thomson, W. (1851) 'On a mechanical theory of thermo-electric currents', *Proc. Roy. Soc. Edinburgh*, Dec. 1851; reprinted in *Collected Papers* I, pp 316–323, Cambridge University Press, 1882.

Thomson, W. (1852) 'On a universal tendency in nature to the dissipation of mechanical energy', *Proc. Roy. Soc. Edinburgh*, 19 April 1852; *Phil. Mag.* Oct. 1852.

Thomson, W. (1854) 'Thermo-electric currents', *Trans. Roy. Soc Edinburgh* **21**, part I, read 1st May 1854; *Collected Papers* I, pp 237–41, Cambridge University Press, 1882.

Thomson, W. and Tait, P. G. (1867) *Treatise on Natural Philosophy*, Oxford University Press.

Thomson, W. (1870) *Nature* **1** 551, cited by Brush (1976).

Boltzmann, L. (1872) 'Weitere Studien ueber das Waermegleichgewicht unter Gasmolekulen'. *Sitzungsberichte der Akademie der Wissenschaften, Wien*. II **66**, 275; English translation in Brush (1966), pp 88–175.

Thomson, W. (1874) 'The kinetic theory of the dissipation of energy', *Proc Roy Soc Edinburgh*; reprinted in Brush (1966).

Boltzmann, L. (1877a) 'Bemerkungen ueber einige Probleme der mechanischen Waermetheorie', *Wien. Ber.* **75**, 62 cited in Brush (1976).

Boltzmann, L. (1877b) 'Ueber die Beziehung eines allgemeinen mechanischen Satzes zum zweiten Hauptsatze der Waermetheorie', *Sitzungsberichte der Akademie der Wissenschaften, Wien*, II, **75**, 67–73 (1877); English translation in Brush (1966).

Boltzmann, L. (1877c) 'Ueber die Beziehung zwischen dem zweiten Hauptsatze der mechanicshen Waermetheorie und der Wahrscheinlichkeitsrechnung, respective den Saetzen ueber das Waermegleichgewicht', *Wien Ber.* **76** 373 (1877), *Phil. Mag.* [5] **5**, 236 (1878).

Helmholtz H. (1878) 'Ueber galvanische Stroeme, verursacht durch Concentrationsunterschiede; folgerungern aus der Mechanischen Waermetheorie', *Ann. der Physik und Chemie,* **239**, issue 2 pp 201–216.

Culverwell, E. P. (1890) 'Note on Boltzmann's kinetic theory of gases, and on Sir W. Thomson's address to the British Association, 1884'. *Phil. Mag.* [5] **30**, 95.

Boltzmann, L. (1884) 'Ueber die Moeglichkeit der Begrundung einer kinetischen Gastheorie auf anziehende Kraefte allein.' *Wiener Berichte*, **89**, 714, cited by Cercignani (1998) p 134.

Burbury, S. H. (1895) *Nature* **51**, 78, 175, 320, and **52**, 104 , cited in Brush (1976).

Boltzmann, L. (1895) 'On certain questions in the theory of gases', *Nature* **51**, 413–315.

Zermelo, E. 1896 *Ann. Phys.* **57**, 485, and **59**, 793.

Kelvin, Baron (1900) 'Nineteenth century clouds over the dynamical theory of heat and light', *Proc. Roy. Inst.* **16**, 363; *Phil. Mag.* [6] **2**, 1 (1901), cited in Brush (1976).

Gibbs, J. W. (1902) *Elementary Principles in Statistical Mechanics, Developed with Special Reference to the Rational Foundations of Thermodynamics*, Yale University Press; Dover reprint 1960.

Planck, M. (1901) 'On the law of distribution of energy in the normal spectrum', *Annalen der Physik*, **4**, 553.

Einstein, A. (1905) 'Ueber die von der molekularkinetischen Theorie der Waerme geforderte Bewegung von in ruhenden Fluessigkeiten suspendierten Teilchen.' *Ann. Phys.* **17**, [2], 549; *Investigations on the Theory of Brownian Movement*. New York: Dover, 1956. ISBN 0–486–60304–0

Einstein A. (1910) 'Theorie der Opaleszenz von homogenen Fluessigkeit und Fluessigkeitsgemischen in der Naehe des Kritischen Zustand', *Ann d. Physik* **33**, 1275–1298.

Johnson, J. B. (1928) 'Thermal agitation of electricity in a conductor', *Phys. Rev.* **32**, 97–109.

Nyquist, H. (1928) 'Thermal agitation of electric charge in conductors', *Phys. Rev.* **32**, 110–113.

Onsager, L. (1931) 'Reciprocal relations in irreversible processes', *Phys. Rev.* **37**, 405–426 and **38**, 2265–2279.

Denbigh, K. G. (1951) *Thermodynamics of the Steady State*, Methuen.

Kubo, R. Kubo (1957), *J. Phys. Soc. Japan* **12**, 570.

D. A. Greenwood (1958), *Proc. Phys. Soc.* London **71**, 585.

Coleman, B. D. and Truesdell, C. (1960), 'On the reciprocal relations of Onsager', *J. Chem. Phys.* **33**, 28–31.

de Groot, S. R. and Mazur, P. (1961), *Non-Equilibrium Thermodynamics*, Dover reprint 1983.

Brush, S. G. (1965), *Kinetic Theory*, vol 1, Pergamon.

Brush, S. G. (1966), *Kinetic Theory*, vol 2, Pergamon.

Lanford, O. E. III (1975), Time evolution of large classical systems, pp 1–111 of *Dynamical Systems: Theory and Applications: Battelle Seattle Rencontres 1974*, ed. J. Moser (Springer Lecture Notes in Physics **38**).

Brush, S. G. (1976) The kind of motion we call heat, *Studies in Statistical Mechanics*, ed. E. W. Montroll and J. L. Lebowitz, vol VI (North-Holland, Amsterdam).

Truesdell, C. (1984) *Rational thermodynamics*, Springer, New York.

Cercignani, C. (1998), *Ludwig Boltzmann, The Man who Trusted Atoms*, Oxford University Press.

Ball, J. M. and James, R. D. (2002) 'The scientific life and influence of Clifford Ambrose Truesdell II', *Arch. Rat. Mech.* **161**, 1–26.

CHAPTER 16

References

1. Elisabeth Crawford, J. L.Heilbron, and Rebecca Ullrich, *The Nobel Population: A Census of the Nominators and Nominees for the Prizes in Physics and Chemistry* (University of California, Berkeley, 1987).
2. Silvanus P. Thompson, *The Life of William Thomson: Baron Kelvin of Largs* (Macmillan, London, 1910) [SPT], p. 905–9; Crosbie Smith and M. Norton Wise, *Energy and Empire: A Biographical Study of Lord Kelvin* (Cambridge University Press, Cambridge, 1989) [SW], pp. 801–11.
3. Harold I. Sharlin, *Lord Kelvin: The Dynamic Victorian* (Pennsylvania State University Press, University Park, 1979) [HIS], p. 238.
4. Dipankar Home and Andrew Whitaker, *Einstein's Struggles with Quantum Theory* (Springer, New York, 2007).
5. SPT, p. 510.
6. Lord Kelvin, *Baltimore Lectures on Molecular Dynamics and the Wave Theory of Light* (C. J. Clay and Cambridge University Press, Cambridge, 1904).
7. SPT, p. 819.
8. SW, p. 489.
9. SW, p. 91
10. Joe. D. Burchfield, *Lord Kelvin and the Age of the Earth* (Macmillan, London, 1975).
11. David B. Wilson, *Kelvin and Stokes: A Comparative Study in Victorian Physics* (Adam Hilger, Bristol, 1987).
12. Marie Boas Hall, *All Scientists How: The Royal Society in the Nineteenth Century* (Cambridge University Press, Cambridge, 1984) [MBH], p. 145.
13. MBH, pp. 124–5.
14. Lytton Strachey, *Eminent Victorians* (Chatto and Windus, London, 1918).
15. Thomas S. Kuhn, Energy conservation as an example of simultaneous discovery, in: *Critical Problems in the History of Science* (Marshall Clagett, ed.) (University of Wisconsin Press, Madison, 1959), pp. 321–56; reprinted in *The Essential Tension* (University of Chicago Press, Chicago, 1977), pp. 66–104.
16. Yehuda Elkana, *The Discovery of the Conservation of Energy* (Hutchinson, London, 1974) [YE], p. vi.
17. Clifford A. Truesdell, *The Tragicomical History of Thermodynamics 1822–1854* (Springer-Verlag, New York, 1980) [CAT], p. 185.

18. W. Thomson, 'On the dynamical theory of heat, with numerical results deduced from Mr. Joule's equivalent of a thermal unit, and Mr. Regnault's observations on steam' *Proceedings of the Royal Society of Edinburgh* **3**, 48–52 (1851); reprinted in *Mathematical and Physical Papers* (Cambridge University Press, Cambridge, 1882) [MPP], Vol. 1, pp. 174–210.
19. For a full list of Kelvin's publications, see SPT, pp. 1223–74.
20. SPT, pp. 169, 173–4.
21. SPT, pp. 175–6.
22. W. Thomson [signed P. Q. R.], 'On the uniform motion of heat in homogeneous dolid bodies, and its connection with the mathematical theory of electricity', *Cambridge Mathematical Journal,* **3** (14), 71–84 (1842).
23. W. Thomson, 'On electrical images', *British Association Report,* **17**, 6–7 (1847).
24. SPT, pp. 113–21.
25. SW, p. 281.
26. SPT, p. 50; SW, p. 213.
27. SPT, p. 19.
28. SW, p. 203.
29. B. G. Doran, 'Origins and consolidation of field theory in nineteenth century Britain: from the mechanical to the electromagnetic point of view', *Historical Studies in the Physical Sciences* **6**, 132–260 (1975).
30. David Gooding, 'Faraday, Thomson and the concept of the magnetic field', *British Journal for the History of Science* **13**, 91–120 (1980).
31. Matin Durrani and Peter Rodgers, 'Physics: past, present and future', *Physics World,* **12** (12), 7–13 (1999).
32. HIS, p. 120.
33. Donald Franklin Moyer, 'Continuum mechanics and field theory: Thomson and Maxwell', *Studies in the History and Philosophy of Science,* **9**, 35–50 (1978).
34. HIS, p. 122.
35. M. Norton Wise, 'The flow analogy to electricity and magnetism, Part I: William Thomson's reformulation of action at a distance', *Archive for the History of the Exact Sciences,* **25**, 19–70 (1981).
36. SPT, p. 1065.
37. Translated by John Tyndall as; 'On the motive power of heat, and the laws regarding the nature of heat itself which are deducible therefrom', *Philosophical Magazine,* **2**, 1–21, 102–19 (1851).
38. D. S. L. Cardwell, *From Watt to Clausius* (Heinemann, London, 1971) [DSLC], p. 244.
39. W. Thomson, 'An account of Carnot's theory of the motive power of heat; with numerical results deduced from Regnault's experiments on steam, *Transactions of the Royal Society of Edinburgh,* **16**, 541–74 (1849); reprinted in MPP, Vol. 1, pp. 133–55.
40. Crosbie Smith, 'Natural philosophy and thermodynamics: William Thomson and the "the dynamical theory of heat"', *British Journal for the History of Science,* **9**, 293–319 (1976).
41. Sir Joseph Larmor, 'William Thomson, Baron Kelvin of Largs 1824–1907', *Proceedings of the Royal Society A,* **81**, I-lxxvi (1908).

42. SW, pp. 89–91.
43. SW, p. 306.
44. SW, pp. 91–9.
45. CAT, p. 34.
46. DSLC, Ch. 3.
47. SPT, pp. 275–6.
48. W. Thomson, 'On an absolute thermometric scale founded on Carnot's theory of the motive power of heat, and calculated from Regnault's observations, *Philosophical Magazine,* **33**, 313–7 (1848); reprinted in MPP, Vol. 1, pp. 100–6.
49. SPT, p. 285; CAT, section 9D
50. Translated by John Tyndall as: On the conservation of force: a physical memoir, in *Scientific Memoirs, Natural Philosophy* (J. Tyndall and W. Francis, eds.) (Taylor and Francis, London, 1853), pp. 114–62.
51. YE, pp. 79–82.
52. W. Thomson and P. G. Tait, *Treatise on Natural Philosophy* (Cambridge University Press, Cambridge, 1883).
53. SW, pp. 352–3.
54. YE, pp. 47–9.
55. SW, pp. 544–51.
56. SPT, pp. 228–9.
57. SPT, pp. 338–9.
58. SPT, pp. 354–5.
59. SPT, p. 370.
60. SPT, pp. 654–5.
61. SPT, p. 697.
62. SPT, p. 735.
63. Paul Tunbridge, *Lord Kelvin: His Influence on Electrical Measurements and Units* (Peter Peregrinus, Stevenage, 1992).
64. SPT, p. 805.
65. SPT, p. 297.
66. SPT, p. 845.

INDEX

Note: page numbers in *italics* refer to Figures.